301.31
F18t

81756

DATE DUE			
Jul 5 '74			
Dec 15 '75			
Feb 23 '76			
Dec 15 '76			
May 16 77			
Dec 16 77			
May 22 78			

BOOKS BY RICHARD A. FALK

*Law, Morality, and War in
the Contemporary World*

*The Role of Domestic Courts
in the International Legal Order*

Security in Disarmament
(editor, with R. J. Barnet)

The Strategy of World Order, 4 volumes
(editor, with Saul H. Mendlovitz)

Legal Order in a Violent World

Neutralization and World Politics
(with Cyril E. Black, Klaus Knorr, and Oran Young)

International Law and Organization
(editor, with Wolfram Hanrieder)

*The Vietnam War and International
Law,* 2 volumes (editor)

The Status of Law in International Society

The Future of the International Legal Order,
2 volumes to date (editor, with Cyril E. Black)

This Endangered Planet

THIS
ENDANGERED
PLANET

Prospects and

 RANDOM HOUSE   NEW YORK

THIS ENDANGERED PLANET

Proposals for Human Survival

Richard A. Falk

Acknowledgment is extended to W. H. Freeman and Company for permission to reprint from *Resources and Man: A
Study and Recommendations* by the Committee on Resources
and Man of the Division of Earth Sciences, National Academy
of Sciences—National Research Council, with the cooperation
of the Division of Biology and Agriculture. Copyright © 1969
by W. H. Freeman and Company.

An excerpt from "Toward a Critique of Economics" by Paul
Sweezy is reprinted by permission from *Monthly Review,* January 1970. Copyright © 1970 by Monthly Review, Inc.

FOR

Florence/Dimitri

Lili/Erich

four vital horizons
of love
and survival

Note of
Acknowledgment

I have been the grateful beneficiary of many kinds of help in the course of preparing this book. Especially, I have benefited from friendly counsel and stimulating discussion covering many of the issues. Among those who have been particularly helpful I would mention Pierre Noyes, John G. Sperling, Saul H. Mendlovitz, Gidon Gottlieb, Harold Feiveson, Irwin Mann, Harry Hollins, Scott McVay, Lester R. Brown, and Charles G. Westoff. Indeed, my conversations with Pierre Noyes led me to abandon earlier plans and embark upon this project.

The Center for Advanced Study in the Behavioral Sciences at Palo Alto, California, provided me with a year of highly provocative leisure (1968–69) within which I developed the underlying plan of the book. I was much helped at the Center by its excellent and affable staff, and I would especially thank Irene Bickenback who performed with grace and efficiency numerous "special requests."

After I returned to Princeton, the Center of International Studies made available its facilities to permit the completion of the book. Once again, I am beholden to Jane G. McDowall and her talented secretarial staff for typing several drafts of this manuscript.

The World Law Fund has also been a strong and consistent supporter of my work in this area, including this book. My participation in the World Order Models Project of the Fund has greatly influenced the direction of my thinking on world order problems.

Claudia Cords has served with great excellence as my research assistant and I am indebted to her for many contributions of detail to the final product.

My association with Random House has been a valuable experience, itself very much of an ingredient of the writing process. I am, in this sense, particularly grateful to my editor, Albert Erskine, for his patience and counsel.

Finally, my wife Florence participated at every stage of this long adventure. Her enthusiasm for the undertaking often provided me with the energy and will to persevere during periods of discouragement. Also, as with earlier books, her excellent suggestions as to style and content are reflected in the final draft.

With so many expressions of gratitude, it seems necessary to confirm my own responsibility for persisting errors, confusions, and insufficiencies.

Richard A. Falk

Princeton, New Jersey

Contents

THIS
ENDANGERED
PLANET

I. An Endangered Planet

To be a man is to be limited and mortal. To be on earth is to live within a finite and restricted environment. Life is sustained by a thin belt of atmosphere above a skin of earth crust. The life-support system based on air, earth, and water is delicate, subtly intertwined, and remarkably intricate.

The Need for Limits. The rise of the industrial state, and with it, science and technology, has led us to overlook these conditions of finitude and fragility. We have come to accept theories of progress and of inevitable development that look toward an indefinite improvement of the human condition by continuous economic growth made possible by an endless sequence of technological improvements. We have identified growth and expansion with progress, and we have not acknowledged the existence of any limitations on progress. The decline of an active religious consciousness in our century has reinforced this habit of inattentiveness toward the limits and contingencies that surround our individual and collective presence on earth. In earlier periods of history the active presence of religious thought helped keep alive the distinction between the finite and the infinite.

A storm presages its coming by a variety of signals. This book attempts an interpretation of these storm warnings, giving special attention to their political significance. To take proper precautions requires that, first of all, we rediscover the reality and significance of our finitude. We need to identify and clarify the limits of our planetary existence and plan to live within those limits. The task is urgent. We may not have more than a few years to make fundamental adjustments; we certainly do not have more than a few decades. Unless we can adapt our behavior to the carrying capacity of the planet, the future of mankind will be indeed bleak—assured short-run deterioration of life circumstances, and a growing prospect of human extinction. The stakes of this struggle are nothing less than the habitability of the planet.

Alongside the need for finding and setting limits is the growing gap between the *unity* of the *world of facts* and the *fragmentation* of the *world of authority and power.* The scale of modern technology is overflowing every political boundary. Whether it is a matter of radioactive fallout, carbon monoxide or other poison effluents, multinational corporations, computer technology, satellite broadcasting, or air and space travel, we constantly witness a drive toward operations on a planetary scale. And yet most political behavior continues to be dominated by the territorial state. States compete with one another for power, wealth, and prestige, and jealously guard their sovereign prerogatives. This competitive pattern generates conflict, waste, and distrust. Huge amounts of resources are devoted to national defense, collective violence is persistent and pervasive, and wars occur at many points on the planet. An organizational crisis appears emergent. The governmental machinery that exists in the world is mostly located on or below the national level and cannot be expected to cope with problems of global and near-global proportions. We need to develop new political forms responsive to the new problems, directed toward planetary ideals and endowed with funds, respect, and authority. Many formidable obstacles dim the prospects for fulfilling these needs, not least of which is the persisting vitality of the myth and traditions of national sovereignty. There is little evidence that any change in the structure of world order is likely to come about in the years ahead, unless it is provoked by a catastrophe of awesome proportion. But let us not be deceived. World federalism is no panacea. The task of dealing with the problems of human survival is far more complicated than the advocacy of a world government or a world state. It is desirable to preserve political autonomy and cultural diversity and to prevent tyranny at the global level. We also do not pretend that the mere creation of a central world government can solve the problems of this endangered planet. Governments at all levels have not demonstrated a great capacity to solve the most urgent problems of human society. Violence and misery persist in most national societies of the world.

The first need is to awaken the consciousness of men and women everywhere to the causes of their distress. Above all else, men need to understand their vital dependence upon nature. The political society of the future must work to establish and maintain harmony between man and nature, as well as between man and man. Perhaps the achievement of harmony is too ambitious and unrealistic to serve as a goal, but at least there is needed some form of détente between man and nature, some better understanding of the dangers and difficulties that arise from ill-advised human interventions in the processes of nature. The rediscovery of finitude is above all else an understanding of what it means for men to live within a limited natural environment. In the future, perhaps, the burden of justification will shift against those who propose new interferences with natural design; initiative will have to be filtered through a knowledgeable screen of prudence if life on earth is long to endure.

In this sense, the situation is simple. The precarious situation of the peregrine falcon may be an omen for mankind. In fact, George H. Lowrey, Jr., a zoologist, reviews a book on the prospects for survival of peregrine falcons by writing that "it might well be a classic—the handwriting on the wall for all mankind." The peregrine falcon is a species mainly endangered by passing DDT, DDE, and other chlorinated hydrocarbons up the food chains. These chemicals disrupt the calcium metabolism of these raptorial birds, leading to reduction of egg-shell thickness that apparently induces egg breakage and egg eating by parents, resulting in so drastic a cut in population that extinction is now forecast.

The human danger is brought on by human activity. Unlike the peregrine falcon that is being threatened with extinction by changes in the environment brought about by human intervention, man is fouling his own nest. In this sense, even an ecological view of life tends toward putting man at the center of the stage, as both protagonist and villain.

It is hardly novel to assert that population pressure is growing, that the danger of nuclear war persists and is probably increasing, that various forms of pollution are impinging upon our environment, and that resources upon which our affluence depends will not be able to satisfy the demands of future generations. It has become commonplace to point out that these four circumstances add up to a situation of unprecedented danger for the human race. For some time now, ever since the birth of the nuclear age at Hiroshima, numerous prophecies of doom have been delivered. The crisis is provoked by the explosion of people and technology past the point of overload. Man is being confronted with the grim actuality of his contingent existence in a limited environment.

The rhetoric of apocalypse has not changed basic patterns of feelings and behavior of those who live within these circumstances of acknowledged danger. People rarely move away from the village that lies below the active volcano, or relocate to avoid a predictable earthquake hazard. To grasp a new reality requires a combination of experience, understanding, and an organized effort at positive response. There needs to be hope as well as fear. A posture of fear immobilizes rather than transforms. Man has an alarming capacity to adapt to conditions of persisting hazard without taking available steps to reduce or eliminate them.

The problem of change is associated with the existence of inequality among groups and nations. The powerful, the privileged, and the rich generally oppose drastic change as it shakes the foundations upon which their advantages depend, whereas the poor tend to accept, and are often induced to believe in, the permanence of their deprivation. Even if the retention of the status quo poses a common danger, there is rarely evident a human capacity to meet the greater danger by assuming the lesser, but perhaps more definite, risk. Who among us would give up summer vacations or consumer luxuries to improve the prospects for enduring peace? And besides, how can we ever be sure that our individual sacrifices will turn out to be improvements upon what now exists? Until the crisis validates itself by catastrophe, the whole concern is an abstraction, in the critical sense of not entering actively into our consciousness, its dreams, fears, fantasies. The great danger of an apocalyptic argument is that to the extent it persuades, it also immobilizes.

We are dealing with vast and vague imponderables in a setting of rapid and largely unplanned change. The future cannot be demonstrated, but only conjectured in more or less convincing terms at the very outer edges of human imagination. There are disagreements about how great specific dangers are; there are special-interest groups that use their resources and talents to obscure or suppress relevant information and perceptions. There is no consensus about how to deal with principal issues. Societies distrust one another's pretensions of global concern, and look for selfish motivations. Such a context of perception produces a sense of bewilderment and immobility that tends to produce indifference, and to assure the persistence of the old ways. Occasionally, the evidence becomes overwhelming on a very specific issue—say, mine safety or the damage done by DDT—and then a burst of curative effort is visible. But usually such an awareness arises in a situation where the evidence is conclusive and confined in scope; normally, a specific disaster or a series of disasters is needed to trigger public indignation—the offshore oil blowout of 1968 has made Santa Barbara into a community with a radical approach to environmental issues. Sometimes, impartial studies, if issued under august auspices, can command enough respect to stimulate legislative action that can provide some standards. The problem of responding to danger in a domestic context is different from responding to danger in a planetary context.

For one thing, there exist only the weakest instruments for the identification of planetary interests. Even within the organs of the United Nations, state sovereignty prevails, and many states distrust the Organization altogether. The UN is appreciated and used by national governments as an instrument for the *promotion* of their interests in a competitive world system, in which the main governments are constantly preparing for a possible war against their rivals. As yet these governments are unwilling to subordinate their interests to some ideal of planetary welfare. National governments continue to devote their energies to building a strong and prosperous domestic society that measures achievement by Gross National Product levels, welfare by GNP per capita, and security by armed forces and military allies. Peace is "enforced" by national defense budgets and alliances that enable an exchange of effective threats. The principal states devote their foreign policy and huge resources to sustaining the credibility of these threats by military, economic, and political means. International history is dominated by warfare and conflict.

The rise and fall of every civilization is, finally, a reflection of the outcome on the battlefield, which in turn expresses the facts of cohesion, capability, and geographical position. For centuries the outcome of these struggles for international ascendancy, although often accomplished at great human and environmental cost, did not imperil basic modes of human existence. Struggles for regional or global primacy, even if carried on in self-destructive fashion, did not threaten the basic structure of human existence. In fact, with the benefit of Darwinian or Calvinistic logic it became possible to believe that even bloody conflict contributed to the all-pervasive selection process that elevated the most dynamic and worthwhile of contending groups and eliminated those who were weaker and less well endowed. Changes in the status and values of dominant groups brought about by competition, however beastly, were believed generally to reward the best and punish the worst; a mystical confidence associated victory with the eventual promotion of secular, even divine, justice. The history of mankind, in its modern presentation by most Western historians, is characterized by an upward linear slope, rather than either a circle of repetition or a movement toward an abrupt and final collapse.

Nuclear weapons appeared at first to cast grave doubt on this violent process of adjusting conflict, at least in the relations among more powerful states. The changed character of war is the most visible expression of the new planetary setting. A world contaminated by radioactive fallout largely negates not only the distinction between victory and defeat, but also that between belligerent and neutral. There is created a new community of danger that embraces the whole of mankind. "One teaspoonful of strontium 90 would be enough, it is said, to kill the entire human race." The existence of this community of danger is verbally acknowledged and enters into defense planning in the form of elaborate precautions, but the risks of nuclear war have not yet reshaped the methods and structures of forming policies, making decisions, allocating resources, and organizing power within the world. Certainly, powerful governments now incline toward a more cautious diplomacy so as to avoid unintended conflict, but the basic patterns of international thought and action remain frozen in their prenuclear cast. The *old* political consciousness has become a *false* political consciousness because it has not adapted to the new *objective* circumstances of the present situation.

My early chapters seek to demonstrate why certain critical changes in objective circumstances make national governments increasingly incapable of solving the principal problems of mankind. The point can be developed in its most dramatic form: the present framework for problem-solving in international society increasingly imperils human survival. Such a statement can be put in a more positive form: the conditions of human existence could become safer and the quality of human life could be improved by inducing certain changes in the organization of world society. Whether to guard against extinction or to struggle for a better world, there is a need to show why and in what respects the present structures of power and authority cannot be expected to deal with the emerging problems of the planet. I will emphasize four interconnected threats to planetary welfare, each of which interacts and reinforces the others. It is an awareness of these threats and a sense that their character will grow rapidly more grave and less manageable in the years ahead that create the special crisis of our time. We are living in a period of constantly increasing risk and diminishing opportunity. To illustrate: as the atmosphere grows more contaminated by a variety of poisons, it becomes ever more difficult to restore conditions of purity. A situation of irreversibility threatens to arise in which no amount of feasible effort can counteract the process of contamination or temperature change. Irreversible change has led specialists to pronounce Lake Erie "dead," that is, no amount of effort at reasonable costs can be undertaken to restore life systems to that large body of fresh water. The body of water has become virtually an inert mass, useful for transportation, dangerous for health, useless for life-support. To rehabilitate Lake Erie, if possible at all, might cost as much as $40 billion.

In countless respects, we are proceeding along a collision course, with little serious appreciation of the degree of danger that arises from the environmental deterioration that surrounds our lives. We have very little understanding of how much pollution of various kinds the atmosphere or oceans can absorb, or the lethal consequences to air, land, and water of certain mixtures of chemicals. The vastness of the oceans and the skies makes it almost impossible to grasp the prospect of their contamination. We inherit attitudes based on several centuries of human existence that make the earth, the oceans, and the skies a dumping-ground for man. Pollution problems have arisen from time to time, but only as matters of local concern, compelling at worst an abandonment of a certain area of land. At no time has man had to deal with the possibility that the scale and character of human existence may be imperiling the conditions of life on earth altogether.

The Distinctive Character of the Planetary Crisis. We are living now in the first stages of a planetary crisis. It is the first such known crisis in the history of the planet. Even earlier ice ages imperiled life only on those portions of the earth covered by glacial formations. The crisis is of planetary scope because the danger is not confined to any part of the planet; the patterns of behavior that generate the crisis are created by the scale of production and life-style in the most advanced industrial societies, and therefore especially by the principal nuclear and space states. An adequate response eventually requires a new pattern of organization and coordination that needs to encompass the entire planet.

Prior international disorders—wars, famines, plagues, or economic dislocations—have had negative effects in many, perhaps in all, parts of the world. But *survival* and *recovery* never seemed to be an issue. Today the deterioration of the environment through everyday technology creates a variety of hazards that may so upset the ecological balance of water, earth, and air as to make the planet uninhabitable or the human species extinct. We do not know at what exact date such an outcome will occur if present trends are maintained, but the evidence now available provides a rational basis for urgent concern. We know that the nuclear arsenals of the Soviet Union and the United States have the capability to destroy all life on this planet. At the present time, defense planning in the United States calls for equipping submarines with the Poseidon weapons systems which have Multiple Independently-targeted Reentry Vehicles (MIRV) that will enable each Polaris submarine to destroy 175 or more cities. A single submarine commander, if able to overcome command and control safeguards, is thereby empowered to destroy almost all major urban life on the planet. Given human fallibility and the alienation of many citizens of the modern world, we are making ourselves ever more vulnerable to the systems of security we so artfully devise.

Where *survival* is at stake, the consequences of breakdown may be *irreversible*. Past history includes many examples of societies or civilizations that did not respond to challenges directed at their survival. Arnold Toynbee, in fact, has written of all human history as the continuous saga of unsuccessful response to fundamental challenge. But these challenges were normally an interaction between adversary groups: a challenger and a respondent. The outcome of the interplay, it is true, could often be explained more profoundly by internal factors of organization, leadership, technology, and morale than by the balance of forces between opponents or by a mere account of battles and wars. Planetary history has been the story of the changing tides of power, the rise and fall of groups, civilizations, and regions.

The nature of the crisis is difficult to describe in its totality. The danger of nuclear war can be contemplated by the imagination in vivid terms as an event that might occur at a particular time under various sets of circumstances. A whole genre of books and films depict some of the more plausible settings and effects of nuclear war; among the best examples are *On the Beach, A Canticle for Liebowitz, Red Alert (Dr. Strangelove), The War Game,* and *Seven Days in May.* But the problems of ecological disarray attributable to modern technology, population growth, and resource depletion undermine the foundations of human existence by imperceptible, yet steady, processes of erosion. It is a matter of termites eating through the foundations. We do not know exactly when the edifice will collapse but we do know the event will occur if the termites go on eating.

The eating away at the ecological foundations of human existence is a useful way to envision the planetary crisis. Such a metaphor emphasizes the importance of time. To call a situation a crisis implies its urgency. We do not know and cannot know how much time is available to bring about the changes in behavior and organization needed to remove or moderate the threats.

To label a situation a crisis also implies the possibility of a favorable outcome. A crisis is a time for decision and response. The planetary crisis of survival can be dealt with, but only in the event that certain fundamental changes in attitude and organization take place. This book is mostly concerned with the international setting of the planetary crisis and concentrates upon the changes in the structure of world society that are needed. Such a focus arises from a further characteristic of the crisis: in order to be effective, change must be planetary in scope. No national society, however enlightened its government, can meet the challenge directed at its survival by independent action. National policy may hasten or defer the impact of certain disorders, but it cannot control the dynamics of the crisis. The planet is now a whole as a consequence of technological development and population increase. A coordinated response presupposes a common, or at least a convergent, interpretation of the situation by all principal governments, ideologies, cultures, and religions. At the same time we live in a world organized into contending units, where boundaries are critical organizing parameters and where power and ideology continue to be largely controlled by national governments and by institutions closely allied with such governments.

The first signs of awakening are likely to be within national societies. Established elites need to be immediately confronted with simple and compelling descriptions of the common danger. Let us suppose that we were lucky enough to discover that another planet was preparing to attack and destroy the earth. The discovery would have to be verified and assimilated in the various main capitals of the world. A strategy of planetary defense would then have to be devised. This strategy would depend upon a unified effort that mobilized the capabilities of the planet against a powerful extraterrestrial adversary. A war-consciousness might be generated out of such a situation. The normal limits on what kinds of international cooperation are feasible might disappear for the duration of the war and the requirements of military necessity might overcome entrenched interests in the economic and social status quo.

As a first step we need to develop a clear awareness of the present situation, not only with respect to its dangers or problems (as symptoms) but with regard to the underlying disorder (as cause). Only then does it become possible to take suitable action. Seriousness gives rise to hope, and hope to action. There is no known barrier that prevents a solution, although there are, of course, numerous obstacles. The inventiveness of man is not necessarily precluded from working out the conditions for his own survival. Other animals, it is true, have not often been successful in overcoming adaptive crises that placed their species in jeopardy. The balance sheet of evolution is largely negative. Most species of mammal do not survive longer than 600,000 years. Man has already been on earth for more than 1,000,000 years. Of the 500,000,000 forms of life that have existed on earth, 498,000,000 are extinct. Unless man can make some extraordinary changes in his patterns of social and political behavior, he may well soon follow the dodo and the dinosaur down the path of extinction; but when man goes, advanced forms of life on earth are also likely to go.

The argument, then, is very simple. Mankind is endangered by a crisis of planetary proportions. This crisis has emerged mainly out of the interplay between a machine technology and a rising population. The dynamics of this interplay have continued for over a century and are pushing up against the carrying capacity of the earth in a variety of ways.

The political life of man is mainly organized into competitive units called states. These states do manage to cooperate for limited purposes, such as trade or tourism or to wage war against a common enemy, but the basic mode of behavior is competitive and the fundamental unit of organization is limited in its scope. An endangered planet calls for stronger cooperative patterns of behavior and for more embracing forms of organization.

How can we mobilize the kind of social action needed to create the sorts of institutions that are now necessary to assure the welfare, and possibly the survival, of mankind? This question expresses the great political challenge of our time, and the quality of the planet's future—and even whether there is to be a future—will depend upon our capacity to generate an adequate response. It is a task of such magnitude and complexity that it requires the participation of men of good will everywhere.

Skeptics will not rally in support of such a proposal. It sounds almost as absurd as putting a man on the moon.

The Primacy of Politics. We live in an age of scientific vanity. Part of our complacency arises from a widespread belief that when the situation deteriorates sufficiently, scientists, engineers, and technocrats will somehow manage to save the day with some ingenious new device. Arthur C. Clarke, a widely read writer on science, holds the view that "science will dominate the future even more than it dominates the present." Mr. Clarke makes explicit what remains a widely shared bit of contemporary folk-wisdom:

> I also believe—and hope—that politics and economics will cease to be as important in the future as they have been in the past; the time will come when most of our present controversies on these matters will seem as trivial, or as meaningless, as the theological debates in which the keenest minds of the Middle Ages dissipated their energies. Politics and economics are concerned with power and wealth, neither of which should be the primary, still less the exclusive, concern of full-grown men.

The task of building a new social and political order on a world scale is regenerative in essence. It requires a positive vision that can arouse men to action and can

reshape the values and attitudes that give vitality to
political movements. Science can help with the work of
implementation, but it cannot possibly provide a substitute for politics.

There are several steps that need to be taken: first,
we need to understand the inability of the sovereign state
to resolve the endangered-planet crisis; second, we need
a model of world order that provides a positive vision of
the future and is able to resolve this crisis; third, we need
a strategy that will transform human attitudes and institutions so as to make it politically possible to bring a
new system of world order into being; fourth, we need
specific programs to initiate the process, as with learning
to walk—we need to learn to walk into the future. The
first steps are the hardest for a baby, and their occurrence, the most mysterious and precarious. The pressure of the situation suggests that *time* is itself a scarce
resource, making it appropriate to counsel a sense of
urgency. The national governments of large sovereign
states act now as the major power/wealth clot, obstructing
the birth of healthier forms of life circulation. These
governments need to be reoriented toward an appreciation of the reality of the endangered-planet crisis.

Finally, politics on both the national and world level
has been heretofore dominated by the concern with expanding or managing power and by the problems of *man-
in-society;* as later chapters will make clear, a politics
designed for planetary welfare and survival must quickly
begin to incorporate *man-in-nature* into its essence. The
presence of nature as a life-support system can no longer
be taken for granted. Human activities are disclosing in
numerous and alarming ways the vulnerability of this
system to destruction and deterioration. It will become a
prime political task to safeguard this life-support system,
and the prospects for doing so depend, in part, on the
responsiveness of existing political institutions throughout the world. The longer the delay, the greater the costs

and dangers of response, including, of course, the over-arching danger that the effort may come too late, that a threshold of irreversible deterioration has been crossed, dooming mankind and other forms of earth life to a lunar destiny. Furthermore, if the response arises as a reaction to the occurrence of a series of preliminary or partial disasters, then it is more likely to be highly coercive in quality, endangering in countless respects areas of individual and collective autonomy now revered by men throughout the world. Perhaps this may not occur; but it is the actualities of power and wealth that assure misery for most of mankind, that greatly intensify the dangerous conditions under which we live, and that harness science itself to the causes of domination and profit. The autonomy of science is a foolish myth and is hardly subscribed to even by scientists. Science sets neither the goals nor the priorities of any polity; it does not have much to contribute to the identity of nations around which states have been organized. The marvels of science can be used for good or bad social purposes, depending on the political track that guides the leaders of human affairs. The fundamental challenge of the future is not, of course, to land a man on Mars or even to grasp the mysteries of the universe.

Americans, in particular, have been undergoing a series of culture shocks in recent years. The Vietnam war, in its ferocity and persistence, has exposed the destructive potentialities of American foreign policy—potentialities that are particularly shocking because they collide with the pretensions of our statesmen and the self-righteous, although misleading, myth of our own past. It was in reaction to Vietnam that young Americans first began in a widespread way to question the premises of our foreign policy and, beyond that, the pernicious influence of the military-industrial-legislative complex upon the institutions of government and upon their own

educational institutions, and even beyond that, the legitimacy of a government that would serve such a destructive set of beliefs and interest groups.

This process of questioning initiated by Vietnam carried over into trans-political issues of personal identity.
More and more young people raised doubts about
whether they wanted to be programmed into a nine-to-
five IBM world of gray flannel, martinis, and suburban
living. These doubts supported the development of what
Theodore Roszak has called a "counter-culture," based on
a new life-style that shuns careerism, that searches for
communal forms of being together, that affirms more
mystical modes of being, and that has turned against the
rational-scientific traditions of objective consciousness
associated with the rise of the West. In essence, this
counter-culture is striving to protect human personality
against the destructive tendencies at work in advanced
industrial societies.

The explosion of concern among the young about the
environment is linked to these other principal developments. Within their own society the destructive potentialities of the economic and political system are exposed
by the official toleration of forces that are destroying the
very basis of life on earth. It does not require a Marxist
critique of society to grasp the extent to which private
greed and public opportunism underlie the steady deterioration of the quality of life. In this sense, the counter-
culture repudiates, almost instinctively, many of the
attributes of modernity, such as the automobile and the
general tendency to interpret economic growth as equivalent to human progress. The basic point here is that
mounting a life-revering resistance to destructive energies
provides the counter-culture with its own potential, as
yet largely unrealized, for moral and, eventually, political coherence. The interpretation of danger is correct,
although the understanding of its character is incom-

plete and often confused and hostile, and the prescriptions for action by the young are themselves often little more than a blind, reactive inversion of the destructivity that stimulates their own behavior. The consequence is to let loose a dangerous nihilistic dialectic between the forces of change and reaction, and to give credibility to the demands by rightist forces of reaction for repression now, disguised by the label "law and order." The counter-culture needs to organize its critique of modern society into a positive program for political action.

Arnold Toynbee has written that "nationalism is by far the most powerful of all living religions. It is the common religion of people of all races, civilizations, and degrees of economic development." And Mr. Toynbee might have added ". . . of all major ideologies." In this respect, the counter-culture has emerged within the parochial traditions of nationalism, although there are certain tendencies for youth to identify *across* boundaries, at least with each other. Even transnationalism is not enough. The tendencies toward the destruction of life cannot be dealt with until there emerges a much stronger sense of the reality of wholeness and oneness, of the wholeness of the earth and of the oneness of the human family. Such a reality is grounded now in a last-ditch struggle for survival. In this sense, the ecological imperative points both to the limitedness of the earth's capacities to sustain life and accommodate technology and to the necessity to unify the control over these capacities. For this reason we move on now to an exploration of the ecological imperative as the basis for a new politics, a new ethics, and a new culture.

We need to work out a whole new world view, based on a timely renewal of more primitive conceptions of man and nature, based on the idea of the earth as whole and limited, and as sacred and worthy of cherishing. Claude Lévi-Strauss, the French anthropologist, has said:

It would take a spiritual revolution as great as that which led to the advent of Christianity. It would require that man, who since the Renaissance has been brought up to adore himself, acquire modesty, and that he learn the lesson of all the atrocities we have experienced for thirty or forty years. He would do well to learn that if one thinks only man is respectable among living beings, well then, the frontier is placed much too close to mankind and he can no longer be protected. One must first consider that it is as a living being that man is worthy of respect, and hence one must extend that respect to all living beings—at that point, the frontier is pushed back, and mankind finds itself better protected.

Let us heed the mind-stretching counsel of Lévi-Strauss and struggle to push back the frontiers of our concern from ourselves. This is a large demand to make of men who are not yet often able to identify with their fellow men across frontiers of nations, race, religion, gender, and ideology, but perhaps in developing solicitude for nature, we will also almost automatically come to understand and accept the shared danger and glory of humanity, what unifies as well as what fragments.

The secret of beginning is to begin somewhere, with a concrete and personal act, perhaps located in the mind, in one's daily routine, or in one's neighborhood. The early acts should be close at hand, and gradually the circles of effect should be expanded. Otherwise the magnitude and awesomeness of the apocalyptic warnings will, however well taken, induce psychic numbness and will not encourage an awakening of man to his new setting.

II. The Ecological Imperative

The idea of an ecological perspective involves studying the relation between an animal or plant and its total environment as systematically and comprehensively as possible. Ecological interpretations have developed as a way of thinking about scientific and social problems in many disciplines, occasionally under a distinct and apparently independent vocabulary. For instance, "the field of theory" in sociology associated with the work of Kurt Lewin or the Gestalt approach in psychology are variations on the ecological theme. Similarly, systems theory, arising initially out of wartime operations research and extended to planning for all large-scale organizations, has been evolving in an ecological direction. At the present time, there is more and more interest in formulating an ecological basis of human existence.

In this book I am proposing the development of *ecological politics,* whose essence is a political embodiment of man-in-nature, as the ideological underpinning for an adequate conception of world order. Such an ideology, arising out of an ecological understanding of the endangered-planet crisis, will necessarily push the center of inquiry beyond the level of national societies. The American Indians and many other cultures, especially in the non-Western world, intuitively achieved a sense of mutuality as the basis of man's presence in the world. An Iroquois saying summarizes this sense of mutuality: "The Earth is our Mother, and we cannot sell our Mother." The reality of mother earth is the essence of an ecological outlook, achieved without the benefit of evidence and reasoning designed to show that the adoption of such an outlook is essential for human survival. A new politics must correspond in its scope with the principal problems facing mankind, and offer solutions for problems that spill over national boundaries and cope with dangers that arise from prevailing presuppositions that the earth has an infinite capacity to satisfy the ever-increasing human wants of a growing population merely by taking advantage of a constantly improving technology. The pressure to increase production is intensified by social demands for more equitable distribution and by security demands for ever more elaborate weapons systems.

There are some misconceptions about the espousal of ecological issues by naturalists and conservationists. Ecology is usually discussed in the public press—as it is more and more often—when it becomes evident that dam or power plant proposals overlook detrimental side effects that damage the natural environment. For instance, a jetport that is located so near a natural park as to damage its plants and animals, or a nuclear power plant that will raise the temperature of a river so much as to destroy much of its marine life. In this sense ecology has become the hardheaded and scientific ally of nature-lovers. As such, to invoke ecological considerations is often a way of arguing against technology and engineering solutions to human problems. It is certainly true that builders and real-estate developers have done great damage in modern societies by their failure or refusal to take into account the wider environment of man. The business-engineering mentality in charge of economic projects has been, hitherto, especially insensitive to the web of interactive processes and relationships that sustain the ecological balance between man and nature, but other groups in society also, with rare exceptions, have not been very protective of the environment or willing to place obstacles of any kind in the path of economic growth.

The ecologist demands that a greater effort be made to understand all the effects that would probably result from a proposed interference with a natural ecosystem. The ecologist also demands that the engineer plan his project to mitigate these detrimental effects and that interferences with the environment are also costly if some part of nature's diversity or beauty is lost. But ecologists are not professionally against "progress" or somehow wedded to preserving as much of nature as possible in a pristine state.

We are concerned with ecology in two distinct respects: first, as a way of enlarging the political frame of reference to include environmental issues; second, as an orientation toward action that offsets the tendency of the engineer and real-estate developer to rely upon short-term economic criteria to decide on what is and what is not socially valuable. The ecological outlook is also needed in government where regulatory and monitoring schemes should be immediately undertaken to protect the public interest in securing a livable world. It is essential to identify ecological dangers early enough to prevent and moderate human damage—as from various kinds of pollution—or damage of an irreversible kind to the main life-support systems of nature.

For centuries a combination of famine, disease, and war maintained a tolerable, if often harsh and austere, condition of equilibrium between man and his environment. In many world civilizations, especially in the Orient, the existence of such an equilibrium was an ethical ideal as well as a brutal reflection of a short life of hardship for the masses. Changes took place, disease and hunger were prevalent, local disasters occurred, but the popular imagination conceived of history as a series of cycles, time as a turning wheel, and there was no expectation of basic improvement in the lot of most men. In the West, from as early as the Old Testament era, there was a linear sense of history as unfolding and developmental. The Christian religion developed these ideas in more conclusive fashion by placing stress upon the Last Judgment and the Second Coming, and by depicting in vivid terms the prospect of a glorious end of history. Notions of human reason and confidence in science nurtured confidence in the view that mankind was forever moving forward. The great intellectual achievements of Greek civilization also must have contributed greatly to the separation of modern man from nature. In Greek philosophy, the idea of using reason to attain knowledge was of fundamental importance, whether it took the form of the merciless logic of a Socratic interrogation in a dialogue by Plato or was set forth in the Aristotelian idea that man's highest fulfillment consisted of thinking about thought. The mind as separate from and transcendent over nature lies at the basis of the whole development of modern science with its stress on "objective consciousness." Through the use of this objective consciousness it becomes possible to *observe* nature clearly and carefully, gradually learning how to master and control it, but also forgetting that the observer is also a *participant,* dependent for life on his own equilibrium with nature. When we look out at something as an observer—for instance, even when we sit watching people in a room—we have no way to observe the observer. The ecological outlook depends on rehabilitating our sense of man as *participant,* an awareness that has been lost through the ages by the ability of Western man to make such effective use of the mind to acquire knowledge, power, and riches and to become the *apparent* master of the earth.

The historian of science, Lynn White, Jr., traces the modern ecological crisis back to the insistence in the Middle Ages upon human dominance *over* nature. He contrasts Irish legends showing how man was able to dominate animals with a story about Saint Francis: "The land around Gubbio in the Apennines was being ravaged by a fierce wolf. Saint Francis, says the legend, talked to the wolf and persuaded him of the error of his ways. The wolf repented, died in the odor of sanctity, and was buried in consecrated ground." In Franciscan thought man was regarded as a harmonious part of a sanctified environment, rather than as a being that stood apart from nature and was, by virtue of his soul, alone sacred. The idea of man's superiority in relation to other forms of life underlay the ethics of appropriating nature for human benefit and gain. Nature, like the black man, came in the West to be regarded as subject to enslavement. As Professor White shows, the Franciscan view of animals and nature was quickly and decisively repudiated by the Church, and this decision initiated the process of ecological disruption, made gradually worse by the improvement of technology and the expansion of human numbers, that we are only recently beginning to perceive as a challenge to our very existence on earth. The triumph of science over spiritual authority—the ultimate victory of Galileo over the Catholic Church—developed these tendencies further. Science as a cumulative body of knowledge that could support an ever-growing mastery of man over nature induced the rise of a technology-based industry. Man became intoxicated with the power of reason, and conceived of an idea of progress that was compatible with a linear view of history. This idea of progress was expanded in Hegelian and Marxist thought, and embodied the expectation of a steady, if periodically convulsive, improvement of the human condition. In economic terms this orientation culminated in the Protestant Ethic with its subtle links between the rewards of work and service to God. These beliefs tended to identify successful exploitation of the environment with human fulfillment. The growth of modern science and industry in the West took place against this background of supportive ideas and attitudes. Immunology, hygiene, and better farming enabled more people to live longer. By the end of the seventeenth century the long period of slow population growth gave way to more and more rapid population increases. Applied science upset the equilibrium that had persisted for centuries between man and nature. Agriculture and industry assured that more people could be amply supported per acre and there seemed to be every reason for pride in the medical ability to extend the period of the average human life. The prospects for increases in food supply in relation to the potentialities for population growth led Thomas Robert Malthus at the end of the eighteenth century to set forth a pessimistic account of the future of mankind. But Malthus underestimated the capabilities of applied science to expand food supplies. Even today, when a widely recognized population crisis is upon us and there are many neo-Malthusian voices to be heard, some experts continue to believe that increases in food supply will outpace increases in population during the decades ahead. The early returns from the Green Revolution in Asia suggest that food may not be the first limiting condition that fixes a ceiling on human survival, but rather the human output of waste and poison. That is, we may bury ourselves in garbage and contaminate our atmosphere before we run out of enough to eat. Of course, large-scale hunger and malnutrition currently afflict large numbers of people even in a country as affluent as the United States, but these food shortages are a consequence of poverty and failures of distribution and do not indicate an overburdened agricultural capacity. As is well known, the United States Government each year pays its farmers several billion dollars to withhold acreage from productive use.

Mankind is passing through the early stages of its first planetary crisis. The interrelated dimensions of this crisis are population pressure, multiple forms of pollution, resource depletion, and the danger of wars of mass destruction. It is the technological character of contemporary society that gives the planetary crisis its apocalyptic character.

The crisis is the outcome of processes associated with deeply entrenched behavior: the tendency toward families of more than two children, the success of mass medicine, the continuing pursuit of a production-oriented technology, and the identification of progress with a larger per capita GNP. Such patterns of behavior suited the situation of man under "pioneer" or "frontier" conditions, although the costs for displaced peoples were always high, sometimes total. The environment could absorb more people, the establishment of consumer industry brought a more comfortable life to many, the consequences of environmental abuse were local or could be absorbed in time by nature, the supply of resources seemed limitless. For a pioneer society of abundance the encouragement of individual initiative and competition seems to produce good social results: a dynamic and powerful economic order that rewards its members with individual wealth and collective power. Qualities that are virtues under pioneer or primitive conditions become vices under contemporary circumstances where technological modes of operation impinge on the fundamental capacity of the environment to provide the oxygen, space, tranquillity, and food to keep mankind at a decent level of existence, and perhaps in being at all.

The technology of the present, especially its ultra-toxic dimensions, also threatens to produce accidents of great scale and scope. A shipment of deadly nerve gas across the country by train to be dumped into the oceans symbolizes the hazards that surround our lives. The weaponry of man now threatens to overwhelm the planetary environment once and for all. But beyond this the effects of heat and noise and contamination are such that conditions of human habitation are being subjected to greater burdens and risks, many of them unknown or imperfectly known, all the time. It is not possible to reestablish the conditions for human survival without transforming, at the same time, some very basic human attitudes toward life and nature. At present, we contaminate the world by our morality as well as by engine fumes. This interconnectedness of spiritual and physical survival has been vividly expressed by Claude Lévi-Strauss's comparison between the behavior of men and the behavior of maggots in a sack of flour: "When the population of these worms increases, . . . even before they meet, they become conscious of one another, they secrete certain toxins that kill at a distance—that is, they poison the flour they are in, and they die. I think what's happening on a human scale is a little the same sort of thing. We are secreting psychological and moral toxins."

In these circumstances, we need to establish the overall conditions of planetary survival and amelioration, including the establishment of upper limits and optimum numbers for GNP and population. It is not enough to prevent Armageddon. We must reverse the processes that are consistently operating to impoverish our existence in the name of progress.

To reverse these processes requires a momentous effort of education and persuasion. At the present time the poorer, most crowded countries of the world are organizing their economies around the paramount goal of achieving as rapid GNP growth as is possible. Such growth can only be achieved by industrialization in the Western style. The poor and populous countries of Asia, Africa, and Latin America look toward the advanced technological societies as providing the model of how to lift themselves out of misery and impotence. But it is just this model of man-over-nature that has contributed so greatly to the creation of a planetary crisis. Jean Mayer, President Nixon's top adviser on food problems, has made this point in striking fashion: "It might be bad in China with 700 million poor people, but 700 million very rich Chinese would wreck China in no time. It's the spread of wealth that threatens the environment." Even in the advanced countries the model of development continues to be a growth model measured in GNP per capita terms. Very few countries publish distribution figures that would alone enable an appraisal of which groups benefit from GNP growth. Such a model entails escalating demands for resources, more waste, more sources of pollution, and even greater consequences attaching to the hazards of technological accident.

These problems are accentuated by the competitive dynamics that underlie the organization of the national economic system. Whether the yardstick is profits, as it is in capitalist societies, or quotas, as it is in socialist societies, the reward system is structured around self-sustained growth and maximum production. A market economy may induce more attention to quality, whereas a planned economy may avoid the creation and satisfaction of spurious wants by advertising and promotion, but in both situations it is the fulfillment of production standards as measured by output statistics that shapes the evaluation of economic performance. Conservation considerations may enter into the web of relationships where one set of uses tends to upset another set or where the particular use is causally associated with a particular situation of distress. In this spirit, it has been proposed that a large investment in the culture of oysters for the Atlantic coast would not only contribute to food supply but would "provide the best possible deterrent against pollution, since the first threat of damage to the pollution-sensitive oyster industry would be immediately translated into political action!" Such a strategy of development involves creating interest groups whose profits depend on protecting the quality of the environment. Stocks of companies manufacturing anti-pollution equipment were featured for growth potential by the end of 1969. Moody's Investment Service has issued a booklet advising potential investors of $100 billion worth of business opportunities in pollution control that lie ahead. The political logic of making the system responsive to survival considerations is, of course, a practical way of guarding against some of the most manifest forms of environmental abuse.

But it is not possible, especially in the international context, to come to terms with the challenge by such superficial forms of responses. A large portion of world society is faced with immediate problems created by poverty and population growth in a political atmosphere dominated by a growing sense of deprivation and desperation. Just as public health has managed to achieve a steady decline in the death rates in these countries, so modern communications have created an awareness of the poverty gap between poorer and richer societies. National leaders, to acquire and retain political power, are led to promise their populations an improving economic future. To carry out this promise the goal of the leaders of these countries has been almost invariably to follow as rapidly as possible the path of "development" and "modernization." The ideal has been to achieve a high rate of annual growth in GNP at any cost, with the eventual objective of reaching "take-off," a point where an economy generates enough surplus to sustain its own growth indefinitely. This goal takes precedence over any competing considerations, except possibly a concern with military security and international prestige. To preach ecological prudence to an Asian or an African leader would be tantamount to advising Mao Tse-tung to invest in the stock market.

Rich countries with a per capita income 10 to 50 times as great as that prevailing in the poor countries are in no position, even if so inclined, to preach economic self-sacrifice to achieve a better equilibrium between mankind and the environment. For one thing, the productive success of the rich countries has followed from the full development of a modern economy of scale, forever improving the ratio between human input and material output. For another, this success has been associated with an ideology or world view that identified human progress largely with uninhibited mastery over the natural environment. Such a conception of progress has been thought to have universal application as the natural means by which a government provides its population with the fruits of the modern world. Besides, the long and humiliating servitude experienced by Afro-Asian peoples has been correctly interpreted by their leaders as a consequence of a failure to achieve the industrial base needed to provide these societies with adequate fighting capabilities. The rise of the Soviet Union to world eminence has also been comprehended as a consequence of its ability to build up a base of heavy industry in a short period of time. Even China's breakthrough to a position of world power has been generally understood to result from the success of its nuclear weapons program, which also demonstrated a competence in the most modern techniques of science and industry. There have been exceptions—that is, groups who were thriving until displaced by the avalanche of industrial progress. The American Indians are among the most poignant victims of "progress." As expressed by Black Elk, an Oglala holy man, we can grasp the plight of these dispossessed people everywhere: "Once we were happy in our own country and we were seldom hungry, for then the two-leggeds and the four-leggeds lived together like relatives, and there was plenty for them and for us. . . . But the Wasichus [white men] came, and they have made little islands for us and always these islands are becoming smaller, for around them surges the gnawing flood of the Wasichu; and it is dirty with lies and greed. . . ."

Throughout the world, then, there is a general acceptance of the Western creed that national goals should be measured in the material terms of *developmental thinking*. Benevolent Western advice to the poorer countries tends to reinforce GNP-mindedness. For instance, an economist and India specialist, John P. Lewis, analyzes India's future almost totally in terms of whether its annual rate of economic growth is 4½ percent, 5½ percent, or 7 percent; "a relevant radicalism" for India involves adopting the policies and developing the organizing skills needed to achieve a 7 percent annual rate of GNP increase. Although Professor Lewis is sensitive to the question "growth for what," nevertheless the whole style of analysis is oblivious to the ecological chaos that would result if this kind of radicalism succeeded in India and elsewhere. There are certain isolated resistances evident —for instance, the effort of Tanzania under its President, Julius Nyerere, to carry out the postulates of the Arusha Doctrine, which calls for economic goals supportive of Africa's traditions and the quest for self-reliance and autonomy, emphasizing agriculture, economic austerity, national service by the elite, and small socialist communities, rather than seeking the buildup of heavy industry and big cities.

But the main momentum throughout the poorer countries is to achieve an increasing rate of growth in GNP by creating an underlying base of heavy industry. This strategy calls for emulating the richer countries and is already beginning to produce the same kind of environmental deterioration as is associated with the advanced industrial societies of the West. Jean Mayer has made the point effectively: "The ecology of the earth—its streams, woods, animals—can accommodate itself better to a rising *poor* population than to a rising *rich* population." The very goals of the poorer countries—those where the population increases are the greatest—involve a tremendous effort to enter the modern world, such entry being signaled by geometric aggravation of the conditions that cause the planetary crisis in the first place. The irony of development is that to the extent that it succeeds, the world situation worsens.

Primitive conditions of poverty and preindustrial development may cause great hardship for the social groups involved, but their effects on the planetary crisis are of marginal significance. Overpopulation becomes a world problem most alarmingly to the extent that poverty and famine are averted. The pressure on the environment increases at a rate that is directly related to, but more rapid than, increases in the level of industrial development and in the standard of living of the population.

On further reflection, however, such a view is not entirely correct. The failure to achieve developmental progress in a situation of aroused public demand might be likely to produce more extremist kinds of political strategies. A specialist on the causes of violence, Ted Gurr, has convincingly demonstrated the correlation between *perceived relative deprivation* and the propensity of a society to experience civil disorder. When this correlation is considered in a world of increasing ease of access to nuclear technology and weapons of mass destruction, then one consequence of remaining poor and primitive is to augment the dangers of large-scale war.

We are confronted, then, with a fundamental dilemma: *success* in industrializing the poor countries is likely to result in less poverty and turmoil, but in a rapidly and possibly decisive worsening of the ecological situation, whereas the *failure* to industrialize these countries is likely to generate political behavior that would be likely to increase risks of general war.

In one sense, then, a major political challenge is to discover the means by which misery and poverty can be

alleviated without, at the same time, provoking an ecological collapse. It would appear clear that the present
structure of international society is totally unable to deal
with this dilemma. Principal sovereign states are, themselves, unable to satisfy the rising demand schedules of
their own populations except by continuing down the path
toward ecological disaster. An economy is called "stag-
nant" as soon as it loses the capacity for growth. A stag-
nant economy is a sign of weakness and distress in every
economic system, and there is everywhere in the world
an insistence upon overall growth. The preoccupation of
affluent societies with improving their own output of
goods and services indicates the very limited extent to
which transfer of wealth and resources can be expected to
alleviate mass poverty. Even the one percent of GNP, the
level of foreign aid recommended by the UN, has proved
to be an unrealistically heavy burden to impose on rich
countries for assistance to the poorer parts of the world.
Socialist societies refuse to give this level of aid, often
arguing that there is no reason why they should be made
to pay for the economic distress that was sustained in Asia
and Africa by the colonial system of export capitalism.
The needs of minimum world community welfare cannot
be met by external capabilities, although in circumstances
of emergency, food and aid can diminish or even avert
disaster.

Ecological considerations call for a concerted effort to
establish an equilibrium between man and nature. The
minimum requirement of such an equilibrium is some
kind of ratio between technology and society enabling
enough goods to be produced to satisfy the needs of
everyone without resorting to a scale or modes of production that deteriorate the quality of the environment, including its resource base. This requirement cannot be met
without rigorous planning, including a crash program in
population control. Also, the Gross Planetary Product
(GPP) must be *constituted* in such a way as to reduce

greatly the *waste* of resources through *superfluous private consumption*, through *failures to recycle* used materials, and through *public expenditure* unrelated to the satisfaction of *human needs*. A needs-conception entails the elimination of national security as the basis of world order and the imposition of restraints upon production in free-enterprise economies. In addition, major *distributive* reforms will have to be introduced so that the basic needs of poorer populations can be satisfied without requiring further expansions of the world industrial base. And greater proportions of resources will have to be devoted to restoring an ecological balance between man and nature with regard to both resource management and technological innovation.

Such an ecological imperative cannot be realized without far-reaching political changes in the structure of world society. It seems naive to suppose that the ecological pressures of the affluent societies can be reduced without relieving the social and economic pressures of the impoverished societies, and vice versa. There is a kind of interrelatedness that points toward unified guidance for the benefit of all men as the only basis upon which life on earth can safely endure. Such guidance need not take the form of drastic centralization of power and authority. In fact, pluralism, zones of autonomy, diversity of life-style, culture, and ideology are all values that could be emphasized in a system of world order suitable for the ecological age. The essential modification of the world system as it operates today would involve the elimination of the national boundary as the basic organizing idea for purposes of security, wealth, and loyalty. If the boundaries between sovereign states could become as inconsequential as the boundaries between Connecticut and Massachusetts, Hunan and Hupeh, Rajasthan and Uttar Pradesh, and the Uzbek and Kazakh, then the new era will have emerged. There is no reason to work toward the elimination of the nation or state. The basic need is to find ways to deprive the national boundary of its mystique, and thereby of its capacity to cut mankind off from lines of action that could assure human survival and even promote human welfare. Perhaps the ecological angle of vision—the earth as seen in the pictures sent back by astronauts on their way to the moon—can serve to create a new political consciousness in the rising generation of man. Perhaps, even, Gene Marine, a provocative journalist, is correct when he writes that "the first public figure with a national audience who seizes on the ecological conscience as his subject and refuses to let go, who insists on the demonstration that *all* our other problems are related to it, who refuses to avoid its implications (for he will have to attack some mighty profit makers, among other things), will find, to the astonishment of many, that there are millions waiting for him."

III. Underlying Causes of Planetary Danger

The planetary crisis is an outgrowth of certain basic patterns of human behavior and political organization. These patterns need to be clearly understood before we can even begin to work out a better future for the human race. At this stage, groping for understanding is the necessary prelude to constructive action. Our point of departure is not a happy one. Despite the marvels of the modern world, there is an increasing realization that mankind has strayed from the pursuit of its own welfare, perhaps decisively, with little prospect of recovery and a growing appreciation of urgency. The French poet, Paul Valéry, put the situation very clearly some years ago: ". . . we are blind and impotent, yet armed with knowledge and power, in a world we have organized and equipped, and whose inextricable complexity we now dread." We seem entrapped in this central paradox of unprecedented knowledge and power on the one side and of unprecedented helplessness and danger on the other side.

The State System: A world of sovereign states is unable

to cope with endangered-planet problems. Each government is mainly concerned with the pursuit of national goals. These goals are defined in relation to economic growth, political stability, and international prestige. The political logic of nationalism generates a system of international relations that is dominated by conflict and competition. Such a system exhibits only a modest capacity for international cooperation and coordination. The distribution of power and authority, as well as the organization of human effort, is overwhelmingly guided by the selfish drives of nations.

The significance of nationalism is clearly revealed by the inability even of states closely aligned by ideology and objective to coordinate their behavior sufficiently to evolve common policies that promote the relevant community of states. The Sino-Soviet split is one prominent example of the persisting precedence of nationalism in world affairs. The slow progress toward regional integration in various parts of the world, despite the considerable benefits in the form of economies of scale, more effective security communities, and enlarged areas of free trade, is a further indication of the strength of the sovereign state. Nationalism is even more evident in the Third World, where the effort to overcome subnational tribal and ethnic divisions so as to build a modern state depends on the potency of symbols of nationhood and sovereignty.

Sovereign states can sometimes cooperate in situations where *perceived* common interests exist. The idea of an alliance against a common enemy is by far the most successful form of international cooperation. We need only to compare the great effort and resources invested by the United States in the North Atlantic Treaty Organization or the Soviet Union in the Warsaw Pact with the trivial level of attention given to the work of the United Nations to grasp the extent to which international cooperation is itself dominated by ideas of competition and conflict. The vitality of international cooperation against a common enemy has long prompted science-fiction writers to invent stories about world government arising in response to a credible threat of aggression from a distant planet. Writing in 1928, André Maurois even went so far as to conjure up a conspiracy among world leaders consisting of a world-government movement. The essence of the conspiracy was to use the mass media to convince people everywhere that the earth was threatened by an invasion from the moon. In the Age of Apollo it would probably be necessary to move the enemy farther off in space than the moon, but the point holds. One can imagine the countries of the earth banding together in a single political unit to stave off extraterrestrial aggression, but under almost no other circumstances. Maurois ended his story on an ironic note, with the discovery that there really was an extraterrestrial enemy ready and willing to attack the earth from the moon. For the first time the people of the earth are confronted by a common danger, but it remains doubtful as to whether a situation as diffuse and abstract as the ecological crisis can arouse the political imagination sufficiently to serve as the equivalent of a personified enemy aggressor. Somehow the history of man is so bloody and internecine that only a personified enemy seems able to produce fear and a sense of emergency. In this respect, the earlier mentality of man in which the forces of nature were identified with (and personified by) gods was much better suited to waging a war of planetary survival. The rational consciousness of modern man has emptied nature of personality, and has deprived man of any sense of personal kinship with nature. The rejection of pantheism, animism, and polytheism makes it more difficult to engage the survival instincts of man in this struggle to rehabilitate life on earth.

Except for cooperation against an enemy, states have been able to cooperate effectively only on matters of nonvital concern, such as postal service, safety at sea, telecommunications, the extension of commercial credit, and the like. In many areas of international life international law has played a positive and important role by creating an orderly framework for routine competition and rivalry, but such a role is to be distinguished from achieving common policies on common problems. In the area of endangered-planet concerns there has been some effort to formulate a world interest in relation to the prevention of war, the resistance to population growth, and the curtailment of environmental pollution. But the world interest to date has been formulated only in highly general terms designed neither to prohibit dangerous behavior nor to implement positive goals. The formation of a world moral consensus and awareness is part of a move toward understanding, but by itself it is a feeble development considering the seriousness of the situation and the urgency of the need to begin curative action now.

And here is where sovereignty as the organizational basis of world society burdens the struggle for human survival. States have priorities distinct from one another that lead them to perceive the issues of the endangered planet in very diverse ways. As a consequence, it is virtually impossible to obtain agreement even on an agenda of concerns. National governments formulate *planetary priorities* to reflect the ranking and character of *national priorities.* Diversities of power, wealth, ideology, and history create the basic diversity of outlook on the part of national governments. For most governments, especially those with mass poverty, the primary concern is to raise GNP at a satisfactory rate and to secure internal security in relation to rebellion and external security in relation to potential aggressors. Any other concern seems remote and may be viewed with the suspicion that it is nothing but a malicious distraction from the business of the day. I recall a conversation in late 1969 with some African intellectuals and community leaders in Uganda who listened to me in polite amazement while I tried to discuss with them the dangers of planetary survival. Even to intelligent Africans and Asians, the concerns of this book reduce to problems of pollution which are the special and somewhat exotic and ephemeral concerns of the industrial societies and have no relevance whatsoever to the Third World emphasis upon securing a tolerable basis of existence for their various ill-housed, ill-fed, and ill-treated populations. There is, of course, some appreciation of the fact that a rising population may be a national detriment, but not because of the ecological hazard; rather increases in population prevent increases in GNP from being translated into higher standards of living. Even here, however, the appreciation is far from universal, as many poor countries view an expanding population as a source of pride and potency. The Government of Thailand, for instance, seeks as a matter of national policy to double its present population within a decade.

In addition, many states are either generally or periodically absorbed in intense conflicts that involve struggles with neighboring states for control over disputed territory. A special problem of the present world concerns the struggle for the reunification of the divided countries of China, Germany, Korea, and Vietnam; African states are concerned with the elimination of colonialism and racism from southern Africa. These immediate issues of foreign policy have a concreteness and relevance for national politics that the more abstract and intangible issues of the endangered planet do not yet possess. Try talking to an Arab or Israeli about the importance of world order!

The danger of nuclear war has, it is true, generated widespread apprehension throughout the world and a strongly perceived common interest in its prevention. But again it is a reflection of the state system to compare the resources and energies invested in the arms race with those invested in arms control, much less disarmament. Each year the Arms Control and Disarmament Agency has to fight for its budget of a few million dollars in a closely divided Congress, while the defense budget in the United States has risen at one point to over $80 billion per year during a period of "world peace"; in other words, in a society that correlates relative expenditure closely with relative value, it is worth noting that military competition receives thousands of times more resources than does the work of international cooperation. The ratio is even wider if personnel or man-hours or media coverage are compared. We live in a world organized around competitive pressures, with success associated with competitive superiority. And yet the objective condition is such that we need to be able to shift the resources used for competitive purposes to secure cooperative ends if we are to have any hope of dealing with the problems of an endangered planet. These problems call for central planning and coordination, for the sort of effective implementation that both signals and enables a real shift in patterns of behavior. Neither by disposition, habit, nor organization is a world of sovereign states constituted to deal successfully with the agenda of the endangered planet.

The role of sovereign states is rendered more dangerous by sheer size. Leopold Kohr, a rather eccentric but creative political scientist, has argued that many of mankind's most severe problems arise from the creation of large-sized political units; his position has an ancient heritage, going back to the writings of Plato and Aristotle. The requirements of technology now make it impossible for any but the largest national economies to make the latest weapons systems, to participate in the exploitation of deep-sea resources, or investigate outer space. Technological determinism combines with the ideology of sovereignty to produce very tightly controlled large-sized states that are the dominant actors on the world scene. It is these few huge states, because of the bureaucratic impersonalism that almost necessarily results, and the heterogeneity of internal population groups that are likely to do the most harm to their own societies and to the rest of the world. Unity is artificially or coercively sustained and tends to produce a continuous atmosphere of danger and threat. There is no natural base of governmental legitimacy, and so abstract goals of power, wealth, and welfare are emphasized to secure a semblance of popular support. All large states are multinational in makeup, with some nationalities subordinated and others dominant. Such a potentiality for internal breakdown is averted only by a combination of an exceedingly strong internal police control and by invoking the advantages of scale that may be decisive if progress is measured by technological criteria of GNP, military prowess, and space activity. The United States, the Soviet Union, and China each in a distinct way manifests this internal vulnerability to crisis and collapse. The leadership needs to mobilize some mass base of support through exaggerating external dangers, imposing rigid internal controls, and adhering to a simplistic ideology that explains away problems as combat between forces of good and evil. The United States, by virtue of its advantages of position and wealth and by the considerable energy of its economy, has managed to engender a relatively high degree of loyalty from its population without often resorting to overt coercion. But the strains of technological society are beginning to widen the fissures in the social and political fabric, which in turn cause widespread alienation from the values of the ruling groups and stimulate revolutionary and counterrevolutionary pressures to mount. The People's Park episode in Berkeley, California, the 1968 Democratic Party Convention in Chicago, huge rock music festivals, ubiquitous displays of the flag on cars and homes and lapels, and the rallying of forces for and against the Black Panther Party are typical battlegrounds of our times in the United States.

Smaller states that have dangerous and irresponsible governments can do harm to their own populations, but the results of this harm are likely to be of shorter duration and more restricted effect. Also the foreign policies of such governments are not likely to jeopardize world peace or human survival, except to the extent that big states intervene in their affairs. Francois Duvalier in Haiti and Ian Smith in Rhodesia are governing their populations in a manner that most of the human race finds reprehensible, but such states are hardly dangerous outside their own boundaries. The same insulation of external society does not occur if destructive dreams of expanding power dominate the actions of leaders in more powerful states. It is the *scale* of statehood, as well as its primacy in the organization of the world, that makes the persistence of sovereignty as ideology and organizational form so dangerous for the future of mankind. In Kurt Vonnegut's novelistic account of the destruction of Dresden, *Slaughterhouse-Five,* Billy Pilgrim, the book's hero, travels his imaginary path from New York to Illinois to give a speech on flying saucers: "He has to cross three international boundaries in order to reach Chicago. The United States of America has been Balkanized, has been divided into twenty petty nations so that it will never again be a threat to world peace." Such a fantasy of a Balkanized world does not tell us how it may be achieved, but it does dramatize the relevance of *size* to the extent and depth of the planetary crisis. Size and technological capacity are directly correlated and the scale of technology is very directly related to the seriousness of the ecological imbalance. Size is not a sufficient explanation, but it is a critical factor that is often neglected in discussions of the state system. We might be able to survive indefinitely in a world constituted only by Haiti- or Denmark-type states, but in a world that includes one or more United States- or Soviet-type states the survival of life on the planet is imperiled. As World War II suggested, Germany and Japan were of sufficient size to embroil the entire world in a destructive war. Given the increasing ease with which weapons of mass destruction may be produced, the image of tolerable size may have to be diminished considerably, and may almost grow obsolete. Note also that *distance* has diminished as a *size-limiting* factor because missiles, supersonic aircraft, and satellites can move their destructive capabilities over great distances without much loss of strength; communications links can be maintained at remote places by computerized patterns of administrative control. The Intercontinental Ballistic Missile (ICBM) is, perhaps, the prime expression of the shrinkage of distance and the contraction of time in the modern world of political conflict.

The situation is made even more complicated by the extent to which national governments are unable to pursue even their own self-interest as a consequence of distortions of perception and policy caused by domestic economic, political, and social forces. We have become familiar with the extent to which military spending creates a domestic momentum of its own, partly sustained by the matching behavior of adversary states, but essentially detached from an analysis of the best and most reliable defense system operated at the lowest risk and with the minimum investment. Governments allocate a sector of their budget to military preparedness, and as the GNP increases so does the level of military spending; labor, business, professional military, and governmental interests are all tied into the maintenance of the arms race. Images of the world are adopted to support a conception of domestic fulfillment: power, prestige, and wealth. The makers of foreign policy are severely constrained by the belief-systems and vested interests of the domestic political leadership. These constraints are not very often coordinated with real needs, as these needs would be understood by disinterested third parties. The rationality of the arms race is a reflection of the domestic underpinning of principal governments to a far greater extent than it is an expression of the competitive dynamics of the state system. It is essential to grasp these domestic sources of foreign policy so as to understand the conditions of its change.

The paradox of aggregation arises out of the conflict between individual benefit and collective benefit. This conflict is present in many critical areas of human choice and has a peculiar relevance to the circumstances of the endangered planet. The paradox of aggregation is relevant whenever an actor has a schedule of priorities that is inconsistent with the priority schedule of the community. In circumstances of abundance—of space, time, resources—the operation of the paradox of aggregation is of far less social and political consequence.

A corporate manager or shareholder is primarily interested in securing a maximum profit from his industrial operations. To achieve maximum profitability often depends on having a cheap way to get rid of waste by discharging it either into the atmosphere or into water. For a long period of time no sense of conflict emerges. The environment seems able to absorb the waste products without suffering general deterioration. Local pollution hazards may result, and may lead to a struggle between the community and the industrial firm. Already at the end of the nineteenth century Henrik Ibsen, the Norwegian playwright, explored the ferocious dynamics of this kind of conflict in *An Enemy of the People,* showing the extent to which the politics and perceived "welfare" of a community become identified with the selfish interests of its richest members, but also with those of poorer people whose livelihood was dependent on the profitable operation of the polluting tannery. It is very difficult to persuade those who destroy the environment to take serious protective action if profits are diminished, especially if anti-pollution methods of production put a particular firm at a disadvantage vis-à-vis its competitors. Such difficulty has been illustrated on the American scene by Ralph Nader's campaign to induce the auto industry to install safety devices in cars and by the inability of the Santa Barbara, California, community to prevent offshore drilling for oil. A material culture generally gives priority to considerations of profitability and attendant disasters tend to be viewed as the unavoidable costs of industrial progress. It is worth noticing the contrast between the *cost-efficiency* approach to safety that is applied to factory or mine conditions and the *absolute* approach to safety as it is applied to the Apollo Program; no cost is too great if it is justified in terms of protecting or rescuing astronauts. Progress is measured, as has been noted, in terms of GNP and GNP per capita. A typical defense of this outlook has been made by an economist, C. E. Ayers:

> Industrial society is the most successful way of life mankind has ever known. Not only do our people eat better, sleep better, live in more comfortable dwellings, get around more and in far greater comfort, and—notwithstanding all the manifold dangers in the industrial way of life—live longer than ever done before. Our people are better informed than ever before. At the height of the technological revolution we are now living in a golden age of scientific enlightenment and artistic achievement.

Such an outlook is resistant to information about the mounting dangers to the planet that arise out of the very processes described as beneficial, and tends to be confident about the capacity of technology to overcome the hazards and harms of technology. I suppose there is reason to be encouraged that doubts about economic progress have even penetrated the White House. In his 1970 State of the Union Address President Nixon said, "Now I realize that the argument is often made that there is a fundamental contradiction between economic growth and the quality of life, so that to have one we must forsake the other. *The answer is not to abandon growth but to redirect it*" [my emphasis]. But note the persisting belief that nothing needs to be given up, no sacrifices endured, no interruption of the constantly increasing flow of goods even to the affluent. And, of course, there is no specific analysis of who has caused the decay of the environment and for what reasons and who shall bear the main burdens of its protection and rehabilitation.

The paradox of aggregation is a subtle and fundamental social process. An imaginative and provocative biologist, Garrett Hardin, who happens to live in Santa Barbara, has evolved an effective metaphor of this process from a historical experience, the destruction of the common pastures of English country towns in the 1700's and 1800's through overgrazing herds. "The tragedy of the commons" occurs because each farmer calculates his own advantage by reference to the enlargement of his own herd. The gain that results from each additional animal added to the herd is a definite increment to the farmer's wealth and profits, whereas the decision to forgo expansions of the herd in view of the limited grazing capacities of the commons is ineffectual to prevent other less conscientious farmers from adding to their herds. When the sum of the separate herds pushes up against the carrying capacity of the land, then the logic of the paradox of aggregation takes over. It does not help much even to crystallize the community interest and appeal to the conscience of farmers or to demonstrate that only by population control can their collective longer-term interests be protected. Unless the appeal is uniformly successful, a very unlikely outcome as it requires farmers of unequal wealth to forgo immediate private gain, then its results are merely to reward the less socially responsible members of the community at the expense of the more socially responsible. In this kind of setting, the temptation to maintain or enlarge one's own share of the pie is difficult to resist. Inequality is important, as it tends to vindicate a position of less restraint by poorer farmers and a consequent lessening of restraint by all relevant landholders. The typical outcome in such a situation then is, as with the commons, the exhaustion of the common resource by overuse to the detriment of *all* users. Only a common plan of conservation that allocates quotas and effectively punishes violations can hope to protect the collective interests under conditions of scarcity.

As Hardin argues, the deepest problems of the endangered planet arise out of similar conflicts between the pursuit of private gain and the protection of the public good. Decisions as to family size, or as to the use of insecticides or detergents, or as to resource extraction, all involve the exercise of unguided discretion in a context where the public good requires restraint and community control. Even where the public good has been carefully and clearly defined there exists a great deal of pressure to neglect its relevance in the definition of private good. My decision to have two, three, or six children is so insignificant in relation to the total world population as to have no bearing, nor does my use or nonuse of DDT have any discernible effect on the prospects for the survival of hawks and pelicans. Under such circumstances why shouldn't I have as many children as I want to have or protect my garden from the annoying presence of insect pests? To add or subtract a gram of sand from the beach is a decision without social importance. The paradox of aggregation, especially in societies whose mythology exalts a cult of individual initiative, gives an ethical reinforcement to patterns of action that involve the satisfaction of private wants.

To protect public concerns is the task of government. Within the national setting it is exceedingly difficult to use coercive means to uphold community standards of welfare. In a country as generally sensitive to issues of public policy as is the United States there is no agreed-upon optimum population for the society, much less a strategy for its attainment. There is not even much appreciation of the fact that a society may exceed its optimum population, even when it can feed everyone, or that most ecologically minded experts are in agreement that the present American population is too large for the United States and that the present world population is too large, given the capacities of the global ecosystem to sustain life. The notion of "the invisible hand" guiding private action toward the automatic fulfillment of the public good and the belief in a laissez-faire role for government inhibit collective action that regulates individual choice on issues as intimate as family size. In the area of population policy the role of regulation has been further inhibited by the opposition of the Catholic Church to any artificial interference with the human reproductive process. More recently, in the United States, blacks have even asserted that efforts to limit population increase amount to genocide. Underlying such an assertion is the fact that a group's degree of power is often associated with the proportion of the total population that it represents. An increasing birthrate for blacks relative to whites assures a greater power base and a lower relative birthrate would lead to less power for blacks. Therefore, if access to power is the dominant objective, encouragement of a high birthrate for blacks would be the most rational kind of population policy to advocate. Obviously, a similar logic would apply to ethnic groups eager to deny blacks any greater influence within the society.

Matters of pollution and resource conservation exhibit the same general conflict between private and public interests. The conflict is obscured so long as the pollutants are absorbed at a rapid enough rate so that environmental deterioration does not visibly occur, except as some sort of local aberration—as might result from the negligent discharge of some very toxic waste substance. Similarly, if resources are used at a rate that is slower than the rate of discovery or replacement, then it becomes difficult to make a case for a stringent conservation policy that would hamper economic growth. Basically, the individual actor is concerned with private gain and is not often inclined to sacrifice private gain for the sake of public good, or to defer benefits so as to uphold the patrimony of future generations.

There are many kinds of rationalizations for widespread indifference to these issues, even aside from callousness and greed. In discussions with socially sensitive friends I have noticed a general disposition to assume that the problems are not so serious, that the dangers are being exaggerated, and that the issues are too vast to think about. If pressed, many people seem to assume that "something will turn up," that technology can solve the problems that technology creates, that earlier warnings about resource shortages have always been exaggerated. One sophisticated exponent of this view is Alvin Weinberg, the Director of the Oak Ridge Laboratory, a major center of research and development for nuclear energy, who has written as follows:

> To me then, the job and purpose of science and technology remain overwhelming: to create a more livable world, to restore man to a state of balance with his environment, to resolve the remaining elementary and primitive suffering of man—hunger, disease, poverty, and war. These are not small tasks, nor are they new ones; that in science and technology we have the possibility of dealing with them is an article of faith of all who have committed themselves to the scientific way of life. . . . It is up to us, members of the older scientific-technological establishment, to persuade our younger impatient scientific nihilists that ours is the course of reason, and that in our arduously built scientific-technological tradition lies our best chance of ultimate survival.

Such a view is now repudiated by many scientists who are alarmed themselves by the *hubris* and insensitivity of such scientism, which is dazzled by the achievements of science without being mindful of the extent to which the sum of these achievements may equal disaster, and may, in any event, produce a mind-clenching impersonalism that robs human beings of their dignity and sense of individuality.

A political variant of this confidence in the capacity of science to turn up solutions at the brink of disaster—that is, when the problems grow really serious—is the feeling that the government will step in effectively to uphold the public welfare before things get completely out of hand and a breakdown occurs. We feel so helpless in the face of the awesomeness of these issues that we base our sense of well-being on the capacity and willingness of our governors to rescue us before the situation deteriorates beyond recall. But why? The history of government is one long record of a failure by political leaders to take curative action in time to safeguard an endangered society or civilization. Political leaders are often beholden to the wrongdoers and are certainly hemmed in by an assortment of pressures, interests, and traditions; resistance to change is fierce, especially on the part of those who expect to bear its main burdens. Today, the scale of danger is greater and the failure by principal industrial societies to make adjustments imperils not only themselves but all forms of life everywhere on the globe.

Technological solutions are also not very effective in circumstances where the imbalance between private interests and public good is considerable. Birth control technology, even if freely distributed, is an extension of an appeal to conscience and to an enlightened form of self-interest. It works only to the extent that public welfare standards have been internalized in the minds of the people. So long as rural Indian women want, on the average, between four and five children, there is no hope of solving that country's population problem by technical apparatus. Improved technique for birth control and its widespread, free distribution might slow down increases in the population but it will not produce a stable or declining population, at least not for several generations. If population levels are already close to their sustainable limits such an approach is clearly deficient. This deficiency is accentuated, not alleviated, if it is probable, as seems now to be the case, that food supply is not likely to be the limiting condition on population growth during the decades ahead.

Here again a common pattern of reasoning and behavior is illustrated that expresses a facet of the paradox of aggregation: if gain is renounced to prevent public detriment—for instance, arms races among secondary and tertiary states—then the public goal is not served, but the private loss is sustained. Ocean resources provide still another illustration. As long as some mariners kill the endangered blue whales for gain, or some other endangered species, then there is little incentive for others to forgo profit by respecting world conservation policies. Unless virtually *all* whalers are effectively regulated the paradox of aggregation induces even the more responsible whalers to ignore standards that would alone protect the long-term whaling interest. Otherwise restraint results in enriching the least scrupulous competitors. To prevent this result tends to make *all* arms suppliers or whalers or polluters irresponsible, and it is this tendency that makes it so difficult to reverse ecological gears.

Appeals to governments based on the UN Charter appear about as futile as appeals to individuals to forgo children. A government's insistence upon discretion to pursue its own interests is deep-seated and is not normally responsive to rational criticisms or suggestion by others. Occasionally, one sees a glimmer of an ethical imperative that qualifies, to some extent, the hopelessness of the situation. For instance, Stephanie Mills caused something of a temporary and local stir when as valedictorian of Mills College in June of 1969 she announced her decision not to have children of her own to emphasize her convictions about overpopulation. Such examples of personal action may be very significant, not in showing us a model solution, but in demonstrating the gravity of the problem and in initiating a wider chain of thought and action about the issue. But such exemplary behavior tends to be absorbed in the larger continuities that maintain the patterns of the past and, in any event, at best ignites a very slow-burning fire.

As Secretary of Defense, Robert McNamara, in the Posture Statement of the Department of Defense for 1968-69, explained that the United States was earning valuable foreign exchange through its program of arms sales to foreign countries. Besides helping to correct an imbalance of payments, Secretary McNamara pointed out that if the United States were to give up this lucrative trade, other, even less responsible states, would take over our share of the market.

The international setting is even more vulnerable to the impacts of the paradox of aggregation than are national societies. Within international society there are almost no effective mechanisms to implement or even to perceive world interests. The one effective mechanism is probably war, but war itself is a major source of harm and danger. Besides, war as an instrument of national policy arises out of conflicting interests among nations rather than out of some struggle between those undermining and those supporting planetary welfare. Wars arise over conflicting national claims about territory, resources, or political control or over efforts to maintain or establish some kind of "balance" of opposed power groupings in some sector of international society. Efforts to build up a community interest in the maintenance of peace and the renunciation of force have not been successful, even in light of the moral consensus that emerged out of the experience of World Wars I and II and out of the understanding that nuclear war would be a mutually destructive and intolerable means to resolve future conflicts between nations. The League of Nations and the United Nations, although operating within a framework of rules and procedures calling for collective action by states in opposition to any international aggressor, have themselves exhibited the persisting dominance of rivalry, alliances, and the competitive ethic. In these circumstances, the community norms and rhetoric embodied in the Charter of the United Nations strike most people as a sham and the Organization is held in disrepute as little more than "a debating society." I have heard the experienced diplomat George Ball, who actually served as the American Ambassador to the UN for a short time, say that big governments played only "background music" at the United Nations while the real action took place elsewhere, namely, through the hard government-to-government bargaining of traditional diplomacy.

In conclusion, then, there are powerful motivations underlying exploitative activity. The aggregate effect of this activity is to generate a multi-faceted ecological crisis of planetary proportions. A response to this crisis requires more than a formulation of world interests and an appeal for voluntary compliance. An adequate response requires a public policy that takes into account the varied interests of different sectors of the community and provides effective mechanisms for prompt and uniform enforcement. The paradox of aggregation generates a call for governmental planning and enforcement on a planetary scale. But even should such a call be answered, there is no reason for confidence. All our experience with government in the twentieth century suggests that it is no panacea. More specifically, experience with government regulation of business interests should make us wary of placing too much faith in the capacity of government to carry out its mandate to uphold the public good. In case after case, the regulatory agency becomes subverted by the industry or activity that it is supposed to regulate and ends up as the exponent of the industry rather than as its overseer. The Interstate Commerce Commission has become the spokesman for the railroads, the Federal Communications Commission for broadcast industries, the Federal Aviation Administration for the aviation industry; the list could go on and on.

The Efficacy of Violence. Both social change and social control depend heavily on the effective use of instruments of violence. The effective threat of violent behavior by groups with grievances may induce accommodating gestures from those in control of the machinery of government. All large-scale political systems rely upon police methods to assure domestic peace and upon defense establishments to provide national security. Governments are, above all else, constituted by institutions that claim the right to use violence for the benefit of society. Structural violence of this kind is employed to sustain existing patterns of internal and external domination. Principal governments have had to repress rebellious and intimidate dissatisfied sections of their own population and subordinate foreign societies deemed to fall within their sphere of influence. The role of the Soviet Union in Eastern Europe and that of the United States in the Caribbean area illustrate contemporary zones of political domination. In many countries the government maintains its control by the systematic suppression of the majority of its population. South Africa is an extreme example, where the 20-percent white minority enjoys a fantastic superiority in wealth, power, and liberty at the gruesome expense of the other 80 percent of the population. Under these circumstances methods of violent control are essential to assure the stability of the political order.

In disputes among countries it is rare to find a government prepared to accept nonviolent procedures of settlement if its position would be seriously worsened by an unfavorable outcome or if the issue involves disputed territory or vital rights. Usually the threat of war, if not actual violence, is a prelude to diplomatic efforts at peaceful settlement. If the issue is of a fundamental character, as is the dispute between Israel and the Arab countries, then violence and the balance of military capabilities play a decisive role. The side that is able to implement its claims through the effective use of force will prevail. The community of states "deplores" the recurrence of war in the Middle East, but continues to supply the contending belligerent states with a steady flow of modern arms.

Violence continues to pervade the collective existence of modern man. The establishment of effective, and even popular, government does not eliminate political violence and its threat from the domestic sphere of national life. As American experience highlights, even prosperity does not assure domestic tranquillity. The United States Constitution guarantees its citizens "the right to bear arms." Millions of weapons are held by citizens. Efforts to enact gun control legislation have encountered fierce opposition, especially from such well-financed lobbying groups as the National Rifle Association. The skepticism of the society is summarized by the slogan "when guns are outlawed only outlaws will have guns." Such a slogan extends the paradox of aggregation to regulation itself, implying that the imperfections of government will remove guns from the responsible law-abiding citizenry and leave the lawless minority with a virtual monopoly over the private stockpile of weaponry.

The few relatively peaceful and nonviolent domestic societies that exist are to be found in relatively small states with homogeneous populations where little extreme hardship exists and where a relatively popular government operates in accordance with a welfare ideology. The smaller countries of Northern Europe (Norway, Denmark, Luxembourg) are, perhaps, the best contemporary examples.

In general, however, intergroup violence persists within modern states, especially if the violence used by totalitarian and authoritarian governments to keep population in a state of perpetual intimidation is included. The domain of government is certainly not synonymous with the maintenance of peace. In fact, the growth of national governments has been accompanied by a steady increase in the scale and modalities of violence.

It is true that rule of law prohibits recourse to violent strategies of change, and constitutional procedures normally provide the citizenry with some kind of protection even against unauthorized violence by the government. In practice, however, social change of any significance has been achieved by the effective use or threat of violence, and the maintenance of social control has usually depended heavily on the threat or use of violence by government. Within international society there is a consensus that supports a prohibition on nondefensive uses of violence, but this consensus is not yet implemented by serious structures of control. States with deeply felt grievances do not feel obliged to forgo violence in the pursuit of national goals. Despite the destructiveness of nuclear weapons, violence and warfare remain the main legislative instruments in the world today.

The efficacy and pervasiveness of violence in human affairs bear heavily on the problems of the endangered planet. So long as national governments rely on structures of domination to sustain internal power or external position they will oppose serious forms of disarmament, as well as moves toward more centralized approaches to issues of world order. The continuing emphasis on national military power to achieve national purposes reinforces a *competitive approach* to political survival and severely restricts the prospects for the growth and development of a more *communal approach.* Almost every government in the world devotes about the same percentage of its GNP to military spending as do its principal rivals. There are no significant dropouts from the world competition in armaments. Japan is the closest thing to a dropout, a feature of Japanese policy that is a direct result of its defeat in World War II and of the availability of a firm American military alliance. In any event, Japan has been gradually building up its military capability and seems likely in the near future to rejoin the world system in the full sense, that is, by devoting 5 to 6 percent of its GNP, rather than a mere 1 percent, to military expenditures. The persistence of these enormous investments in arms continues without respite, despite the great demands on these funds arising from the welfare sectors of society. The momentum of arms spending seems also to preclude any awareness that a new kind of political order is needed to safeguard group interests, given the destructiveness of war, the decay of the environment, and the dwindling stockpile of critical renewable (wood, food, air, water) and nonrenewable (minerals) resources.

There is as yet no firm evidence that human nature is violent by genetic disposition, although violence is habitual almost everywhere in the world. Man, ants, and rats are among the few principal species that engage in organized aggressive patterns of behavior toward their own kind. At the same time, the territoriality of political units induces a certain tendency toward the defense of homeland, especially given conditions of scarcity and of *inequality*. That is, under world conditions of insufficient resources to satisfy total demand there is a natural tendency for those with less to seek a larger share. This tendency induces those with a larger share to organize their defenses against those with less and to use their superiority to obtain still more. The rich get richer, the powerful grow more so. But *inequality* in the face of *insufficiency* often induces a guilty conscience together with fear, and a policy of repression. Those who suffer from inequality, depending on their degree of deprivation, their level of awareness, and their belief system, are potential sources of revolutionary action. We live now at a time when the ideologies and belief systems developed to vindicate social and economic privilege and inequality have by and large been discredited and replaced by the virtually universal endorsement of the values of equality. To give the values of equality some embodiment in material and political terms is a major issue within and between principal states. But the creed of equality has not yet produced the kinds of social reforms and resource transfers that would be needed to diminish inequality. The rhetoric of equality accompanies the reality of hierarchy. In actuality, gaps between rich and poor countries are continuing to grow and are expected to grow further in the years ahead. In these circumstances where awareness of the growing disparity is increasingly associated with injustice and exploitation, there is created a latent revolutionary base, and hence, the prospect of a counterrevolutionary response. This base has given rise to various theories of change that glorify the role of violence in producing a sense of dignity for agents of radical change. The ideas of Mao Tse-tung, Frantz Fanon, Che Guevara, and other revolutionary leaders have had a profound influence on oppressed people by romanticizing violence. This radical potentiality for violent seizure of power is activated, of course, by the vitality of revolutionary communism and the counterrevolutionary ideology it has inspired. These political patterns of our time illustrate the centrality of violence in sustaining power and promoting change.

The main forms of political order in world society rest on either *domination* or a *balance of forces* provided by *countervailing* power. Deterrence and domination are the principal forms of coping with the external world. Both forms rest on the *sufficiency of violence* for purposes of the extension of control and defense. A perception of *insufficiency* is generally understood as endangering national survival, except in those few sectors of international society where there is at present no serious security risk, as in North America and Northern Europe. Recourse to alliances is part of the effort of a particular government to overcome a situation of insufficiency. In the nuclear age there is a strong incentive for rival states to secure a *mutual sufficiency of violence* at a tolerable cost and in a setting of reasonable stability. Therefore, we witness the search by superpowers for agreements to curtail the arms race either to save money or to avoid a "provocative" situation in which each side might find itself vulnerable to attack. And yet, as principal issues of arms policy such as underground testing, MIRV, and ABM demonstrate, unrestrained *competitive energies* still enjoy a considerable margin of advantage over *stabilizing energies* within national societies. Note that even advocates of stability are not calling the efficacy of violence into question, but are merely making out "a case" for a safer world at a lower level of expenditure. Such prominent exponents of arms control as Herbert York, George Rathjens, or George Kistiakowski are not overt or even covert advocates of an abandonment of the war system. In fact, anyone who advocates really drastic changes in the security system is automatically read out of the main debate and discredited as "a one-worlder" or "a unilateral disarmer." The "responsible" debate is in the 1 to 5 percent range where the issues are not the fundamental questions about the national security system best adapted to the nuclear age, but rather the marginal questions as to whether a particular weapons system or even just the design of a weapons system is the best way to maintain "stability" — that is, to maintain a defense posture in which the adversary can never have an incentive to strike first because the target state will be in a position to strike back with such devastating impact. The limit of reform is more prudent principles of management; to call for a cooperative system of world security is to step beyond the limit, however cogent or persuasive one makes the argument. Therefore, all major governments, for the present time at least, seem locked into patterns of action based on largely obsolete assumptions about the basis of national security.

In fact, those who seek a stable world system resting on

a safe balance of nuclear capabilities are carrying the logic of a *competitive* system to its proper conclusion under present world conditions. Automobile manufacturers learned that the survival of an unregulated system depended on eliminating price wars and developing more benign forms of competition. In a similar way, the advocates of stabilizing measures of arms control are working to save the system by adapting it to new conditions. These efforts are being resisted by those who by disposition, ideology, bureaucratic position, or vested economic interest are so situated in national society as to resist all efforts to regulate arms competition. These largely internal "compulsions" introduce a measure of irrationality into the arms debate, or rather it is irrationality at the level of general policy. To recall the discussion of the paradox of aggregation, from the point of view of a particular defense supplier or bureaucrat, the expansion of a certain sector of defense spending may be tantamount to the survival of a particular set of jobs or source of profits. Hence, from such an individual perspective it is rational (i.e., maximizes self-interest) to advocate what it is irrational for the community as a whole to do. The guidance system of society is in trouble, as it is in the United States, whenever the irrational special interests grow so influential as to shape the adoption of community policy. It is in this sense that the problems of militarism and the military-industrial complex need to be understood. Comparable problems exist in the Soviet Union and other foreign societies.

Must the efficacy of violence be repudiated to resolve the planetary crisis? It would appear that the paradox of aggregation can be dealt with only by relying on central guidance backed up by community mechanisms of coercion to secure compliance with planetary standards of welfare. At the same time the pressure of competing wants and the actuality of resource shortage cannot sustain the continued misappropriation of resources for the weapons of war. These weapons consume immense quantities of scarce resources and satisfy no constructive human needs. As such, the war system is the most spectacular example of man's inability to put the earth's resources to positive use. The magnitude of wasted resources is one of the *most imperiling* of human patterns.

The endangered-planet argument asserts that mankind is mismanaging its existence by destroying the environment that sustains it. Part of this destruction involves the diversion of resources from needs to superfluities, not only in arms, but in many other features of the modern life-style. The efficacy of violence and "defensive" efforts to neutralize this efficacy underlie what is perhaps the most significant of all the cycles of human waste. In a planetary setting of limited resources and space it is essential to allocate what is available on a basis that sustains life and provides individuals and groups with minimum necessities. Such allocation must be able to meet the needs of people for food, housing, health, education, and dignity. At present the destructive procedures of the world leave most people of the world with their minimum needs unsatisfied. One writer estimated that at least a third of mankind lives below the poverty line, and to attain subsistence is itself far below the demands for a tolerable level of human existence. Even in a country as rich as the United States a substantial sector of the population, in excess of 26 million, continues to lack enough food to satisfy minimum nutritional requirements.

Where some people do not have enough and others have a huge surplus there is created a situation where the deprived groups have every reason to want and demand drastic changes. Considerable evidence exists to demonstrate that recourse to political violence is directly correlated with the intensity of perceived deprivation. The systematic elimination of poverty—of individuals, groups, regions, and nations—is an essential part of any effort to downgrade the efficacy of violence in the world community. Such downgrading is itself essential as part of the struggle to use the limited resources of the world in a more survival-oriented fashion.

The Cycle of Destruction. The mortality of man, the power of death, the disbelief in immortality all may contribute to a reckless attitude toward the future and may cause resentment that expresses itself through destructive and self-destructive forms of behavior. Men may act as they do in part because of an uncomprehending rage in the face of their own prospect of extinction, a condition of absurdity that may distort much of what they do. Certain dangerous social attitudes militate against a constructive response to the endangered-planet crisis. The very conditions of self-preservation and of the maintenance of the species and its habitat (earth) over time do not seem widely appreciated. It seems difficult to induce survival-oriented behavior except against "an enemy." The evidence has long existed that heavy cigarette smoking induces cancer, worsens health, and tends to shorten life expectancy. Fairly easy ways exist to end smoking habits, and yet statistical studies of smoking behavior show that relatively few people have so far decided to stop smoking. Well-verified prospects of natural calamities, such as floods, volcanoes, or earthquakes also do not induce people to move away from the threatened area, even if their lives remain in jeopardy by staying where they are. It remains difficult to raise taxes to combat pollution or remove blight, despite their manifest intrusion upon the quality of human life. Socialist societies, despite their central planning and idealistic traditions, also have not yet devoted a high proportion of their resources to securing the conditions for beneficial human survival.

Fear and greed appear to be the animating psychic forces with respect to resource use. It seems easy to frighten societies into devoting virtually limitless resources to police or military functions if the government relies on the dangers of crime in the streets or aggression from abroad. Fear becomes an effective political force only if it can be convincingly associated with the vivid actuality of "the other" who can be treated as an enemy and destroyed without any feeling of remorse. Greed relates to fear in the sense that where the prospects for profits reinforce patterns of fear, then it is exceedingly difficult to readjust behavior to altered political circumstances.

The presence of greed is also revealed whenever a specific industry is charged with inflicting unacceptable social costs upon the society. The oil industry spends millions on public relations to undo the damage done its image by ocean pollution, the cigarette industry fights to discredit studies that demonstrate the high incidence of cancer among heavy smokers, the chemical industry contends that conservationists exaggerate the damage done by DDT, the auto industry lobbies to defeat compulsory safety and anti-pollution legislation, mining and construction interests oppose costly safety measures, and so on virtually without exception. Vested governmental interests also tend to exhibit a greed for power and influence that displaces all other considerations. The Atomic Energy Commission issues overly reassuring reports about the consequences of radioactivity; the armed services subject national communities to needless hazards by testing nerve gases, and use thier vast lobbying apparatus to organize opposition to arms control and disarmament by any possible means. Most legislators vote for defense appropriations if their districts benefit financially, almost regardless of whether the expenditure is warranted or not. Of course, there are conscience-minded exceptions throughout industry and government that blur the clarity of these assertions about the priority of greed over other considerations. But the basic point holds: Men are by and large animated by selfish motives of greed and fear, and social and political patterns reflect this outcome.

It may yet be possible to link up the energies of fear

and greed with the planetary crisis in a constructive fashion. Such an effort is held back because it is difficult to induce action unless the threat can be *personalized* in a fairly *concrete form.* Organized religion evolved the imagery of the devil and witches to satisfy this demand for tangibility. The threats to the planet appear to arise from impersonal and indefinite forces. The stage of human sensibility has passed beyond an animistic frame of mind—which provided primitive man with concreteness by endowing the principal environmental forces with godheads. The sun, rain, earth, and disease were all forces that could be appeased, angered, cherished. Such a man-milieu relationship is highly realistic in its insistence that human welfare depends on man's ability to make peace with—as distinct from imposing peace upon—his environment. The illusion of the technological mind is that it can *master* the environment rather than live within its constraints. Scientific rationality, by exhibiting the capacity of the mind for *autonomous thought,* misleads man to the extent that it suggests the grounds for *autonomous living;* reason detached from the limits of the earth is certainly part of the confusion that has simultaneously induced the achievements of scientific civilization in the West since the time of the Greeks and produced a breakdown of the ecological balance that allows man to persist as a dependent part of an overall life-support system, itself a tiny unit in an immense universe largely inimical to life as we know it. But the inertia of the past is almost impossible to overcome without some kind of traumatic breakdown. All major historical efforts to secure world peace have come after destructive wars: the Peace of Westphalia after the Thirty Years' War, the Congress of Vienna after the Napoleonic Wars, the League of Nations after World War I, the United Nations after World War II. The structure of world society has been taut in interwar periods, and relatively flexible in the immediate postwar atmosphere.

The unwillingness of men to act for their own preservation, except in circumstances of fear and greed, is accompanied by a callousness toward the suffering and destruction of others. The world has grown accustomed to watching and reading about mass suffering without displaying any deep compassion for the victims of war or natural catastrophe. The willingness to destroy ourselves and our environment is connected with our indifference to the destruction of others. The acceptance of this almost casual attitude toward destruction is a measure of the weakness of the survival impulse, and helps to explain how hard it is to get people agitated about threats to their existence that do not arise from some tangible enemy. Such analysis is admittedly at variance with standard assumptions about the vitality of survival instincts. It may be that the weakness of the human survival instinct expresses a failure of imagination more than anything else, man being usually unable to appreciate danger in advance or, at least, until the danger is incorporated into an argument for action against "an enemy."

Hazards of Velocity. We live in a world of accelerating change. The rate of change produces pressure upon available procedures of adjustment. A single basic statistical projection conveys some sense of the magnitude of change in our world: "In the 30 years from 1970 to the year 2000 there will be more construction than came to pass from 3000 B.C. to date." The endangered-planet crisis in part arises because our technological abilities are evolving so much more rapidly than are our abilities to solve social and political problems. Tensions have always resulted from the inability of political man to cope with the changes wrought by technological man, but these inabilities now threaten irreversible disaster on a *planetary* scale. The political challenge created by the phenomenon of acceleration concerns whether governments can manage the processes of change so as to assure the survival and betterment of their own societies and of human society in general. A certain pessimism arises from the contrast between the rapid mastery of space and the persistent reliance of governments on the most primitive and traditional means to resolve conflicts between sovereign states. On the one side of our experience we have the elaborate computerized world of NASAland and, on the other, the persisting cruelty of torture and warfare being carried on in many parts of the world, often for no reason more evident than "pride" and "honor."

The velocity of change has been unsettling, and has induced a kind of insensitivity to the lessons of the past and the challenges of the future. Because the patterns of change alter the whole human environment, it is important to plan ahead to accommodate their impact. There are numerous manifestations of the strains imposed by change. The nightmare of urban traffic throughout the world partly reflects the failure to redesign the city to accommodate the impact of the automobile or to restrict the use of cars in the city, or even perhaps to design systems of mass transit that solve the need for movement without clogging the streets and poisoning the air of our cities. It also exhibits a failure to restrict population and its excessive migration to large cities.

The art of accommodating change requires an adequate sense of societal priorities and managerial needs as well as a grasp of the social importance of technological change. It also presupposes the capacity to put rational policies into action, a presupposition that we know to be unrealistic given the vigor and power of pressure groups and vested interests. The public interest is diffuse and rarely can assert itself, nor can it often mobilize a strong constituency in the general community, which is a crazy-quilt of selfish, private interests that sustain one another by a variety of subtle, and often tacit, bargains.

A recent study shows that the length of time between an initial discovery of a technological innovation and the discovery of its commercial significance has decreased from 30 years (1880-1919) to 16 years (World War I period) to 9 years (in the period since World War II). In the same vein, the diffusion of technological innovation was twice as great after World War II as after World War I; the recent rate is four times as great as it was for innovations introduced in the years before World War I. David Braybrooke writes: "Within a decade, computer technology has already advanced far enough to threaten computer programmers with obsolescence." The sociologist and futurologist, Daniel Bell, commenting on the increasing velocity of technological change has said that "Perhaps the most important social change of our time is the emergence of a process of direct and deliberate contrivance of change itself."

Computers bring even greater speed to the routines and standard transactions of our world. Each new generation of computers can both do more and do it faster than its predecessor.

In 1946 it generally took 10 seconds or more to multiply two fairly large numbers. This kind of multiplication can be now made 10,000,000 times faster. The increased speed of jet flying over walking is, by way of comparison, only a factor of 100, while radio waves are about 1,000,000 times faster than soundwaves. "These changes in rates of transportation and communication have completely remade the world. The even greater speedups in information processing are remaking our world again." As well as speedups in the spread of misinformation; rumors, hostile incitements, and outright lies, in addition to other more useful messages, now travel at the speed of light. There is a subtle yet profound identification of accelerating change with progress in those portions of the world where the most advanced technological developments are taking place. This identification has been evident even in the arts. Artists often first notice and depict the most salient features of an altered human situation. The fine arts in the United States and Europe have, first of all, given increasing prominence to the ascendancy of the technological dimension of the sensibility of modern man—in fact, artifacts of sculptors have come very often to resemble machines, and even to be machines. One celebrated artist, Yves Tanguy, carried his negative view of this process to its extreme by creating machines that destroyed themselves, the artistic intention being to stage the destructive event for a participating audience. In general, the modern artist excludes the human form and condition from his work, carrying the distinction between observer and participant into the creative process. As argued in Chapter II, the loss of a participating sensibility underlies man's separation from nature and contributes to the ecological crisis. Classical Chinese landscape painting, with the artist deliberately lost or almost lost in a natural scene, is one of the most authentic embodiments we have of man's participatory dependence upon the world he inhabits. The Chinese painter aspires to portray the wider cultural ideal of harmony between man and nature as the most perfect expression of human repose. We find in Western art no comparable conception of the artist as participant in the world about him. The separation of man from nature in the West can be summarized by noticing that *even* the artist conceives of his relation to nature from the safe distance of observation—the Western artist looks out at the scene he paints. To complete the contrast, the Oriental artist is guided by a spiritual image of man-in-nature that takes precedence over that which he sees with his own eyes (except by theory, or by reflection in a mirror or the surface of a lake, we are excluded from visually witnessing ourselves as participants, and may accordingly be misled by our senses into thinking of ourselves only as observers).

More germane to the present discussion is the infatuation of artists with change for its own sake. The art critic, Harold Rosenberg, coined the ironic phrase "the tradition of the new" to describe this insistence upon newness, above all else, by the characteristic artists of our age. In Rosenberg's words,

> Changes in art have been taking place at such lightning speed that compared with even three years ago the current scene is virtually unrecognizable. . . . It matters not in what mode an artist begins, whether with colored squares, a streak of black, the letter 'D' or the drawing of a nude. All beginnings are clichés and the formal repertory of modern art was fairly complete by 1914. It is finding the obstacle to going ahead that counts— *that* is the discovery and starting point of metamorphosis.

In earlier periods of art history, artists concentrated their talents upon the preservation of the positive elements of the cultural and moral tradition within a civilization, restating ever again what was fundamental and unchanging in the experience of the group and its relationship to the world. Many traditions of great art have been static in view, anti-developmental, maintaining the integrity of symbols by manifesting their immutability through endless repetition. Islamic and Hindu art embody in their purest form the identification of the good with the unchanging, rather than with the changing aspects of experience. In such a cultural setting the idea of progress is itself deeply suspect and, if it exists at all, tends to be associated with a deepening of insight into what is permanent rather than with cumulating achievements through change and growth. Human fulfillment depends on the escape from change rather than on its mastery, and salvation is portrayed as liberation from the turning wheel of life.

But in the modern scientific state we live according to a scale of values that pays great attention and gives acclaim to the attainments of technology, especially to those innovations that alter our daily experiences of living or feeling. The idea of change is also very much embedded in the arms race where the psychological essence of national security is to keep ahead, to change first, and to avoid, at all costs, falling behind. Military planning is oriented toward the future; weapons systems take at least 5 to 10 years to develop; the lead time between research and development (R&D) and deployment has expanded and the rate of obsolescence has increased. Weapons systems are sometimes obsolete, as with various missile defense systems, by the time they are ready for deployment. The waste of resources is magnified by this feature of "the tradition of the new," as is the sense of impermanence and conjecture that surrounds debates about security issues.

And, of course, planned obsolescence underlies the operation of a free enterprise profit system. There is, hence, a built-in incentive to glorify change so as to substitute what is new for what existed earlier. Advertising is carried on in "the tradition of the new," exaggerating the benefits of change and associating the latest model of car, TV set, washing machine, or detergent with such intangible experiences as happiness and vitality, and even with sexual prowess. In writing about space exploits, Kurt Vonnegut records that to reach the moon we spent $33 billion and that "Somewhere cash registers ring." After all, "He who pays the fiddler calls the tune." To obtain such funds for these purposes means taxing the poor and downgrading their claim on our resources and energies. Vonnegut believes that "we should have spent" the $33 billion "cleaning up our filthy colonies here on earth. There is no urgency whatsoever about getting somewhere in space, much as Arthur C. Clarke wants to discover the source of terrific radio signals from Jupiter." Space heroics bear witness to the reinforcing pressures that make technological progress, scientific discovery, and economic profit part of a single seamless web. The entire activity is given credibility by the imagery of "the race" against rivals, whether arms race or space race. We are members of a political system with a priority schedule that puts landing a man on the moon ahead of feeding the hungry, housing the poor, or caring for the sick. "The cathedrals" of modern science bear about as directly on the alleviation of human suffering as those older gilt edifices built with such effort and sacrifice by the impoverished masses. To emphasize his concern for the problems of social distress, President Lyndon Johnson initiated "a war on poverty"; such labeling draws on the potency of war as an instrument inducing change, but when its use becomes rhetorical and no war consciousness or large appropriation of effort and money ensues, then the hollowness of the gesture accentuates the inattentiveness of man to the misery even of fellow citizens.

A characteristic of the modern industrial state is to conceive of the future in technological terms. As money is associated with power, and as power arises from successful applications of the most modern technology, the characteristic industrial leaders of our time tend to be technocratic in outlook. The ascendancy of the technocrat gives rise to a false sense of optimism about the future. In a recent survey of industry prospects, all 34 board chairmen and company presidents, even the spokesman for the tobacco industry, turned in optimistic reports, foreseeing a period of indefinite prosperity and economic growth in the years ahead for their particular sector of the economy. The awesome feats of astronautics—such as the successful missions of the Apollo Program—help foster this illusion of technological omnipotence. In thinking about the future this kind of technocratic sensibility approaches endangered-planet problems, if at all, by proposing cities on or under the sea, urban communities beneath geodesic domes, new chemical sources of food supply, and the like. The attention is directed toward fantastic schemes for reaching a technological reconciliation between man and nature. No attention is given to the

underlying social processes and human attitudes that
make the modern world so fraught with danger, unhap-
piness, anxiety, and ugliness. The illusion of technological
omnipotence assumes that there exists the capacity to
deal with all fields of human endeavor, as if there is a spillover from the technological domain to the domain of
politics. In fact, getting a man to the moon dramatizes the
gap between our technical aptitudes and our societal in-
eptitudes, and the scale of technological enterprise makes
the cost of human failure so much greater.

Part of the success of technology has been to discredit
prescientific ways of thinking. This process of discrediting
the past has been partly a consequence of science dis-
placing religion as the principal source of human guidance. In both socialist and capitalist spheres of the modern world a kind of scientific objectivity about the status
of truth and evidence has taken over from traditions of
revelation. Charles C. Tillinghast, president of TWA,
says that "It's conceivable in 1975" that a businessman
"may be expected to conduct a staff meeting at 8:00 A.M.
in New York, give a pep talk at 10:00 A.M. to his sales
force in San Francisco, review his 1976 goals over brunch
with his firm's vice president in Tokyo, and trouble-shoot
distribution problems in Rome before noon." The advent
of supersonic and vertical-takeoff-and-landing aircraft
makes such a schedule feasible. And for Mr. Tillinghast,
"All these developments will give new mobility to man and
the products of industry and make his predecessor of the
1960's appear a slowpoke by comparison." The affirmation of acceleration is among the elements of this new
technetronic age that seems so mindless and destructive
unless it is guided by concern with the human problems of
living together in peace and dignity. The failure of Western
religion to prevent the onslaught of the modern world has
stimulated a return to prereligious forms of nature worship,
with a greatly revived interest in magic, astrology, and

prophecy, as well as an unprecedented Western enthusiasm for Eastern religions. The young of post-industrial society are increasingly drawn to the thought-ways and life-style of the pre-industrial world, calling for a return to simplified modes of living as essential to the development of human communities based on love, simplicity, and mutual respect. Such an orientation looks with horror at the values and behavior of the technocratic elite and holds it accountable for the problems of the endangered planet. In fact, the student movement has gradually assembled a bill of indictment against the reign of technology, ranging in concern from the Vietnam War to the problems of environmental decay. Cars with flower decals or "resist" bumper stickers also now may carry other messages: "DDT digs your liver, man" or "Do not Buy Union" (Union Oil Company being the owner of the oil rig that has most polluted the ocean waters around Santa Barbara, California, and also, quite incidentally, the owner of the tanker *Torrey Canyon*).

One of the disruptive ironies of a rapidly changing technology is to cause breakdowns of communication even in an atmosphere of improving techniques of communication. The difference between a rapidly evolving technology of communication and the reality of meaningful communication parallels the difference between technique and substance. Margaret Mead has written on this issue: "In the past there were always some elders who knew more—in terms of experience, of having grown up with a system—than any children. Today there are none. . . . There are no elders who know what those who have been reared in the last 20 years know about what the next 20 years will be." If we measure the rate of significant change by accepting the common contention that "knowledge" (obviously a very crude judgment) is now doubling every ten years we begin to grasp the extent of "the generation gap." Margaret Mead emphasizes the extent to which those who hold power are "strangely isolated" from the young who have reacted against them:

> The elders are separated from the young by the fact that they too are a strangely isolated generation. No generation has ever known, experienced, and incorporated such rapid changes, watched the sources of power, means of communication, the definition of humanity, the limits of their explorable universe, the certainties of a known and limited world, the fundamental imperatives of life and death—all change before their eyes.

In such circumstances the bonds that hold society together can be burst asunder. There is a loss of societal center; those without power, the dispossessed, demand control in the name of justice and a new, radically inverted social order, whereas those with power are threatened, are inclined to appease and repress, call upon the police, and argue their case beneath star-spangled banners of "law and order."

When the young are alienated from society there is a sense in which the future is discredited, because it is today's young who will necessarily be the future. Commenting on the struggles of 1969 at Berkeley over the creation of the People's Park, two sensitive observers write: "The great danger at present is that the established and the respectable are more and more disposed to see all this [student unrest] as chaos and outrage. They seem prepared to follow the most profoundly nihilistic denial possible, which is the denial of the future through the denial of their own children, the bearers of the future." We are locking ourselves into a situation of danger by repudiating a plea for a new future that is being articulated, often in barely audible terms, by the young. Part of this plea involves a rejection of the IBM-world for which higher education has become the programming

vehicle, and part of it is a call for purity and peaceful-
ness in man's relations with other men and with nature.
There is a genuine commitment to the creation of a future
that is attentive to the agenda of the endangered planet,
not at the level of technological adjustment but at the
deeper level of human adaptation. *Adjustment* implies
tinkering with the system to keep it working, whereas
adaptation implies reshaping and recreating the system,
redefining what work means, and clarifying the relations
between ends and means in our appropriation of nature
for human society.

To some extent the political leadership of human
society is coming to realize that the rates of change are
endangering values of prosperity, welfare, and peace.
There is a general acknowledgment at this point that a
continuing expansion of the world's population will raise
very serious problems for the human race. On July 18,
1969, Richard Nixon, as President of the United States,
gave an address to the Congress on World Planning of
Population Growth which exhibited an awareness of the
seriousness of the problem and the extent to which the
rate of change, not just the numbers of people, made the
population problem so severe, on both a national and a
global basis. Mr. Nixon, in commenting on the increase
of American population from about 100 million in 1917
to 200 million in 1967, said that he believed "that many
of our present social problems may be related to the
fact that we have had only 50 years in which to accommodate the second hundred million Americans." In proposing to deal with population problems Mr. Nixon
stressed heavily the expansion of family planning assistance and the need for careful study of future trends to
enable better planning. He pointed out that at present
growth rates the equivalent of a new city of 250,000 (about
like Tulsa, Dayton, or Jersey City) is needed every month
just to take care of expected additions to the American
population. The world increase, per day, results in the

need for a city of almost 6,000,000 each month—that is, a city three times the size of Philadelphia. As is characteristic for the leadership of post-industrial society, the nature of change is conceived in technological terms (birth control, better pills—once a month, even once a year, education, and aid in the use of the technology), but not in terms of the underlying human and social dynamics that give rise to the crisis. Mr. Nixon, after evidencing the depth of his concern, went out of his way to say, "Clearly, under no circumstances will the activities associated with our pursuit of this goal [increased family planning among the poor] be allowed to infringe upon the religious convictions or personal wishes and freedom of any individual, nor will they be allowed to impair the absolute right of all individuals to have such matters of conscience respected by public authorities." We are, in effect, reenacting "the tragedy of the commons." So long as we hold sacred the social patterns that generate the problems our "solutions" will be, even if successful, patchwork, buying-of-time, essentially unrelated to the creation of longer-term balance between man and nature. Population expansion has become a form of environmental pollution, and needs to be comprehended in these harsh terms. If family size is beyond public scrutiny, then it is comparable to saying that sewage disposal is similarly exempt. Of course, special dangers exist that governmental intervention in private decisions made about childbearing will abuse the dignity of men, but unless the relevance of decisions about family size to the aggregate problem of population growth is stressed, there is little prospect of working out a beneficial future for the human race.

We live, then, in an environment of rapid change, wherein even values change rapidly. Some changes are reinforced by the idea of progress embodied in our post-industrial society. GNP accounting is taken as an index of national progress throughout the world, regardless of the character of a political system or its stage of economic development. The provision of public goods, needed to cope with a rapidly changing environment, lags badly behind social requirements, partly because the public in no society is as well organized to protect, or even to perceive, its interests, as are special sectors of society, such as the defense or oil industry, war veterans, minority groups, labor, farmers, and so on, which devote great energy and attention to satisfying their particular claims upon society. The more diffuse the cause, the more difficult it becomes to bring effective pressure to bear. Most endangered-planet issues, unless temporarily highlighted by a disaster such as a nuclear confrontation, a pollution tale, or a famine, are diffuse issues that do not easily capture the political imagination: there is no specialized constituency that organizes to protect the general welfare, but there are specialized constituencies that exist to oppose those groups that do protest. For instance, the Petroleum Institute is a lavishly financed lobby for the oil industry that exerts a variety of specific pressures that cannot be countered by loosely organized, poorly financed citizens' groups. Despite the recent growth of an ecological crusade, the balance of pressure is heavily weighted in favor of the vested-interest position, reinforcing the overall status quo, reflecting present concentrations of power, influence, and money. Changes can be made, but they come about slowly, and only after massive evidence of the social evil has been amassed and considerable allowance for a painless industrial adjustment has been made. The campaign to advise cigarette users of amply documented smoking hazards and to disallow TV advertising that encourages smoking stands in contrast with the criminal sanctions that attach to marijuana use; in the former case there is a well-organized industrial lobby, whereas marijuana use is carried on outside the framework of the market economy. The relationship between the auto industry and air pollution also illustrates the limited ability of government to prevent societal abuse by big business. No pattern of regulation is feasible at all unless it makes adequate provision for the continuing profitability of the enterprise, including the amortization of capital equipment. The main point to be made is that the pattern of vested interests may prevent, and certainly slows down, the appreciation of the requirements of public welfare, especially where these requirements are based on diffused circumstances. Given the *acceleration* of technological change, this adjustment lag is exceedingly dangerous, and in certain areas of human activity—for instance, the release of carbon monoxide, carbon dioxide, lead, and other chemicals into the atmosphere—may produce widespread and irreversible harm that cannot even be connected up with its cause until many years later.

It is, of course, not only acceleration but *proliferation* that is important. More people have cars each year, and the exhaust from cars means that considerable effort at pollution abatement is needed just to maintain the status quo. Improved modes of production and higher living standards are directly associated with patterns of proliferation. This phenomenon is of world-wide significance. Fiat is now building in the Soviet Union a huge factory for the mass production of cars that will each year enable several hundred thousand more Soviet citizens to drive their own car. The desire to acquire a car, to participate fully in the consumptive economy of the affluent society, seems to be virtually universal and to have acquired a glamor of its own that is difficult to remove. It is, of course, a situation that has been partly created by the manipulation of taste and values through large-scale advertising, but it is also a cross-societal acceptance of technological innovation as indicative of progress. Man progresses to the extent he achieves mastery over nature; speed of movement, of performing household operations, of conveying information, confuse us about what is and what is not an improvement in the human condition. This image of progress animates the expectations of people everywhere for a better future. It is a particularly potent image in those societies where mass poverty exists and, inevitably, casts its shadow upon the life experience of most people. The idea of affluence made possible by the achievements of modern technology possesses an almost irresistible attraction for deprived populations. And, in fact, the newer, more radical leadership throughout Asia and Africa often repudiates traditional cultural attitudes about the passive acquiescence of man to his brief passage through the world between birth and death. These traditional attitudes can be correctly held responsible for the greater poverty and hardship—more disease, shorter life expectancy, higher infant mortality, lower literacy, poorer shelter—that is found in Asian and African countries than in the rest of the world, and especially in the achievement-oriented, progress-infatuated societies of Europe and North America. It is reasonable to explain disparities in wealth and power by a comparison of these attitudes; even the period of colonial humiliation can be explained quite convincingly as a consequence of the passive life-style's inability to resist the challenge and claims of the aggressive life-style that has accompanied the rise of the West.

The most profound danger of deception results from the fact that advanced technological society has created a powerful model for the rest of the world at the very point in time at which reliance on this model threatens to crack wide open the entire basis of human existence on earth. In the face of grossly disparate living standards, it is virtually impossible to address an ecological appeal to poor countries to refrain from patterns of action that have brought affluence and dominance to the rich countries. The United States and the Soviet Union engaged in a long series of nuclear tests in the atmosphere, developing their respective nuclear capabilities to the point where no third state could hope to achieve a comparable capability. Only then did these nuclear superpowers negotiate seriously to achieve agreement on a Limited Test Ban in 1963; at such juncture an appeal to other nuclear-aspiring states such as France and China to refrain from atmospheric testing in the interest of mankind carried little moral or political weight. As might be expected, the French and Chinese governments were unmoved by such appeals and have continued to develop their own nuclear weapons program on the basis of atmospheric tests; they could have been prevented from doing so only at great risk, and possibly at the cost of a major war. There is no doubt that to add to the radioactivity of the atmosphere is to engage in destructive behavior, but there is also no doubt that the states that might have organized a protective campaign were deeply discredited by their own records of destructive behavior. The scientific ability to differentiate weapons explosions from earthquakes is sufficient today to enable a test ban on underground testing, but the United States and the Soviet Union remain unwilling to forgo their race for superiority with one another and their struggle to maintain dominance over all others. Where leaders set the rules of the game it is very difficult indeed to counsel good behavior for followers, especially when all players, as in international society, are endowed with autonomy in the form of "sovereignty." This phenomenon of double standards in rule-setting is of central relevance to the whole effort to discover and embody a central guidance mechanism to put the earth back in safe operating condition.

We know that the projected consequences of the critical trends in population growth, resource depletion, pollution, and arms competition will bring about an ever-more dangerous situation. The future will become worse and worse unless these trends can be reversed. Trend rates can be reduced by reformist action but such effort is a matter of gaining time rather than of achieving solutions, and may even be of negative significance if it induces a false sense of an ability to muddle through by taking an effortless series of 3 percent reforms in a setting where 25 to 50 percent alterations are needed. President Nixon, discussing the population problem, has said that "One of the most serious challenges to human destiny in the last third of this century will be the growth of the population. Whether man's response to that challenge will be a cause for pride or for despair in the year 2000 will depend very much on what we do today." The rhetoric is impressive, but the spirit of implementation is still in the 1 percent range—a study commission and an extra $100 million or so in family assistance. What is needed is a conception of optimum population based on a set of societal and ecological objectives and a program for its realization. We find the same dissociation between Nixon's *rhetoric* on world peace and environmental defense and the marginality of *action* proposed. It is almost as if the problems are beyond solution and all that a leader is able to do is to slightly stir his population by endorsing vague sentiments.

We are dealing with fundamental conditions created by processes, by long, slow directions of behavior that have a cumulative and, eventually, a dramatic impact. To build an ABM system, install MIRV warheads, subsidize the supersonic transport (SST) aircraft, involve processes of action and reaction that sustain a direction of competitive behavior. Most behavior is also interactive and interdependent: We are influenced by what the Soviet Union does and does not do, and because of faulty intelligence and presumed hostility we plan to deal with what they *may* do as well as with what they are *likely* to do. We presume a maximum effort by our rival that can only be offset by our maximum effort, leading to an endless chain of actions and reactions based on "getting ahead," "staying even," or "catching up." And vice versa: the defense posture of each country aims, above all else, at avoiding a perceived sense of strategic inferiority. We do not often question this logic of security, although the existence of military superiority on the one side or of nuclear vulnerability on the other does not appear to enable the Soviet Union to secure a favorable outcome of its dispute with China or even to allow the United States "to prevail" in South Vietnam.

What will exist a decade hence depends on what we do now. The pace of change makes the interaction between present and future of increasing importance. It is less possible, because of rapid change, to adjust continuously to an unfolding future. Partly, the difference between the rate of adjustment and the rate of disruptive change is too great, and partly the underlying dynamics of the process need to be transformed, to avoid further deterioration. Such a situation necessitates advance planning and procedures for effective coordination, activities that are difficult within a well-organized national society when transformations are called for and are virtually hopeless within the highly segmented structure of international society. To plan effectively presupposes an effective consensus both as to goals and as to means; such a consensus is hard to achieve if interests and priorities are disparate and tension exists between the principal actors.

Alienation, Manipulation, and False Consciousness. The phenomenon of alienation underlies much of the politics of our world. Man is alienated when he is estranged from understanding the conditions of his own welfare, from other men, and from nature. Marx attributed fundamental importance to alienation as the main expression of the human distortion that resulted in a capitalist economy where a laborer is exploited and the surplus capital attributable to his work becomes the basis by which some men gain wealth at the expense of others. Alienation involves false consciousness, an estrangement so extreme that a person loses the ability to discern his own interests, the conditions of his own fulfillment, or the actuality of his role as a victim or perpetrator of exploitation. In the late nineteenth century, alienation became associated with mental disorder and "an alienist" was the earlier term used for a psychiatrist, an expert on the existence, control, and cure of insanity.

Acute alienation is visible on all levels of society, in all parts of the world, in socialist countries, in welfare states, in the mixed economies of the West, in coercive and relatively permissive societies. Man has lost his way, sunk in sullenness, unable to understand or control the forces that are shaping his life; the experience of estrangement is widespread, and often grows so acute as to be repressed or sublimated through such mind-killing compulsions as TV, alcohol, drugs, tranquilizers. The situation is worsened by the growing depersonalization of work, by the decline of traditional sources of community participation, by the extension of mass media into everyday life, by the substitution of computerized directives for face-to-face relations with those in positions of authority, by the reliance for security on threats to use weapons of mass destruction against innocent populations, and by the gross contradictions between the ideals of equality and dignity and the misery, oppression, and brutality that surround and pervade our collective existences.

The development of a strong interest in environmental issues reflects a deepening of understanding. It has long been evident that one society can justify destroying another for its own purposes. With nuclear weapons we make this justification a persistent feature of our political lives, resting national security on the hostageship of civilian populations in large societies. But we now also are beginning to understand that greed is also turned inward and that it is unmistakably destroying us, that we are quite literally destroying the air we breathe, the water we drink, the land we live on, and that this destruction is unnecessary and interwoven with the profits of those who run the show. We also are witnessing the gestures of official response coordinated with the failures to pay for effective action. Those who are called "alienated," often in the spirit of derision, are those who are aware of and angry about their estrangement from the psychic and material conditions of life-support. Those who support prevailing values and structures of power are often not aware of their estrangement, and are, in a sense, alienated *even* from their alienation. To the extent that we live on an endangered planet and fail to take the kind of drastic action that is needed to reduce the danger, we are bearing witness to the profound hold over our destinies exercised by the destructive and self-destructive energies that pervade our world: "I find myself unable to see anything at the end of the road we are following with such self-assured momentum but Samuel Beckett's two sad tramps forever waiting under that wilted tree for their lives to begin. Except that I think the tree isn't even going to be real, but a plastic counterfeit. In fact, even the tramps may turn out to be automatons . . . though, of course, there will be great, programmed grins on their faces."

To overcome alienation requires some reintegration into the world based on a persuasive diagnosis of the existing causes of estrangement, and also on a new vision of what our individual and collective lives might be like if we could overcome alienation. The search of the young for pre-industrial forms of being, for communal living, for simplicity and love are the affirmative side of the demands to wreck the Establishment, dismantle the national security state, and shut down the university. Given the modern world, however, filled as it is with destructive forces, it is impossible to carry out such a plan of action. As soon as it becomes dangerous it will be trampled upon in its homeland, and even if it should enjoy isolated instances of domestic success, the world system with its teeming, hungry, frustrated populations will not abide islands of isolated contentment. To overcome alienation requires a vision of the earth's wholeness, it requires nothing less than a new system of world order that is associated with the defense of life on earth. Such an orientation requires a series of major changes in values, structures, and behavior; the task of later chapters is to consider what these changes might be and how to bring them about. So long, however, as most men and their leaders remain estranged—and unaware of their estrangement—from the realities of life on an endangered planet, so long is the situation without hope. And so long as it appears hopeless men will turn away from problems and their possible solution, whether such turning away takes the form of endless breast-beating about impending catastrophe or more discreet forms of hedonistic escapism via TV, drugs, sports, or of placing their faith in ideologies that pin the blame somewhere else, on some acceptable enemy at home or abroad.

What Is a Usable Future? We live at a time when the environment of rapid change and the need for effective adjustment are increasingly evident, and have influenced business practices and government operations. It seems clear that part of the reason why the Defense Department has enjoyed so much more influence than the State Department in the United States since the end of World War II has been its capacity to develop a systematic and coherent set of programs for coping with the future. Most of this kind of futurism involves depicting in detailed form the prospects for technological innovation and the sorts of adjustments that will be called for as a consequence. There is little attention given to the kind of national or world society that we would like to create by planned effort. The image of the future is constituted by the interaction between technological momentum and human existence. The goals of human existence are identified with achievement, on an individual and collective level. Landing a man on the moon is a public good for Americans, because it produces a feeling of national pride and exhibits the leadership of the United States.

An imaginative physicist, Gerald Feinberg, has written

an important book, *The Prometheus Project,* in which he points out the need to develop a human consensus on the long-range goals of mankind. Only after an agreement on human goals will it become possible to determine whether and to what extent it will be necessary to supervise certain basic innovations and breakthroughs that are becoming technologically possible. Feinberg makes his main point by reference to genetic engineering that will soon develop to a point where practitioners will be in a position to design the sort of "human" beings that are brought into the world and where artificial intelligence (via computerized activity) will have grown to the point that a machine will be able to perform *all* the functions, including even those associated with artistic creativity, of the human mind. Once introduced, these impinging technological prospects will drastically alter the human situation. Feinberg argues that these prospects should be appraised in terms of a consensus as to the goals of mankind, and that this process of assessment needs to be initiated immediately and from many representative angles of human perspective so that universal attitudes might emerge. Again the intersecting time series create desperate pressure because the rate of technological change is still accelerating whereas the rate of value assessment has remained relatively static.

Goals for mankind are not enough. We need as well some conception of the sort of global village that we would like to inhabit, the processes that might bring it about, and the structures that might sustain it. The celebrated microbiologist, René Dubos, has written that "All great periods of history have created . . . utopian images. Inertia is the only mortal danger. Like the poet, the social planners should break for us the bonds of habit." We need to invent new images, new rituals, new symbols and myths, new models of world societies, and new conceptions of plausibility and fulfillment, or revive old ones that could serve our needs. Such inventions would draw more on the sensibility of the poet than on the social and natural scientist who has emerged in modern technological society, and as such, might correct the mind-distorting separation of man from nature. The whole tendency of the social sciences to emulate the style of marginal analysis developed by professional economists discourages the birth of radical ideas. Economists have become successful policy advisers because they have developed rather precise models of incremental change in which minor adjustments are made to keep the economy in a relative condition of dynamic equilibrium. Rationality has even been identified with *marginal* analysis as distinct from *systemic* analysis. Such a conservative disposition is also encouraged by Anglo-Saxon legal traditions of thought associated with the basic growth patterns of the common law in which law mainly develops through a series of judicial interpretations, rather than by basic adjustments in the legal order. The whole judicial temper of "all due deliberate speed" pervades the bureaucratic policy procedures of major governments and makes it very difficult to give serious attention to any non-gradual approach to problem-solving. Even if the government adopts serious rhetoric to dramatize the importance of a social issue, the tendency is still to propose and adopt only the most minimal program of action. The crescendo of concern in the United States about the deteriorating environment has stimulated both an inflationary spiral of rhetoric and an absence of any correspondingly serious program of curative action. The record of civilization suggests that only war, natural catastrophe, or economic depression can loosen up the political system sufficiently to bring a new scale of priorities for government expenditures into being and to permit a high degree of planning and coordination for the entire society to take place.

At the present time, governmental efforts actually *contradict* the rhetoric of concern. One of the most perceptive analysts of the ecological situation, Barry Weisberg, has shown that there is actually a *declining* (!) proportion of the federal budget (although not less dollars) being devoted to environmental defense and resource management. Weisberg gives the following percentage figures:

1965	2.3%	1968	1.9%
1966	2.2%	1969	1.9% est.
1967	2.0%	1970	1.8% est.

Weisberg also points up the fact that the government is continuing such programs as subsidizing the SST and granting a depletion allowance to offshore oil drilling. The consequence is not only that the new rhetoric of ecological concern is a gigantic exaggeration, but that the government continues to be an active, knowing culprit: "The frightening conclusion . . . is not that the government should do more, for the more it does the worse our ecological systems get."

Given the structure of world society, such planning must first take place within the relatively closed national systems that evolve policies, make decisions, and have the power to shape relevant patterns of human behavior. Therefore, the first step is to alert national societies to the condition of planetary danger and to the sorts of response that are possible and desirable. Such a process of awakening should go on in every part of the world and is almost certain initially to arrive at divergent appraisals of what is happening and why. Existing differences in historical experience, wealth, power, ideology, and status are likely to produce conflicting calls for action. Although it is notable that the Soviet physicist, Andrei Sakharov, has urged the Soviet and American governments to adopt a radical response to world crisis, it is also significant, of course, that Sakharov's document had to be smuggled into the West and has been implicitly repudiated in an official Soviet account of the future. Sakharov's document is radical because it recommends very major transfers of resources from the rich to the poor countries (up to 20 percent of GNP over a ten-year period) and drastic disarmament, and most of all, because it rejects the Marxist-Leninist canon that the world situation will be automatically corrected if socialism triumphs in all principal states.

It is to be expected that poorer countries would demand adjustments in patterns of trade, aid, and investment so as to enable the elimination of mass misery within their countries and also to permit a greater equality of influence on the world stage.

At the same time, any analysis of critical social and economic trends suggests a common peril arising out of uncoordinated national behavior. There is no way to protect renewable food resources in the oceans against depletion by new fishing technology and persistent insecticides, except by some kind of system of resource management that is applied effectively to all nations. So long as the oceans are a commons the basic tendency will be to maximize the national share of the resource; restraint will be interpreted as enriching others rather than as contributing to beneficial conservation; and the cumulative tendency toward over-fishing, pollution, and depletion will follow.

The process of awakening to the situation of world crisis should include a sense of the need for *global planning* with respect to managing resources and intergroup relationships. At this stage it is unclear as to whether existing international institutions, especially those within the United Nations, will be in a position to play a major role in such global planning operations. It is probably significant that President Nixon in the course of outlining the prospects for world peace in the 1970's barely mentioned the United Nations. The fact that this treatment of the UN in the 1970 State of the World Message was not even generally noticed, is a further indication that the UN is not taken very seriously as an element in world society. This situation could change abruptly if major governments begin to appreciate the extent of their need for common institutions to deal with common problems. In the meantime, the UN may be able to play a significant role in disseminating a basic diagnosis of the endangered planet, in facilitating an identification of points of agreement and disagreement, and in exploring various proposals for initiating adaptive change at the global level. It is very important to minimize the impact of political and socioeconomic antagonisms upon the perception of the shared interest in a global approach to the endangered planet. As international behavior is interactive, the hostile perception of other societies tends to reinforce tension, distrust, and suspicion on both sides. The Soviet Government often views proposals for new systems of world order as covert means of subordinating the socialist sector of world society to the larger and richer non-socialist sector. Such a perception needs to be understood and transcended. Nothing short of a voluntary consensus on goals, programs, and modalities of change among principal governments can provide the foundation for common effort that is required. Just as an individual cannot alone, by voluntary adherence to constructive standards, safeguard the community welfare—for instance, to achieve effective and just environmental and demographic planning and priorities—so distinct national governments cannot, by their individual adherence to constructive standards, safeguard planetary welfare. A first step is to establish lines of communication *within* and *between* states to achieve a clearer sense of common danger, and also to foster an early awareness of the directions of change that are needed if mankind is to survive.

IV. The Four Dimensions of Planetary Danger

Politics is concerned with the management of human affairs. The adequacy of political organization depends upon the nature of the tasks facing a particular political community, the scale of values imposed upon those whose job it is to complete these tasks, and the institutions and procedures available. Political communities are very different as a consequence of diversities in size, tradition, wealth, technological development, ideology, military capability, climate, and sense of cohesion. All political communities share an aspiration to survive and to sustain their social identity. Most political communities also seek to achieve stability and to bring about a condition of at least minimum subsistence for loyal members of their citizenry.

On a global scale, the primary tasks of politics have been controlled by governments that operate as centers of territorial power. There has never been any kind of global unification of political authority, although some great empires, religious movements, and political ideologies have been animated by this goal. In recent centuries the predominant unit of political authority has been the sovereign state, although the tasks of management have been distributed among a variety of centers of political authority including the family, tribe, corporation, village, empire, and church. National governments took over, especially, the task of preserving domestic tranquillity and organizing the common defense. In recent decades, both in capitalist and socialist societies the government has also assumed an increasingly active responsibility for maintaining a viable economy, and has sought to provide some kind of assurance of minimum welfare services for its entire population.

The expansion of national governmental structures has accompanied the growth of national societies, and follows from the complexity of these social orders and the intricacy of industrial and postindustrial society. It is evident that strikes by sanitation workers or truckers immediately endanger the urban population of a modern city, and that a postal strike can paralyze the economy in a few days. The total blackout arising from a power failure in the Eastern grid of the United States on November 9, 1965, illustrated the vulnerability of modern society to a breakdown of a single interlinking component in a complex electrical network. A society that relies on candlelight is not nearly so dependent on the skills of management both to prevent failures and to recover from their inevitable occurrence. The vulnerability of modern states to intentional and unintentional disruption is of critical significance. The interconnectedness of the networks of control and organization is a characteristic part of our world situation, perhaps most dramatically evident in the relationship between human decision and recourse to nuclear weapons systems.

Until recently the planet as a whole was not in need of coherent management. The parts did not interconnect with sufficient regularity and significance to require any continuing coordination of human activity over the entire globe. Besides, the orbit of effective administration was restricted by the slowness and reliability of media of transport and communication. Most really large units, especially great empires, were unable to maintain their full territorial extension for very long. Of course, it is true that men have dreamed of unity ever since the dawn of history and have lamented the consequences of dividing mankind into relatively autonomous and often conflicting units of authority. The prevalence of conflict among these separate political units has caused the history of mankind to be dominated by the story of the rise and fall of civilizations, especially through the interrelated experience of warfare and technological change. The struggle for power, prestige, and enrichment has led social groups to rely on violence to defend what they possess and has spawned a series of efforts by aroused social groups to expand their authority and control through a reliance upon their superior fighting capabilities. Throughout the world a struggle for dominance has periodically raged. As a consequence, human groupings have fallen from positions of prominence and most societies of men have eventually disappeared from view altogether. But the human species as a whole has apparently never before been in danger. In fact, steadily improving methods of agriculture and industry have allowed man to extend greater and greater control over other forms of life on earth.

The enormous aptitudes of the human brain have provided the powerful and affluent sections of world society with an increasing prospect of a long, healthy, and well-provided-for life, although this achievement has always depended on the exploitation of the poorer sections. Unlike other species of animals men have had a sense, reinforced by religious teaching and Western moral tradition, of their own omnipotence in relation to the earth.

Wars, strife, famine, and disease provided men with many incentives to improve the management of their group and intergroup existence. The development of medicine, agronomy, industry, and government has created a tolerable degree of order and continuity within the more successful societies of the West. Rivalry among these societies has produced a sequence of wars, the most recent general war, World War II, being of great destructive magnitude. Such wars were certainly major tragedies, as were plagues in earlier ages, but the degree of peril seemed fully compatible with the preservation of the human speices, even with its progressive emancipation from the burdens of earlier ages. It was certainly possible to contemplate a continuously better world, and even such prophets of modern despair as Nietzsche or Dostoevski never contemplated that the issue of survival was at stake, except in the metaphorical tradition of a Last Judgment. Dostoevski, Nietzsche, Kierkegaard, and more recent Existentialist thinkers lamented over the dismal moral and spiritual failures of man and discerned the absurdity of human existence, but never contemplated that the physical basis of life-support was being seriously eroded by modern social, economic, and technological forms of behavior.

The development and use of nuclear weapons changed this dominant expectation of indefinite survival. It became possible, even plausible, to contemplate the extinction of human life through nuclear warfare. In addition, a neo-Malthusian anxiety about population pressure began to cast its shadow across the political imagination. The most concerted efforts of the newly awakened countries of Asia and Africa to provide food and welfare for their populations have disclosed both the immensity of the present task and the virtual hopelessness of meeting the needs of the 1970's, 1980's, and 1990's. In a world of rising expectation and faltering performance the pressures upon the political managers in these countries are likely to grow more intense, even if technological innovations such as miracle cereal grains and large-scale nuclear power plants work out well. Furthermore, we are just beginning to

appreciate the extent to which the affluence of the rich
societies rests upon the poverty of the poor ones. Only
recently have we started to understand that the high-GNP
societies are destroying the environment and depleting
the resource base upon which modern standards of living
depend.

We are gradually becoming conscious of the increasing
vulnerability of earth forms. We are losing confidence in
the prospects of mankind, without yet quite understanding why. C. P. Snow suggests that, as a result of this
widespread anxiety about the future of the human race,
"We are behaving as though we were in a state of siege."

Part of the emerging awareness is a vague appreciation
that our organizational forms are not able to cope with
the new dimensions of experience and challenge. No government can offer its population much illusion that it is
able to provide protection against a nuclear adversary
that is prepared to risk, or accept, the destruction of its
society. The security of our collective lives depends on
adversaries' adopting a benign or at least a rational posture. But the course of international history discloses
many examples of malign and misguided motivation, and
many assumptions of irrational risks by aggressors and defenders alike. As with war, the momentum of trends underlying population growth—pollution, resource depletion—also suggests the presence of forces that are not
subject to control by independent action, however enlightened, on the part of national governments. *Only new
organizational forms with a planetary scope that corresponds to the planetary dimension of the situation can
offer any prospect of a timely, corrective, and adequate
response.*

In this chapter I present and interpret the underlying
facts and forces. The cumulative effect of these facts and
forces is what gives the situation its urgency and suggests
the importance of an organizational focus. It should hardly occasion surprise that the sovereign state—suitable for

a simpler world of more nearly autonomous units—cannot be expected to cope with the tasks of our world. The scope of modern problems clearly overwhelms the jurisdiction of many national governments; but also the nationalist way of doing things is becoming outmoded, given the circumstances of the endangered planet. This generalization remains far less applicable at this time to the situation of less industrialized states in which central governments play a positive role in building the national unity needed to secure domestic tranquillity and social and economic progress.

Most literature on the endangered planet is specialized around a particular dimension of concern, whether it be the arms race, nuclear war, population pressure, resource depletion, or ecological imbalance. Remedial action tends to be associated with the particular phenomenon, and normally involves technological or specialized institutional proposals. The search for a new birth control pill taken once a year that will check population growth, or the discovery of chemical procedures to achieve photosynthesis that will greatly enlarge food supplies, or ocean minerals that will satisfy resource needs and solar power energy needs, or disarmament that will end the arms race and prevent large-scale warfare, are characteristic expressions of this problem-solving mentality. As indicated already, the analysis here presupposes *the interconnectedness* of the threats to human existence, not only in the banal sense that everything is connected with everything else, but in the fundamental respect that these issues cannot be successfully treated as separate and separable. In essence, the threats are all outgrowths of the underlying circumstance of a *mismanaged environment* that is an inevitable result of a defective set of *political institutions*. Furthermore, the threats pose a *cumulative* challenge that exceeds the individual problem areas separately considered. The problem of poverty is deeply associated with disparities in wealth and income and with the growing access of all governments to the technology of nuclear weapons. The construction of nuclear power reactors makes virtually every country an incipient nuclear power, possessing enough weapons-grade fissionable material and know-how to produce on fairly short notice and in secret a stockpile of nuclear bombs; biological and chemical weapons of mass destruction are becoming even easier to assemble, and will become potentially accessible to subnational groups (Mafia, Al Fatah, etc.) and counter-elites. With the level of distress so high in poorer countries the distinction between life and death is generally eroded, and political strategies that assume high risks of defeat and entail great costs in lives become correspondingly more tempting and tenable. Social distress is likely to grow more acute as a sense of hopelessness emerges when it becomes evident that many governments will never be able to provide their populations with a tolerable life, however hard they work and however talented their leaders are. The spirit of hopelessness can easily be transformed into a politics of fury when it becomes better understood that the poor countries are being assigned a dependent status in the world economic system.

Similarly, it will be impossible to persuade countries facing dire poverty and confronted by widespread malnutrition and constant dangers of famine to forgo agricultural or industrial development that increases their productive capacity by demonstrating that such improvements pollute the skies or the oceans. For one thing, although ignorant about the consequences, rich countries achieved their dramatic economic advances without taking steps to safeguard the environment. For another, the priorities of almost all governments are set by the primacy of *immediate* and *domestic* urgencies, other claims being more remote, more conjectural, hence, more suspect and more easily put off. Such a comment applies especially to countries experiencing a constant state of domestic emergency and, quite possibly, confronted by revolutionary domestic groups. Global coordination of policy by voluntary action of national government is thus a naive and sentimental strategy of ecological defense, a strategy that relies on the *language* of urgency but shrinks from taking seriously its own analysis. As such, it is not a serious approach and, because of the modesty of its demands for change, shares with utopian thinking the worst of all vices—proposing a means that is totally insufficient to attain the end demonstrated to be necessary. The popularity of ecological issues in the West, especially in the United States, exhibits this disjunction of means and ends to an extraordinary degree. It is as if a doctor advised a 400-pound man to lose 200 pounds or die within a year, and went on to prescribe a diet to his patient that consisted only of giving up a second dessert on alternate Thursdays. We would expect a medical association to be appalled by the failure of the cure to fit the condition, and not to be at all mollified by being informed by the doctor that he did not want to upset the patient's life-style. But it is exactly this kind of cure that we have so far been given for an endangered planet by our political leaders.

The cumulative challenge of these various threats to human survival and welfare requires the invention and implementation of a response that transforms the *organizational base* of world society. The first step is to put the facts of crisis in clear perspective, and the second is to demonstrate that this crisis cannot be dealt with by existing organizational forms. Putting the facts in focus is made difficult by certain deficiencies in the information available. There is no way to deal with "just the facts"—data must be assembled, arranged, interpreted. If the asserted danger is related to the future, then trends must be projected, intervening tendencies and uncertainties noted or ignored, rectifying prospects endorsed or discounted. The enterprise *seems* speculative, and *seems* to express only a set of private preferences and to vindicate a personal life-style. Those who are tough-minded on matters of Soviet-American rivalry tend to downgrade these transpolitical dangers, and urge continuing military vigilance as the best approach to security. There is no entirely satisfactory way to overcome the individuality of interpretation in a setting as *complex* and *uncertain* as the *condition* and *prospects* of the planet. Disagreements and severe limitations on knowledge exist, but tentative conclusions can be reached, after the conflicting arguments and their supporting evidence have been carefully examined. We need to consider the evidence, decide, and act, or else we risk the fate of the donkey who starved to death because he could not decide whether to eat the hay strapped to his left or to his right side.

On issues of planetary survival the pessimists and the optimists cannot both be altogether wrong; one set of interpretations is basically correct, and it matters terribly which it is. Furthermore, it seriously matters which is believed to be correct by those who make decisions, shape policies and values, and determine priorities in the main centers of world power. It also matters at what point in time which set of views gains acceptance; the earlier the endangered-planet argument is understood, accepted, and acted upon by those who make policy, the more likely it is that a workable, nonviolent and nontraumatic response can be developed. Should the situation of danger have been overstated, on the other hand, relatively little, perhaps nothing at all, is lost. Even if there is no *survival* hazard of great magnitude, there is certainly the prospect of organizing the affairs of mankind to diminish violence, avoid waste, improve health, raise living standards, protect natural beauty, improve urban living conditions, and so on. It is possible that the credible threat of catastrophe will generate the will and energy to overcome some bad features of our human existence that we have taken for granted or accepted as unavoidable. I would argue, in fact, that the precariousness of human survival might at last give mankind the opportunity to create a social, economic, and political order that would allow human groups to live together under conditions of mutual respect and tolerable dignity.

In addition to the inevitable level of controversy that attaches to conjecture about what the present portends and the future will bring, there is a tendency for experts with opposing interpretations to avoid each other's evidence, premises, arguments, and conclusions. We find, for instance, that one group of population specialists amasses statistical data to show that large-scale famines will inevitably occur in the 1970's, affecting millions, perhaps even a billion or more. We find another group who writes about "the end of the population explosion," who thinks optimistically about intensive agriculture, about chemically processed foods, about the vast food potential that can result from ocean farming, and about a declining birth control. We find yet another group who is optimistic about food production, but pessimistic about the secondary effects of population increases upon scarce wood and mineral resources, upon the quality of urban life, upon civic order, upon human personality, and upon the problems of environmental control and the conservation of nature. These various interpreters of the demographic scene rarely discuss one another's data, do not often refute one another's arguments, and often write as if no contending position exists in the literature. Such a failure of confrontation produces a sense of confusion and supports a position of immobility. The result is to maintain the status quo, as fairly unambiguous evidence is needed to bolster demands for change; otherwise the advocacy of change is easily blunted by the counterarguments and evidence produced on behalf of vested interests. Bureaucratic inertia also acts as a strong brake on change in any situation where the evidence is inconclusive. Such inconclusiveness also serves to reinforce official ideologies of optimism—derived from the illusion of technological omnipotence ("The Apollo Complex") and from a society that believes that thinking positively induces positive results. The Vietnam War has provided one long object lesson in the prevailing American tendency to maintain the smile of the Happy Warrior on all official faces; a dazzling array of "objective" indicators has been relied upon year after year to demonstrate over and over again that the United States is doing better and better in the war. Had there been a realistic assessment of the strength and morale of the National Liberation Front and the North Vietnamese, about the weakness of the Saigon regimes, about the limited effectiveness of military superiority in such a jungle terrain, about the intensity of opposition to the war among American youth, then it is less likely that the war effort would have been pushed to such an outer limit of tragedy and destruction before being gradually relinquished as the most damaging failure in American history. The wider point is that the American emphasis on "winning" and its societal repudiation of "losers" as failed individuals tend to produce inflexible attitudes of resistance to demands for change.

In my judgment, a consensus is beginning to build around the more pessimistic set of interpretations about the dangerous condition of the planet. Men of influence and stature such as U Thant, C. P. Snow, Lewis Mumford, Buckminster Fuller, Robert McNamara, and Andrei Sakharov are sounding warnings about the urgency of acting *now*. Confronting the seriousness of the situation may help mobilize and spread this sense of urgency. Once a sense of urgency exists, receptivity to change is likely to increase. At that point it becomes vital to provide programs of action that have some prospect of success and are not so unrelated to existing patterns of interests and values as to be politically and morally unacceptable. At the same time we need to be vigilant about an acceptance of the endangered-planet argument in political language without any corresponding set of commitments to action.

The leaders of the United States Government are obviously becoming sensitive to the existence of dangerous
conditions in our world, but they seem unable to find the
courage or the programs to moderate the dangers by
drastic action.

We turn now to consider the principal outlines of the
fourfold danger. We proceed from the most obvious
danger—the danger of nuclear war—to the more allusive
danger—the danger of rendering the planet uninhabitable
or less habitable by causing changes in the biosphere.
Note that the dangers come from the character of human
activity, its scale and its heedlessness. There is nothing
about the conditions of human survival and of planetary
circumstance that makes this situation of danger either
inevitable or unavoidable. On the contrary: The planet is
fantastically well endowed, so far as we know, to sustain
life for hundreds of centuries in an extraordinary variety
of forms, including a high level of achievement and comfort for humanity over a long period of time. Also man is
very well endowed to create a life of beauty and welfare
for his species on earth. Even mortality, which makes us
aware of the certainty of our own death, although introducing a note of sadness and short-sightedness into human
life, can be mitigated or deferred by further advances in
medical technology. Surely the planet is a dependent
part of the solar system and the solar system a trivial corner of the galaxy and the galaxy of the universe, but the
time scale of astrophysical change is so large that if we
began to remove man-made threats to survival we could
herald and justify the dawn of an era of planetary confidence in the durability and splendor of the future. We
have the techniques available, provided we can tame
human appetites and limit human population to an optimal number—say 1,000,000,000 people—to provide all
men and women with the material and social basis for a
life of dignity and pleasure.

1. THE WAR SYSTEM. What is the war system? A set of social and political relationships in which the members of a social group expect that violence is likely to be used to settle conflicts with other, foreign groups and with hostile factions within their own midst. Within a family there is, by and large, a peace system; disputes between its members do not typically involve the threat or use of violence; many institutions such as school, church, professional society, and tribe normally operate as peace systems. Violence is rarely threatened, or expected, even in an atmosphere of severe conflict about vital issues. Therefore, preparations to employ violence in an effective manner do not often take place. Violence may, in fact, occur, but its occurrence is not part of the normal pattern of behavior.

Most civil societies are intermediate cases between war systems and peace systems. A society that has a democratic tradition, adequate procedures for social change, adequate police capabilities, and no seriously discontented subgroups tends toward being a peace system, although it will probably need some police capability to protect the social order against deviants and criminals. If the population, or a large section of it, is kept in a subordinate or disenfranchised position through techniques of fear and intimidation, then it is one kind of war system; and if there exists a revolutionary situation within that society it tends toward being another kind of war system. The Soviet Union, to the extent that it keeps internal opposition at bay by suppressive tactics, tends toward being a domestic war system of the first type, whereas the United States, to the extent that the role of the police is enhanced and the claims and organization of militant groups with a revolutionary outlook increase, tends toward becoming a domestic war system of the second type.

A "pioneer" or "frontier" society may be a domestic war system of anarchic variety. Members may employ violence, and relate their security to their proficiency in the techniques of violence, within such a society. In this self-reliant setting police institutions and prerogatives have usually not been reliably established; it is a cowboy atmosphere in which individual self-assertion is highly valued and in which the development of an established society with rights and duties of citizens fixed by law and upheld by government is not looked upon as a necessity. As a result, might makes right to a greater extent, warrior virtues are honored, and there is no sense that nonviolent patterns of living are superior. At some point in the experience of a self-help war system too much anguish and uncertainty result and a transition is made to a policed peace system. As the social and economic order grows more complex the unreliability of a self-help system becomes intolerable to an ever larger share of the citizenry. A famous depiction of such a transition is the ending of Shakespeare's *Romeo and Juliet*, where the Prince of Verona insists that no more violence between the Capulet and the Montague families shall occur and that he will henceforth impose a peace system upon the life of the town.

International society is, of course, an extreme example of a war system. Conflicts abound. Vital interests are constantly at stake. Inequalities of resources and power create incentives to acquire what a neighboring state possesses. War has been glorified in the myths and histories of every country. Many of the great human heroes have been conquerors. The autonomy of separate political units was reinforced in earlier decades and centuries by poor systems of transportation and communication, and by a relatively low degree of dependence upon reliable contact with foreign societies. There was no great urge to control developments in the external environment so long as they did not appear to threaten territorial integrity or political independence. Besides, no serious model of a world peace system existed. As the experience of empire demonstrated, even powerful governments at the height of their glory could extend only a tenuous and temporary kind of political control over peoples far removed from the central bases of ethnic, linguistic, and traditional control. World society could not, until very recently, be technically administered as a single unit. Societies are organized along different, often antagonistic, political principles, adhere to different religions and belief systems, speak different languages, believe in their own cause and oppose that of other societies. The role of violence is central to expectations about security, change, and expansion. Preparations for war and success in war remain the central preoccupation of major sovereign states.

Until World Wars I and II there was no concerted effort to create a peace system in world affairs, although a long tradition of world thought going back to earliest times has advocated this goal. Many efforts were made, often with success, to moderate the scope and barbarism of war, but no serious assault was mounted to remove the conditions that cause war, or to introduce the structures that might be capable of maintaining peace. Since World War I the rhetoric of a peace system has been adopted by most statesmen and some of its essential institutions have even been established in skeletal form. However, no police capabilities have been created on a world level nor have national governments diminished to any extent their prerogatives and capabilities with respect to war. Quite the contrary. Increasing funds everywhere are being devoted to improving the capacity of a government to wage war. The character of international society is that of a self-help war system where the occurrence of violence is frequent and the preparation, threat, and expectation of violence are virtually all-pervasive. Such a condition is characteristic of a war system, and the uncertainty of its operation is heightened by the rapid pace and fundamental character of change in the area of arms technology, a situation that is certainly aggravated by the secrecy and fears engendered by the arms race. The awareness of this uncertainty has induced the most powerful governments to take unprecedented steps *to manage* the war system, especially with respect to the risk of nuclear war. A prominent aspect of such a managerial attitude is an emphasis on the maintenance of a communication link between rivals during a period of crisis; the United States and Soviet Union used "the hot-line," which in 1963 linked the White House to the Kremlin by teletype, during the Middle East War of 1967 in an effort to avoid escalation of that conflict through miscalculation of intentions.

Although it is correct to describe international society as dominated by a war system, it is not uniformly correct. The expectations of violence, and hence preparations for it, vary greatly in the different regions of the world and at different periods of world history. Scandinavia is one area where there is virtually no expectation that a dispute between governments will lead to warfare. Western Europe, as a whole, after several centuries of bitter conflict as a war system, now seems to have achieved progress toward becoming a peace system. Some of this progress may be an indirect benefit of the hostile relationships with the states to the East, and of the vulnerability of a divided Western Europe to the exertion of Soviet military power. Thus, Western Europe's provisional peace system may partially depend for its durability upon the credibility of the threat from the East. Scandinavia as a peace subsystem seems to reflect a genuine repudiation of violence as the means to resolve intergroup disputes, made more impressive by the strong separate identity of the Scandinavian countries, as the failure of projects to achieve Nordic confederation has consistently shown.

The most manifest threat of planetary destruction arises from the danger of a general war between the nuclear superpowers. The extent of this danger is in dispute. We know that contingency plans call for a variety of nuclear attack patterns against potential enemy states. We know that the defense of Western Europe continues to depend on a NATO threat of nuclear response. We know that during the Cuban Missile Crisis of 1962 many participants in the American decision-making process urged a principled stand against Soviet missile deployment in Cuba despite their sense that the stand involved a substantial risk of nuclear war. President John F. Kennedy is reported at the time of his decision to prevent Soviet missile deployment to have believed that the risk of nuclear war was as high as 50:50, and yet he was unwilling to reduce that risk by accepting the symbolic price of removing obsolete U.S. Titan missiles (at the time already scheduled for removal) from Turkey. To be a winner is to avoid the impression of trade-off or compromise.

There remains in the West a strong sense that World War II was partly provoked by the appeasement of the German Government during the 1930's, especially at Munich in 1939. The lesson of Munich has been uncritically applied to the present world situation, and any unfavorable adjustment of the status quo is resisted by our leadership as war-provoking "appeasement." Former President Lyndon Johnson repeatedly vindicated America's involvement in Vietnam as the price of averting a greater war in subsequent years. Should the Soviet Government interpret *history* in the *same* fashion, but appreciate the *facts* in a *contradictory* manner, then the stage is set for a major confrontation. China and other Communist Party groups have accused the Soviet Union of "appeasing" the United States in Vietnam, contenting itself with furnishing assistance at a time when the homeland of a socialist ally was being battered by American airpower. Ralph White, a social psychologist, has shown how normal it is for each side in an international conflict, by selective perception and presentation of facts, to arrive at the sincere conclusion that the adversary is the "aggressor." We live in a world that lacks any very authoritative procedures to assess relative responsibility for recourse to violence. In judging an international dispute even the principal organs of the United Nations arrive generally at conclusions that express the will of a *political majority* rather than a judgment reached after an *impartial assessment* of the merits.

Given the persistence of intense political conflict and the absence of genuine means to inhibit recourse to violence, the danger of nuclear war occurring *at some point* seems considerable. Many scenarios—of a fictional, gaming, and think-tank variety—have been written to depict some *plausible occasions* for nuclear warfare. How plausible depends on a variety of circumstances that cannot be appraised in advance, including the rationality of leadership. The cohesion of bureaucracies, the reliability of command and control, the accuracy of intelligence, the "provocativeness" of adverse military postures, and the unambiguousness of diplomatic intentions are some of the factors that determine whether or not a government is war-prone under varying conditions. As with any fallible mechanism subject to human control, government policy is vulnerable to breakdown, mistake, and misuse. The improvement in "kill" technology, the deployment of ABM systems, the reduction of decision time to a few minutes, the increased number of possible nuclear weapons countries are factors that suggest that the present risk of nuclear war is likely to grow in the future.

If a general nuclear war occurs, the level of destruction could be extraordinary in magnitude. In the last "posture statement" presented to Congress by Robert McNamara, then Secretary of Defense, it was estimated that the Soviet Union or the United States could each inflict 120 million deaths on the other in retaliation. And as George Rathjens has pointed out, "this is probably a lower limit on the actual expected damage, in that indirect and delayed effects of a nuclear attack, such as fatalities from fire storms, maldistribution of resources and fallout, are not generally considered in making such estimates." A more detailed, and more frightening, account of the effects of nuclear attack has been made by Barry Commoner. He shows how utterly devastating it would be for a society to undergo the damage and trauma of a major nuclear attack, and how problematic its prospects for recovery might be. The experience of nuclear attack has been foreshadowed to some slight extent by the atom bombs—miniature (!) explosions by current standards—dropped in 1945 on Hiroshima and Nagasaki; the wounds have remained open for many Japanese, and for those who survived the experience the anxiety and shock of the event have never been removed. Indeed, Garonwy Rees has written that "Since Hiroshima, we are all *hibakusha,* and perhaps the greatest value of *Death in Life* is that it is not the *hibakusha's* experience which it helps us to understand; it is our own." *Hibakusha* is the word invented by the Japanese to describe the survivors of the atomic explosions in August 1945, "a calamity which exceeded men's worst possible imaginings of what the future might hold for them." We must not avert our eyes from the atomic explosions of the past. It is necessary to grasp fully the fact that such weapons have actually been used and are now ready for use on a scale that would make the legacy of Hiroshima and Nagasaki seem in retrospect a trivial prelude.

In the early years of the nuclear age some writers argued that war had become obsolete because nuclear weapons would inflict mutual destruction at such a fantastic level as to make the distinction between victory and defeat irrelevant. But habits and structures of security planning have proved extremely durable. It is exceedingly difficult to transform the behavior of modern governments: cross-cutting vested interests, incremental decisional style, and ingrained patterns of policy-making in large bureaucratic institutions combine to prevent all but minor adjustments in attitude and behavior except under the pressure of an emergency. In other words, the demonstration that recourse to war, or retention of the war system, is irrational under circumstances of the nuclear war is not likely by itself to bring about *fundamental* change. Marginal changes may occur. More effort has been devoted to the avoidance of *accidental war* through misinformation, some effort is being devoted to the implementation of the common interest of the nuclear powers in preventing the proliferation of nuclear states (hence, the Nonproliferation Treaty), and some effort will be made to draw geopolitical boundaries more sharply so as to diminish the likelihood of crisis confrontations (hence, the low level of Soviet response to the U.S. intervention of 1965 in the Dominican Republic and the low level of U.S. response to the Soviet occupation of Czechoslovakia in 1968). Tinkering, caution, prudence, are all signs of adjustment to the nuclear danger, but they do not involve any effort either to pull out the roots of danger (the nuclear capability itself and the absence of effective and mandatory procedures for peaceful settlement), or even to stem the tide of reliance by governments upon military self-help and superiority (the arms race itself continues to maintain its momentum).

The irrationality of nuclear war is not to be confused, as has so often been the case, with the obsolescence of warfare. The threat to inflict nuclear damage may influence an adversary to act or refrain from acting in certain ways. Peace in Europe has often been attributed, although not convincingly, to the deterrent impact of nuclear weapons on Soviet behavior, as have other inhibitions upon risk-taking by revisionist or expansionist states (e.g., West Germany, China, Soviet Union). That is, nuclear weapons are *used* in a latent capacity to shape behavior and to sustain present patterns of dominance even when no nuclear warhead is exploded. Hence, the elimination of nuclear weapons is often alleged to improve the risk-taking situation that faces aggressively minded states. Since nuclear states, especially nuclear superpowers, also are the richest states (except Japan and West Germany, which enjoy firm nuclear guarantees) with the most to lose as the result of a forcible redistribution of wealth and territorial control, there exists a strong incentive to maintain their present position of military dominance.

The early experience of war appears associated with securing food for cities whose inhabitants were not hunters or farmers. Neolithic farming villages were largely undefended, and the best evidence suggests that no general pattern of organized violence existed. But with the development of cities—first of a theocratic variety, later of a more imperial character—it seems clear that warfare arose out of the need of the urban tyrant for an assured food supply from the country. There also existed a drive to expand territory under control so as to provide additional farming land for the children of peasants. Wars also seemed to arise out of conflicts between groups for access to and control over scarce water supplies. The natural size of a political unit was restricted, except under a charismatic leader, to an area relatively close to the urban center of authority. Military strength was degraded by distance, placing a limit on the extent and durability of empire. The impact of such limiting circumstances varied considerably with resource base, energy supplies, geographical position, administrative efficiency, ideology, leadership. In general, however, the possibility existed for the growth of separate, but unequal, sovereign states; and indeed the sovereign state has emerged as the basic organizational unit in world society and continues to possess remarkable vitality.

Given the present situation of mass undernourishment (more than two-thirds of world population), it is worth taking account of the ancient link between war and the control of food surplus, as well as the age-old human practice of protecting positions of political and economic privilege by military means. In 1965, Lin Piao, then Vice-Chairman of the Chinese People's Republic, made a notable analysis of the present world situation in terms of the rural sectors of the world (those countries that are poor and underdeveloped) encircling and eventually overcoming the dominance of the cities of the world (the rich, industrialized countries). Lin Piao tried to demonstrate that the victory in China of the revolutionary forces depended upon their ability to use the countryside as a base area from which to encircle and finally to strangle the forces of reaction and oppression: "The countryside, and the countryside alone, can provide the broad areas in which the revolutionaries can maneuver freely. The countryside, and the countryside alone, can provide the revolutionary bases from which the revolutionaries can go forward to final victory." In a closely linked passage Lin Piao proceeds from the revolutionary struggle within a country to the revolutionary struggle within the world:

> Taking the entire globe, if North America and Western Europe can be called "the cities of the world," then Asia, Africa, and Latin America constitute "the rural areas of the world." Since World War II, the proletarian revolutionary movement has for various reasons been temporarily held back in North America and West European capitalist countries, while the people's revolutionary movement in Asia, Africa, and Latin America has been growing vigorously. In a sense, the contemporary world revolution also presents a picture of the encirclement of cities by rural areas. In the final analysis, the whole cause of world revolution hinges on the revolutionary struggles of the Asian, African, and Latin American peoples who make up the overwhelming majority of the world's population.

It is, of course, fascinating that Lin Piao identified the age-old struggle of "the farmers" to protect their surplus from confiscation by the armed might of the cities as the central feature of the modern struggle among nations for

dominance. We live in a world of all-pervasive exploitation, where the wealth of the rich is based on the primary
commodities of the poor and where the masses in the
poorer countries live under conditions of poverty and
virtually unalleviated misery. A stratified world system is
one in which superior instruments of warfare help sustain
the basic foundation of economic and political *inequality*.

Violence in world affairs persists in the nuclear age.
There is, first of all, the reliance by governments upon
structural violence to sustain positions of dominance in
foreign societies. The Soviet Union *uses* violence *both* to
inhibit insurrection and to suppress it in Eastern Europe,
most prominently displayed so far in Hungary (1956) and
Czechoslovakia (1968). The United States uses violence
both to inhibit insurrection and to suppress it throughout
Latin America and in many parts of Asia, most prominently in Guatemala (1954), the Dominican Republic
(1965), and Vietnam (1964 to the present). Paramilitary
and nonmilitary interference with foreign societies occurs
in a variety of ways, many designed to keep in power a
ruling group that is compatible with the maintenance of
a global position of dominance.

Throughout the world, within and between states, the
politics of force continues to control the course and
outcome of principal international conflicts. Even in the
most prosperous and powerful societies of the world there
are many indications that abrupt and drastic change
depends upon successful recourse to violent techniques
of insurrection and that those who control the system
depend, in turn, on recourse to structural or repressive
violence to discourage, and if necessary, to defeat radical
challenges that arise from within the society. Soviet
society is not peaceful, it is merely efficiently regimented
through the systematic application of the instrumentalities
of structural violence. Those who counsel "law and order"
do not generally seek to move our society away from
violence, but merely support building up the balance of

force and prerogatives in favor of the government. South Africa has few incidents of violent opposition, not because the governed are satisfied or content with existing grievance procedures, but because the institutions of repression have been granted such wide discretion to use violence, intimidation, and brainwashing to throttle potential or perceived enemies of the state. Laurence Gandar, the editor of the *Rand Daily Mail,* Johannesburg's leading daily newspaper, was convicted in July 1969 of *a crime* under the Prisons Act because he exposed—in a circumstance where the evidence was not even seriously contested—the *routine* practice of torture in South African prisons. The prosecution was initiated and he was convicted, although the South African Government clearly knew that what Gandar had published was *true;* a demonstration to all South Africans that *the truth of criticism* is no defense against a prosecution for *its falseness.* Such is the extreme logic of the police state: structural violence is carried to the point where violent opposition appears to vanish, and the mass of inhabitants turned into helpless captives of the governing group. The white regime in South Africa relies, and thinks it must rely, on its military capabilities to assure its continuing domination of the black majority living in the country. The link between violence at the disposal of the white regime and the inequality between the white and black races seems plain.

In such a world structure we find two central political phenomena: first, a stratified world society in which great disparities of wealth correspond to disparities in power; second, highly stratified social and economic structures in a number of key states in which power disparities roughly correspond with disparities in wealth. The maintenance of stratification seems to depend on the threat and use of force. It also depends on convincing the population that the existing hierarchy of wealth and influence is just or inevitable; therefore, direct and indirect forms of thought control are coordinated with direct and indirect methods of police control. The coordination is explicit in totalitarian societies and tends to be more tacit in "open" societies with a democratic tradition. In democratic societies the media are directly and indirectly operating under the control of those who are at the top of the socioeconomic pyramid, although this control is to some extent obscured by the toleration and indulgence of a measure of dissent, an indulgence that disappears at a time of stress, namely, whenever the dissent appears to shake the foundations on which the structure of stratification rests. Files on dissenters are now being kept by security agencies in the United States, making it possible to convert the society quickly and efficiently into a highly repressive one. It appears, therefore, to be very difficult to eliminate violence from the collective life of mankind without removing or, at least diminishing, the most exploitative forms of stratification.

In addition to the use of naked power, many belief systems have been evolved to vindicate stratification or to deny its existence. Theocratic societies invoked the sanction of divine forces, as did early monarchies, claiming that the legitimacy of kingship involved a matter of "divine right." Liberal democratic societies espoused a creed of citizen equality that obscured the realities of inequalities that included in the United States the systematic denial of the rights of man to black and brown people and other minorities. The societies in the communist world attempted by ideological means to eliminate the issue of stratification by associating its presence with the exploitative relationship of entrepreneur and worker, a relationship that was to be eliminated by the revolution. But most socialist societies have built up bureaucratic elites that have held their own populations under severe constraints while their governments engaged in a set of power-seeking policies imposed from above. In addition, a new managerial class emerged, reestablishing social and economic conditions of inequality, although operating under a new myth of socialist community.

International society since the Peace of Westphalia in 1648 has been dominated by the idea of sovereign equality. In actuality, the history of statecraft has mainly exhibited the various consequences of inequalities among states. Notions of empire, imperialism, and spheres of influence express the systematic nature of stratification. Wars often arise to challenge outmoded or deteriorating systems of stratification. The international legal structure embodied unequal relationships by creating the status of colony, dependency, protectorate. The international treaty has always served as a flexible instrument by which to give validity to the consequences of inequality. A peace treaty giving territorial title to the conqueror enjoys a good legal standing. The law ratifies the outcome of a war, might makes right, and as with any self-help system, a winner is assured glory as well as dominance.

In addition to the reliance upon violence to maintain or to disrupt a structure of stratification, considerable energy and resources are devoted to the competitive and hostile relationships that exist among elites in control of distinct structures of stratification. There is a constant struggle to expand, defend, alter the domain of dominant influence in world affairs. The principal expression of this struggle is war and even more so in the nuclear age, constant preparation for war—the arms race that proceeds at different levels depending on the nature of the competition and conflict. The arms race is likely to develop in two principal directions: *vertical* and *horizontal.* The vertical development involves the continuation of the rivalry among principal states, especially the nuclear superpowers.

A series of domestic factors helps to sustain arms competition at a high level of expenditure independent of any assessment of external security needs. The entire economy and governmental structure are geared to large defense industries and a massive military establishment. The discussion of security policy is based on a fairly constant share of the federal budget and there is little disposition to take seriously, whether by way of arms control negotiations or unilateral decision, any proposal for a military cutback of more than a marginal sort. Even when American society is faced by growing demands for more spending for welfare purposes and to protect the environment against further decay, it becomes almost impossible to resist the domestic determinants of a high investment in defense. The decision to go ahead with the ABM Safeguard System was a watershed test of strength of the domestic defense constituency because the case for going ahead was so weak. In such a context the external security arguments operate so as to mystify the public and the leadership, offering elaborate explanations for taking action that is being largely forced upon the government by domestic pressures. As with most effective forms of mystification, the rationale has some measure of tenability, and it is treated by its creators as real; and self-mystification results. Therefore, the tenets of the full logic of deterrence may be professed with great sincerity even when their basic function is to provide an acceptable cover story for the assertion of the power of the Pentagon. My basic contention is that we cannot begin to understand what emerges as security policy without examining domestic structures of power and influence. Part of the reason why it is difficult to establish a cheap, safe world security system is that those who make policy in principal governments are, by and large, socialized into an acceptance of the obsolete existing system and suffer from acute forms of self-mystification (that is, they really believe in its rationality and necessity). Those who do not accept the logic of deterrence are just not recruited into the top ranks of government. There is a narrow tolerance range for deviation that allows for revision in the form of small modifications of the present security setup. These small modifications—domestic reforms and arms control measures—are politically possible only to the extent that they can be reconciled with the basic integrity and preeminence of the defense establishment. Even influential critics—self-styled "doves"—confine their attacks to the exposure of waste or excess profits, the internal inconsistency of a deterrence argument, or the assertion that a particular set of actions or the deployment of a given weapons system will stimulate adversaries to take some very dangerous countermeasures. The rivalry among adversary states involves carrying on a search for constantly improved defensive and offensive weapons systems in an atmosphere of secrecy and conflicting judgment about adversary intentions. The defense posture of both the Soviet Union and the United States rests on the minimum requirement of possessing a retaliatory capability that is highly likely under all foreseeable circumstances to survive a first strike. Such a capability must be sufficiently *invulnerable* to a first strike—if an enemy should attack—so that it both survives and itself is able to get through whatever defenses the enemy may be thought to have. The degree of confidence about the adequacy of an assured destruction capability depends on what the probable adversary does and does not do by way of developing its security policy. The nuclear superpowers can cooperate in keeping their defense needs at the lowest level consistent with a capacity to inflict assured destruction on a rival or they can compete in a variety of ways so as to establish more solid assurance on one side and imperil the assurance of the other. There are many subordinate issues, some of which seem caught up in current debates about whether to develop, test, and deploy MIRV weapons systems. If in addition to assured destruction some effort is made to *limit damage* by developing a defense system that will protect cities (to some extent) against a nuclear attack, then the other side may grow worried about whether its basic deterrent remains intact. This worry may be further increased if one side seeks also to discourage provocations other than a direct attack upon itself. If a government grows worried then it will probably take steps to increase its capability to penetrate its probable rival either by increasing the number or size of its nuclear warheads or by improving its system of missile delivery. By augmenting its penetration capability, however, the other side's assured destruction capability may, in turn, grow vulnerable to a first strike. Such a prospect induces a further reaction, the need to build up survivable missiles by either expanding their number, increasing their mobility and indestructibility, or dispersing their location.

If one side is, or thinks that it is, vulnerable, then it may be tempted to adopt a preemptive strategy in the course of a crisis or if it receives an intelligence report that predicts an enemy attack. The loss of an assured second-strike capability induces first strikes, and the awareness of this situation may tempt the invulnerable actor to preempt the potential preemptive strike actor. The logic of deterrence rests on a combination of *perceived credibility* and *selective rationality of behavior.* It is essential that A believes B does or does not possess the assured destruction capability and would or would not retaliate if attacked. A is expected to refrain from attacking or even provoking B, if B has this capability and effectively conveys its intention to retaliate. Under these circumstances an attack by A appears to invite the destruction of both A's and B's basic social order, and it becomes highly irrational for A to take any appreciable risk of its own destruction by inviting a nuclear response.

But suppose A attacks B in any event, destroying, as it might, at least 80 percent of the industrial and population base of B. At this point, is it rational for B to carry out its threat to destroy a comparable portion of A's society? Why? B has a better prospect of recovery if A is in a position to give it assistance, something that might be bargained for in exchange for refraining from carrying out the deterrent threat. Also, much less radioactivity would be placed in the atmosphere. Country A would presumably give a great deal to dissuade country B from firing a salvo of nuclear missiles at its principal cities. It would seem to be almost always *irrational* to carry out the threat that underlies the logic of *assured destruction.*

If such an assessment is correct, then, there is no reason why A should be deterred except for the possibility that B would go ahead and retaliate despite the *irrationality* of the action in the wake of a nuclear attack. The paradoxical rationality of nuclear deterrence rests on the *irrationality* of the victim of an attack that could be viewed as rational in terms of predicting the outcome. B's retaliatory response could be made almost automatic, that is, engineered into the weapons system so as to be independent of human judgment. Such a posture would serve to convince A that B would, *in fact,* retaliate. But if B puts itself in this position, then it is committed to irrational behavior of a highly self-destructive form, perhaps decisively, should deterrence fail to prevent a nuclear attack.

This kind of analysis of strategic nuclear policies can be carried to ever-greater degrees of refinement. The logic of deterrence is not at all persuasive, at any given point, but its cumulative effect is to convince military planners that the other side would attack first unless at every point we can convince ourselves and our adversary that we can and will retaliate, that we will, in that ultimate moment of decision, pursue the irrational option, despite the existence of many powerful reasons for not doing so. We can understand the *fragility* of our whole security concept by exposing and apprehending its root commitment to *irrationality.* A further point is that human behavior tends to be compulsive in conflict situations. It seems rather likely, therefore, that should the contingency arise, then the retaliatory threat would be carried out *regardless* of consequences. Man is often an irrational animal, perhaps never more so than when using the rhetoric and equipment of hyper-rationality. Computers, systems analysis, the detached reasoning of deterrence theory contribute an elaborate facade of objective analysis to a set of arms policies that are plundering the valuable resources of the earth and placing the entire edifice of human civilization into a situation of extreme and permanent jeopardy.

Given the evident determination of Soviet and American leaders to maintain this kind of security system there are more and less *dangerous* and *costly* ways to maintain it. Danger is increased by security policies that call into question the defensive intentions of a nuclear capability. Any weapons development that seems to place an adversary in a vulnerable position is especially dangerous. It is not likely to result in achieving any military advantage that can be put to political use, but the other side is likely to take countermeasures that involve more spending and give even greater dynamism to the arms race. The logic of deterrence depends on beliefs about the probable intentions of adversaries. Therefore, to influence these intentions it is necessary to manifest a military advantage, thereby further deterring a potential adversary. If the objectives of a particular military acquisition were mixed, and included some interest in being in a position to attack or defend against an attack, then it might be important to enshroud military capabilities with a mantle of secrecy. Cost for all actors participating in an arms race is also enhanced by pursuing more dangerous deterrence routes, especially as arms races aggravate international tensions and tensions rise after evidence that a nuclear rival is seeking to achieve advantages that undermine the reliability of the deterrent.

Such a dangerous and costly "balance" of deterrence has been proceeding through various phases during the past two decades, and seems likely to continue. The competitive ethos dominates, and more effort continues to be devoted to securing relative advantages in military posture than to stabilizing or reversing the arms competition. The cost/resources spiral causes some pressures even on a country as rich as the United States, leading to a call for less defense spending, for a more stable security ideal, and for greater emphasis on domestic social issues.

A significant side effect of international arms competition is to build up a powerful military establishment within domestic society. The virtual militarization of national governments may result, causing a profound influence upon foreign policy-making and the character of political leadership. Such a situation fosters a certain kind of bureaucratic outlook and is connected by strong links with the industrial sectors of the economy. In the United States, we have long heard—since President Eisenhower's farewell remarks in 1961—about the danger of the military-industrial complex, and more recently, critics of government spending for security purposes who enjoy fairly strong backing in the Senate have talked of dismantling "the national security state." The evidence suggests that the arms race has also had a major impact upon patterns of influence within the Soviet Government, although far less is known about its form and effect.

The *vertical* arms race also serves to maintain the gap between the superpowers and all other states. At the present time most, but apparently not all, leaders in the Soviet Union and the United States have adjusted themselves to the prospect of their vulnerability to one another, and they share the common security objective of making every other state in the world vulnerable to attack without any hope for significant retaliation. The existence of a common objective on the part of the United States and the Soviet Union seems particularly evident in their hostile relations to China. The French reaction under De Gaulle to the American effort to sustain its exclusive control of the nuclear deterrent, even in relation to allied states, suggests the same insistence of shared nuclear primacy with the Soviet Union. The Limited Test Ban, and even more the Nonproliferation Treaty, can be appreciated, in large part, as representing an effort by the nuclear superpowers to deny additional states the prospect of becoming significant challengers of their dominance on the world scene. Why, otherwise, concentrate on "nonproliferation" rather than on the renunciation of the right to use nuclear weapons or even on their elimination? The attitude of nuclear powers toward arms control discloses the desire to improve the safety of international life provided it does not imperil the *politics of stratification,* those political, economic, and psychological relationships that ultimately seem to require *inequalities of power.*

At the same time bipolar competition between the United States and the Soviet Union induces a *horizontal* arms race at the subnuclear level in many parts of the world. To sell and supply arms has many incentives, the most obvious being the attainment or retention of political influence and an improved balance of payments for the supplier country. The rich countries supply an increasing quantity of arms to the poorer countries, and these armaments reflect the rapidly improving technology of war. As a consequence, wars are carried on in all parts of the world with more and more destructive weaponry, and states that find themselves confronted by serious threats to their security are undoubtedly tempted to acquire weapons of mass destruction. These more modern weapons are so complex that the recipient country needs to receive training and guidance in their proper use, which itself creates a relationship of dependence upon the supplier country. Perhaps even more significant, as a weapons system grows more complex the recipient country is almost completely dependent on the supplier for spare parts.

Israel and South Africa are obvious examples of states surrounded by a hostile regional environment; at some point the military balance could shift against them to the point where some kind of ultimate weapon that confronted their enemy with destruction could alone safeguard their security. Note that principal states have been willing to rely on nuclear weapons in response to far less central threats directed at their political survival and independence, and their governments have been unwilling to conceive of any serious alternative. Since World War II no serious disarmament negotiations have ever taken place, despite frequent affirmations of the goal of a disarming and disarmed world. The United States Arms Control and Disarmament Agency even went so far as to pay more than $200,000 to a group of known opponents of disarmament to prepare a study as to whether general and complete disarmament would serve the national interest of the United States. The highly negative results were published in a book, and no comparable effort has ever been made by ACDA, the supposed disarmament constituency within the U.S. Government, to solicit a comparable analysis from scholars more sympathetic with the prospects for disarmament and with the creation of a more unified and less statist system of world order.

Vast expenditures are devoted to the development of weapons systems that will never be used under battle conditions, or if used, will represent a breakdown of the security system shaped by deterrence thinking. In view of the insufficiency of world resources to meet human needs, the enormous withdrawal of resources from productive application as a consequence of the war system is a prime social fact of the present human situation. So much more ingenuity goes into the design of better weapons than into finding ways to diminish the reliance on huge military establishments. Compare, for instance, the huge R&D budget of the Department of Defense with the feeble long-range planning efforts of the State Department.

According to the most recently available figures, global expenditures for military purposes were $132 billion in 1964, $138 billion in 1965, $159 billion in 1966, and an estimated $182 billion for 1967. Since 1962, when expenditures were assessed at $120 billion, there has been a 50 percent rise in military spending. The prospect is for further increases, and for a less stable and less secure system of mutual deterrence at the strategic level. "Military spending today exceeds that of any prior period except the peak fighting years of World War II. Global military expenditures now take more than 7 percent of the world's gross product. In money terms they are equivalent to the total annual income produced by the one billion people living in Latin America, South Asia and the Near East." The report goes on: "Very rough estimates indicate that since 1900 more than $4,000 billion have been spent on wars and military preparedness. If the current level of military spending should continue, this total will be doubled in only 20 years. If the recent rate of increase in military spending continues, the arms race will consume another $4,000 billion in only 10 years."

NATO and the Warsaw Pact countries accounted for nearly nine-tenths of the rise in expenditures in 1966 and 1967, with the Soviet Union and the United States being responsible for $35 billion of the $44 billion increase in military spending. The rise of military spending is proceeding more rapidly than is the increase in the rate of population or GNP growth. Therefore, throughout the world military spending takes up a greater per capita and proportional share of national income each year. The developed countries with less than 30 percent of the world's population are responsible for over 85 percent of total military expenditures. The United States and the Soviet Union alone, representing together only about 12 percent of the world's population, spent over $110 billion on arms, or more than two-thirds of the world's total. This immense level of arms expenditure is a shameful ingredient of a world confronted by mass poverty and limited resources. It is true that a reduction in arms spending will not lead to an automatic increase in welfare spending or foreign aid, but it seems evident that current expenditure of funds, talents, and resources on armaments reflects budgetary and political priorities that do not begin to correspond with the needs of mankind, or with the planetary discipline that will be required to create a tolerable future.

Surely, part of the defense effort of the rich is a consequence of economic and political inequality. The rich governments want to be in a position to protect their ranking in the hierarchy of states and to compete with other rich countries for an improved ranking. If nuclear weapons technology spreads, as seems likely, then the temptation grows for disadvantaged states to challenge a world order system that exhibits such grossly unequal per nation and per capita consumptive standards. The danger of this challenge may, in part, sustain the momentum of the arms race, assuring a decisive superiority over poor states even at the price of a less stable equilibrium among the richer states. In this respect, and it may be a disguised or repressed factor in military planning, the Soviet-American military rivalry is a pretext for the joint, yet unacknowledged, interest in managing power to assure their continuing dominance over large sectors of international society and their expropriation of a share of nonreplaceable raw materials far in excess of their proportion of the world's population. The United States population is 6 percent of the world total, its use of resources is estimated at between 40 and 60 percent of the world's annual production, resulting in a degree of inequality that would appear likely to strain any political system. This inequality seems especially likely to strain the world system in which such a great proportion of disadvantaged people live at or below the margin of subsistence at the same time that the visible share of the consumptive life of the richest countries is devoted either to building elaborate machinery for destruction or to indulging consumptive standards of superaffluence for a large segment of its population.

Such circumstances, even in the absence of an endangered-planet situation, would appear to create a revolutionary situation in world society, especially as advances in communication make the populations of the world aware of these economic disparities. The gravity of this situation increases when it becomes evident that the quantum of waste expenditures on the part of rich countries is depleting the resources of all peoples at an alarming rate, is deteriorating the environment to such an extent as to make the planet less inhabitable, and is relying for its security upon nuclear weapons which if used would cause great, perhaps irreparable, harm to all societies and not just to the belligerent societies. The fallout of lethal doses of radioactive material from nuclear explosions would have no respect for national boundaries. It has been estimated, in fact, that a nuclear exchange involving more than a certain megatonnage might well prevent subsequent reproduction anywhere on the planet. In such a setting of shared danger and risk, it hardly seems tolerable to exclude most of world society from the planning and decisional processes.

There are, then, several related sets of issues:

1. The level of military spending on the part of rich countries in a world of deprivation and dwindling resources is itself a potential source of conflict in the years ahead;

2. The degree of socioeconomic inequality that exists within international society would seem likely in the future to generate demands for and efforts to achieve greater equality, or at least to impose on the rich a burden to assure the minimum needs of the poor;

3. Continuing increases in the level of military spending are likely to magnify the dilemma of insufficient resources as well as to highlight the consequences of the present pattern of priorities as between health, education, and welfare and national security sectors of the economy;

4. All states will soon be in a position to achieve access to biological, chemical, and nuclear weapons of mass destruction and will, therefore, be in an improved position to rely on force to press their claims for a less unequal international system;

5. The consequences of a serious breakdown of deterrence between the nuclear superpowers will cause great suffering and hardship for other societies and may even imperil the collective existence of societies in no way involved in the conflict.

It would seem evident, then, that the war system places very serious strains upon the existing international order. The idea of separated centers of governmental authority presupposes the capacity to restrict direct "battlefield" damage to the countries at war. Such spatial restrictions are no longer meaningful if future wars are fought with biological, chemical, and nuclear weapons. In such circumstances the organizational structure of international society seems fundamentally defective. There is *no participation* by most of the world in the decisional processes that control the most basic and vital issues of group welfare and survival. In effect, the entire world has become vulnerable to the particular interests, orientations, and misperceptions that guide governing elites in Moscow and Washington, a vulnerability that seems pervasive, permanent, and profound. The premises of sovereignty rest on the real capacity of national governments to control national destinies, but such control has been lost for all but two governments in the nuclear age. The loss of this control should give most national governments of the world a great interest in working to achieve some kind of new security system based on *central guidance* whereby their interests are taken into account.

The two strongest political tendencies in the world today *within* national societies are the demands for *equality of treatment* and for *participation in processes of political decision.* The war system illustrates the extent to which neither of these demands is being met by the structure of international society. Prospects for the future indicate a worsening of these negative tendencies, abetted by growing access to more lethal weapons, by an increasing number of national governments, and even by subnational groups intent on producing drastic change by violent means.

The war system provided a tolerable, if morally degrading, basis for sustaining group life when the *costs* and *risks* of war bore some proportional relationship to the *stakes* of conflict, when increased investment of resources in national defense *raised* expectations of security against external attack, and when the technology of warfare did not impose such an enormous cost for mistakes and accidents. Under present circumstances, the war system

1. Endangers human survival by risking major nuclear war;

2. Endangers the level of social, economic, and political existence of all countries by risking major nuclear war;

3. Converts huge population groups into hostages of the potential enemy in the name of "security";

4. Withdraws resources and energies from the solution of tasks vital to the quality of life in all societies, especially in poorer countries where the majority of people do not have the minimum requirements of a decent life;

5. Reinforces human separateness in such a way as to prevent creative perception of wholeness, and the formation of a suitable response to the endangered-planet situation;

6. Perpetuates internal and external structures of domination by which people are held in subordination by the use of military means;

7. Provides a costly way to resolve conflicts, achieve and resist changes, and work out the international consequences of new balances of values, expectations, and forces.

2. POPULATION PRESSURE.

There is a Vietnamese folk saying that expresses the mood of rising anxiety about population issues: "When the wind becomes stronger a typhoon is on the way." Most thoughtful people seem now to agree that there are already *too many people* in the world today, and that there will be *many more* in the near future. There is far less agreement on *why* there are too many people and even less on what *should* and *can* be done about it. Population problems touch upon fundamental issues of human existence—the degree to which individuals should be free to determine the size of their own families, regardless of social and political consequences, and the extent to which population growth beyond a certain level imperils the oxygen supply of the planet, leads to very dangerous forms of thermal pollution, or imperils the psychic health of man through overcrowding.

Most people continue to associate the population problem with food supply. Malthus sounded his famous alarm in 1798 by propounding the idea that a disaster was inevitable as a consequence of the tendency of *food supply* to increase at an *arithmetic* rate (2, 4, 6, 8, 10) while population increases at a *geometric* (2, 4, 8, 16, 32) rate; here is the essence of the Malthusian doctrine—"the power of population is indefinitely greater than the power in the earth to produce subsistence for man. Population, when unchecked, increases in a geometrical ratio. Subsistence increases only in arithmetical ratio." This Malthusian emphasis has tended, by virtue of its simple plausibility, to focus population concern on the prospect of famine. A recent revival of Malthusian prophecy has heavily documented the prediction of famine and starvation for millions in the years ahead.

For the public this kind of emphasis causes some confusion. First of all, there is a battle of experts who advance contrary arguments about the prospects of providing enough food for everyone. For instance, Georg Borgstrom, a food specialist at Michigan State University, concluded a scholarly book published in 1968 with the following stark conclusion:

> The fact is that the world in all likelihood, and this on the basis of most available evidence, is on the verge of the biggest famine in history—not, to be sure, the world *we* live in, but the poor world, the countries of Asia, Africa, and Latin America. Such a famine will have massive proportions and affect hundreds of millions, possibly even billions.

At the other extreme is the work of Colin Clark, who urges further population increase as a basis for more general kinds of economic growth and is optimistic about the capacity of the world to feed a growing world population with an improving diet:

> We conclude that the world possesses the equivalent of 77 million sq. km. of good temperate agricultural land. We may take as our standard that of the most productive farmers in Europe, the Dutch, who feed 385 people (at Dutch standards of diet, which give them one of the best health records in the world) per sq. km. of farm land, or 365 if we allow for the land required to produce their timber (in the most economic manner, in warm climates— pulp requirements can be obtained from sugar cane waste). Applying these standards throughout the world, as they could be with adequate skill and use of fertilizers, we find the world capable of supporting 28 billion people, or ten times its present population. This leaves us a very ample margin for land which we wish to set aside for recreation or other purposes.

According to Clark, the world could sustain a world population of 95 billion if Japanese standards of farming and nutrition are used. Clark's analysis seems spectacularly inconsistent with the expectations of agricultural specialists such as the Paddocks or Borgstrom, who foresee very limited increases in the amount of production that can be achieved either by adding acreage or by more intensive methods of agriculture. The growth of population, the expansion of urban and industrial sectors, the construction of highways, airports, and so on, withdraws large quantities of land from agricultural use each year. On a cost-benefit basis agricultural use of land in a heavily populated, industrializing society is far less profitable than almost any competing use, and hence the amount of agricultural acreage tends to decline under market conditions where the bidding is for scarce land.

As might be expected, an intermediate position has been endorsed by most official and quasi-official students of the relationship between expected population increase and expected food supply. One such intermediate assessment has been made by the highly respected Committee on Resources and Man of the National Academy of Sciences in its Final Report:

> Foreseeable increases in food supplies over the long term, therefore, are not likely to exceed about nine times the amount now available. That approaches a limit that seems to place the earth's ultimate carrying capacity at about 30 billion people, *at a level of chronic near-starvation for the great majority* (and with massive immigration to the now less densely populated lands)! A world population of 30 billion is only slightly more than three doublings from the present one, which is now increasing at a doubling time of 35 years. At this rate, there could be 30 billion people by about 2075 in the absence of controls beyond those now in ef-

fect. Hopeful allowance for such controls . . . suggests that populations *may* level off not far above 10 billion people by 2050—and this is close to (if not above) the maximum that an *intensively managed* world might hope to support with some degree of comfort and individual choice, as we estimate such immeasurables. If, in fulfillment of their rising expectations, all people are to be more than merely adequately nourished, effort must be made to stabilize populations at a world total much lower than 10 billion. Indeed, it is our judgment that a human population less than the present one would offer the best hope for comfortable living for our descendants, long duration for the species, and the preservation of environmental quality.

Note that the theoretical possibilities are far in excess of even an optimistic interpretation of *practical* possibilities. The 30 billion ceiling presupposes complete mobility of people and perfectly efficient development of food supplies, unencumbered by ecological interferences or by the shift in land-use patterns from rural to urban and industrial uses. Coping minimally with even the 10 billion prospect presupposes an "intensively managed" world, that is, a world that has been able to evolve an effective central guidance mechanism to govern the whole for the benefit of all mankind.

Even aside from the issue of agricultural capacity is the issue of social structure. Economists express the problem by observing that the schedule of *demands* for goods does not always correspond with the schedule of *human needs;* that is, people are often unable to pay even the production costs—let alone the producer's profits—for goods needed to sustain a decent standard of living. The argument here is that the tendency to specify the theoretical limits of production overlooks the structural problems that currently prevent even food-in-being from reaching those who are hungry. At the present time in the United States the government rents more than 50 million acres from farmers (out of 350 million acres of total cropland) to assure that they will not grow food, although more than 10 million hungry people exist within the population. A study conducted by Iowa State University suggested that a total of 62½ million acres should be withdrawn from production to avoid agricultural surpluses in relation to available demand. Somewhat similar withdrawal schemes are being or will likely be put into operation in Australia, Canada, New Zealand, Argentina, and France to avoid agricultural gluts. We have, then, the anomaly of enormous numbers of people in the world presently suffering from undernourishment and strenuous efforts by a variety of governments to reduce agricultural production. At the *present* time it is estimated that at least 10,000 people per day—or 3.65 million people per year—die from diseases related to malnutrition. It is further estimated that between 65 and 75 percent of the world's population is undernourished if their diets are measured in terms of minimum caloric and protein requirements. Most of the Afro-Asian population is undernourished and malnourished at the present time, entailing disease-proneness and restricted development of human potentiality. We are led, then, to the conclusion that the most troublesome part of the food/population nexus is a consequence of social structure: Many people lack the purchasing power (and often even the knowledge of their nutritional requirements) to obtain the food they need. Even in the United States there are many hungry people because they can't afford to buy, not because the food supply is insufficient.

Jean Mayer has written, "It must be recognized, of course, that many of the worst nutritional scourges of mankind have been historically due as much to ignorance and to callousness as to lack of nutrients as such." Robert G. Lewis has well expressed the *demand* side of the

Malthusian problem in the modern world, especially in societies that adhere to market systems of pricing:

> We have kept our eye on food, instead of on people. We have concentrated on supply, when the real problem has been and still is demand. The result is just about what happens when you try to push a string—frustration and ineffectualism. Throughout the 1950's and 1960's we have groaned about "burdensome surpluses" and paid farmers to "plow under" as much as one acre out of five, only to discover, when at last the U.S. Public Health Service took a look, that we probably have more malnutrition right here in America than at any time since the beginning of World War II. And now U.S. foreign aid policies, built to the specifications of the same conventional agro-wisdom, have started the developing countries down the same road, speeding people off the land to make way for tractors, chemicals, and "efficiency" without a thought as to where the people should go from there—much less what they might eat when they arrive.

In 1968 the Final Report of the 34th American Assembly, distributed under the title "Overcoming World Hunger," paid almost *exclusive* emphasis to increasing the world food supply, setting a target of an annual 4 percent increase in food production in the developing countries and imposing on the advanced countries an obligation to make agricultural assistance available so that such an outcome could occur. In official thought, little attention is given to the *distribution* of existing food supply, and especially how to get food to people who need it and are unable to pay market prices. If the food/population ratio is understood in these terms, then the problem becomes social and political as well as agro-technical in character. The problem of increasing the effective demand for food on the world and national markets becomes a critical element in converting increases in food supply into consumptive patterns that feed hungry people.

Some commentators do not regard the emphasis on food supply as at all warranted. Jean Mayer, perhaps the most persuasive of the non-Malthusian population alarmists, has recently written that "considering the world as a whole, there is no evidence that the food situation is worsening and there is at least a likelihood that food may at some time (20 to 30 years from now) be removed altogether as a limiting factor to population." Like Colin Clark, Mayer believes that much more intensive agriculture can be developed on farming acreage through double-cropping, fertilizers, and pesticides, and that there is a fair chance of a large-scale breakthrough in exotic methods of research going on in food chemistry, involving an effort to harness photosynthesis to permit algae production, and investigating a variety of possibilities for ocean farming. Mayer's pessimism is related to the secondary effects of an increasing global population, and in this sense "the population problem" tends to be more serious in the advanced countries than in the poorer portions of the world. "I am concerned about the areas of the globe where people are rapidly becoming richer. For rich people occupy more space, consume more of each natural resource, disturb the ecology more, and create more land, air, water, chemical, thermal, and radioactive pollution than poor people. So it can be argued that from many viewpoints it is even more urgent to control the numbers of the rich than it is to control the numbers of the poor." In these terms, the essence of population pressure is associated with ecological disturbance, in relation to both environmental quality and resource base. The United States is expected to add almost another 100,000,000 people to its population by the end of the century, or an increase of almost 50 percent of its present total, and this entire population is expected to be living at a far higher per capita GNP by the year 2000. The futurologists, Herman Kahn and Anthony Wiener, offer the following projections of U.S. population and per capita GNP:

	1965	1975	1985	2000	2020
Population					
(in millions)	195	222	258	318	421
Per capita GNP					
(in dollars)	$3,500				
Low projection		4,150	5,000	6,850	9,550
High projection		4,800	6,650	11,550	21,250

This image of the future becomes even more discouraging when one realizes that the model of "development" and "progress" is almost everywhere based upon rapid industrialization leading eventually to the high-consumption economy of the United States and Western Europe.

A professor of biology at the University of Kentucky, Wayne H. Davis, has put the non-Malthusian population crisis in a provocative focus. His essential point is the same as that of Jean Mayer, namely, that the ecological dangers come principally from the rich, not from the poor. In terms of garbage production, pollution, land use, and resource depletion each additional person to the United States is equivalent in ecological terms to the addition of at least 25 people to India; Davis suggests that 500 might be a more realistic figure and recommends that we think of our population growth in terms of "Indian equivalents." If we examine the situation in this light the U.S. population expressed in the conservative-measure Indian equivalents is now 5.15 billion (25 x 203 million)—or far more than the total world population of 3.5 billion. By the end of the century the U.S. population will be 7.5 billion in Indian equivalents (as measured by present per capita GNP in India—that is, 25 x 300 million), but given expected increases in American standards of living it would be closer to the higher reading of Indian equivalents (500:1) or 150 billion, far above even the most optimistic view of the theoretical carrying capacity of the earth, as measured in terms of food supply. Professor Davis also shows that the use of Indian equivalents to assess relative population growth changes our sense of where to locate the emergency—India's population is growing at 2.5 percent per year, whereas the U.S. population is growing at only 1 percent per year; but converting Indian equivalents gives the U.S. a 25 percent rate of annual increase or a rate of growth that is 10 times as fast as that of India. Davis makes his overall case in a domestic context, which exhibits in our country the same unevenness of impact that arises from a comparison between India and the United States: "Blessed be the starving blacks of Mississippi with their outdoor privies, for they are ecologically sound, and they shall inherit a nation." We could extend the analysis to the quality of life in the United States, and use a notion of "Mississippi equivalents" to express the degree to which ecological pressure is a result of what rich people in the cities do; again we might use a figure of 25 to emphasize the extent to which population growth among the rich may be socially less desirable than among the poor, although under present conditions neither is desirable.

There is another entire dimension of population policy that concerns the relationship between increases in the numbers of people and the probable capacity of governments to relieve the mass misery of their populations. The larger a national population, other factors being held constant, the larger the aggregate social needs are for food, clothing, shelter, health, and education. The prospects for raising—or even sustaining—the per capita quality of life in the developing world are worsened by the outlook for increasing numbers of people everywhere, the increases being especially concentrated in the cities of the poorer countries.

We have, then, two kinds of non-Malthusian positions

of emphasis in relation to population policies: *first, any further population increases are harmful to the extent that they further deteriorate our environment and deplete our resources; second, population increases are detrimental to the extent that they interfere with efforts to relieve the misery of population groups in Asia, Africa, and Latin America, where an appalling poverty is endured by a great proportion of the population.*

Finally, there are apocalyptic interpretations of population increase that seek to demonstrate the inability of the world to accommodate existing rates of population increase for very much longer. The explosive dimensions of population increase are contained in the simple arithmetic properties of the compound interest formula:

Rate of Population Increase (% per year)	Time Taken to Double Population (number of years)
0.5	139
1.0	70
1.5	47
2.0	35
2.5	28
3.0	23
3.5	20
4.0	18

Applying this table to the population of Mexico, with a rate of growth of 3.5 percent, Karl Sax has shown that in 42 years her present population of 40 million would equal that of the United States today, that in less than 130 years it would exceed the present world population, and that in 200 years it would have increased 1000 times or reached a total of 40 billion. At present rates of world population growth of almost 2.0 percent, which lead to a doubling of world population every 35 years, the following alarming results occur:

1965	3.5	billion
2000	7	billion
2035	14	billion
2070	28	billion
2105	56	billion

If these projections are accurate, the world might even, it has been suggested, begin to run short of oxygen well before 2035, that is if it has not already starved to death, been decimated by pandemics and nuclear war, or perished in an ecological catastrophe. Other projections of continuing population growth have been made to demonstrate overcrowding to such an extent that there would be literally no space for additional people to stand even if they occupied every inch of the earth's surface; in such a scenario depicting the outer limit of population density, the end result is that we squeeze one another to death by overcrowding. The entire world's surface becomes comparable to a subway train during rush-hour in Tokyo or New York, but worse. The point of these sorts of projections is to demonstrate that whatever else, things cannot continue much longer the way they have been going without producing a catastrophe of a decisive and vivid character. Such a demonstration serves some purpose it it can convince people that we need to make as early an adjustment as possible in population increase to avoid an increasing prospect of a demographic Armageddon.

For centuries the population of the world remained approximately stable at about 10 million. By the time of the early Christian era, or after the passage of many thousands of years—possibly a million—the human population had expanded to 250–300 million, by 1650 it had increased to 500 million, by 1830 to 1 billion, by 1950 to 2.5 billion, by 1960 to over 3 billion, by 1975 it is expected to be at least 4 billion, and by the end of the century between 6 and 7 billion. In terms of compound interest increases the rate of growth was 0.3 percent between

GROWTH OF HUMAN NUMBERS

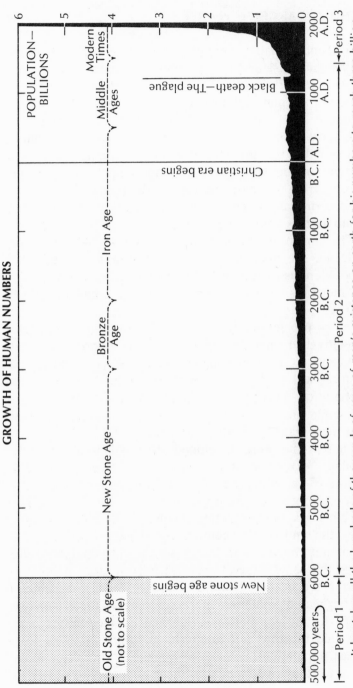

It has taken all the hundreds of thousands of years of man's existence on earth for his numbers to reach three billion. But in only 40 more years population will grow to six billion, if current growth rates remain unchanged. If the Old Stone Age were in scale, its base line would extend 35 feet to the left!

1650 and 1780, 0.9 percent (or three times higher) between 1900 and 1950, and 1.8 percent (or six times higher) since 1950. René Dubos has summed up the dramatic acceleration of population increase in recent times: ". . . it took hundreds of thousands of years for *Homo sapiens* to achieve a population of 3 billion, but his numbers might be doubled in the next 40 years!" By examining these broad trends in population growth we find support for the earlier contention that basic changes in the human situation are occurring at an accelerating *rate,* and that the resulting situation is something genuinely *new* in human history. The current world population growth of 700 million per decade—a number greater than the combined 1960 populations of the entire continents of Africa and Latin America—is equal to the entire world population up to 1650.

The projections for future population growth suggest relatively higher growth rates in the poorer countries. As a result, an increasing proportion of the world's population will be born in countries where low per capita conditions prevail in the future. According to the most reliable projections "the less developed areas will have an aggregate population of about 5.4 billion by 2000, whereas the more developed areas will total about 1.6 billion." Expressed in percentages, this means an increase from 67 to 79 percent of world population living in the poorer countries. As population specialist Philip Hauser points out, this projection involves a 3.4 billion increase by the end of the century for these countries, or an amount equivalent to the present world population; during this same interval only 598 million more people are expected in the advanced countries. Such a distribution of expected population increases is part of the explanation of why the gap between rich and poor countries is expected to grow, not diminish, in future decades; these vast population increases also make it virtually impossible for the governments in these countries to satisfy the demands of their populations for improving standards of living.

One further aspect of the population picture arises from the very much more rapid growth of cities than of rural areas in the current period. Barbara Ward has suggested the universality of this phenomenon by noting that "If world population on the average is growing by 2 percent a year, cities grow by 4 percent, the megalopolitan areas by 8 percent. Moscow grows as quickly as Paris, Rio de Janeiro as Chicago." In terms of aggregate increases, the prospects are for the rural sector to increase at the rate of 1.6 percent, whereas the urban sector is likely to increase at the rate of 6.8 percent; in terms of doubling time, this means that the urban population of the world will double within eleven years. Kingsley Davis, one of the foremost experts on urbanization trends, has made the following comment about this world process of accelerating urbanization: "The rate of change was remarkable. Ten percent of the world's population were already living in million-plus cities. In 1950, 161 million were so living and today the figure is 375 million. On present trends the whole population of the world would be living in cities of a million and over within 76 years—15,000,-000,000—and there would be cities of 1.3 billion population." Obviously such projections cannot and will not be realized, and are in no way to be understood as predictions. As already suggested, adjustments can come about by more or less traumatic processes; those that are more traumatic are likely to involve severe human costs, violence, repression, disease, famine—the occurrence of a disaster sequence. By showing the inevitable necessity of adjustment, given the projection of present trends, a case is built for the early management of human affairs through concerted action. The exponential character of the trends—rates of change of 2 to 8 percent—establishes the urgency of doing something now.

Urbanized societies are more easily mobilized for

political action. Given the modern technology of mass communication—available everywhere—the prospects of totalitarian repression or wildly expansionist political programs promising a cure to the travails of "the people" seem almost certain. All population groups seem very susceptible to well-administered programs of thought control. The most primitive societies in the world have public squares where loudspeaker systems can harangue assembled masses day after day with the most incendiary kind of political appeals. The susceptibility of even sophisticated populations to primitive and orgiastic political appeals has been revealed by the political success of Nazism and Fascism in conditions of societal distress. Such distress will be present everywhere—it is already manifest in many parts of the world—and the extremist forces with their violent tactics and simple solutions will find it ever easier to acquire and hold power, thereby diminishing even further any hope for an increase of human sensitivity in relation to the kind of issues of world order that will determine whether there exists any possibility of achieving a positive political future for mankind.

The population outlook, then, seems fairly clear: there are too many people, given the resources and the social and political structures of the world. Hence, the trends toward rapid growth need to be moderated immediately and reversed as soon as possible. Such an assertion represents a baseline minimum consensus on population policy, although there are certain kinds of resistances from the most fundamentalist of Catholic and Marxist groups, the Catholics adhering to the view that the increase in the number of human souls can only enhance the glory of God, and the Marxists holding to the view that the only source of human misery is class exploitation which, if eliminated, would establish the conditions for a virtually automatic harmony of man and society—at which point even the state is expected to wither away.

Some radical left groups also regard a further expansion of the miseried masses as providing the objective conditions for revolution, and hence, of human and historical benefit. At this point the evidence of population imbalance is so overwhelming that even Catholic and Marxist centers of thought seem to be moving toward the reformulation of their world views to incorporate some more active concern with the importance of limiting further population growth. We are moving toward, although at a rate far slower than the dynamics of population growth itself, a realization that something must be done *now*.

Families with more than seven children although quite unusual in our country are quite common elsewhere. According to the current figures of the Population Council, the average number of children per family is as follows:

Africa	6.1
Asia (without the Soviet Union)	5.1–5.5
Latin America	5.7
U.S. and Canada	3.7
Europe	2.7
U.S.S.R.	2.9

Such figures show the extent to which existing childbearing patterns exceed the 2.14 figure that would be needed to establish a stable population, what has been referred to as "a people-stat" or a zero rate of population growth.

A recent study by two demographers, Larry Bumpass and Charles F. Westoff, does support the significance attached by Nixon to the reduction of unwanted offspring. These authors estimate "that one-fifth of all births and more than one-third of the Negro births between 1960 and 1965 were unwanted" and reports that "the incidence of unwanted births is negatively related to both education and income." According to the findings of Bumpass

and Westoff "the proportion of births reported as unwanted . . . is more than twice as high for families with
incomes of less than $3,000 as for those with incomes
of over $10,000." Obvious political consequences result
from eliminating unwanted children, among the most
apparent of these consequences being the reduction of
the flow of voters from sources that normally identify
with the Democratic Party. The time-deferral is so long
between birth and voting age that such a consideration
would seem to have only symbolic weight.

Bumpass and Westoff developed a useful model of
"the perfect contraceptive society," that is, a society
in which there would be no unwanted children. Applying
such a model to the latest survey of U.S. attitudes would
reduce annual births by 20 percent and cut the fertility
from 3.0 to 2.5 births per woman. Because of the age composition in the United States it would take 70 years to
reach a zero rate of population growth even if the fertility rate fell in 1970 to the level of replacement. An
immediate zero rate of growth for American population
would require a reduction of fertility to 1.1+ children
per family. To achieve an ultimate zero rate of growth
in the United States population depends on reducing
children per married couple to 2.14, but as we have
suggested, such a rate of reproduction would not, even
if soon realized, quickly produce a stable population for
the United States. The goal of a perfect contraceptive
society seems desirable, especially in countries where
there is a relatively high rate of unwanted fertility and
where the average desired family size is not much larger
than 3.0. We do not have very good information as yet as
to what impact a perfect contraceptive model of fertility
would have on population growth in India, Africa, or
Latin America, nor do we know whether to assume that
the mother who is faced with the prospect of an unwanted
child is ready, willing, and able to practice contraception
to prevent it. Nevertheless, the approach is attractive

because it appears to reconcile—in a very humanitarian way (that is, by enabling people to do what they want to do)—values of personal freedom with the attainment of a near-stable population level. As such the thinking that underlies "the perfect contraceptive society" may provide a socially feasible way to mitigate the effects of the paradox of aggregation as applied to matters of population limitation. Practical implementation even in the U.S., however, seems far away. However, the trend toward legalizing abortion in the United States may considerably diminish present prospects of population growth.

We are growing *alert* to the actuality of population pressure, but our idea of a *corrective response* seems far too superficial. Once again we observe the workings of *inertial consciousness* that does not alter basic patterns of behaving and feeling until the pressure of disaster induces a sense of emergency. In the present setting, with the time scale of population increase so much shorter than that of adjustment, it becomes difficult to avoid a pessimistic attitude toward what the future holds in store.

Such pessimism is confirmed by the experience of the past. The problem of population pressure goes back to early man. The dynamics of exponential growth in recent decades suggests that brutal checks on population increase operated in the past, as seems clear, since the birth rate of man has held fairly steady. It is not yet widely appreciated that the dramatic increases in world population reflect a falling death rate rather than rising birth rate. An increasing proportion of children born reach child-bearing age under current conditions of improving public health, especially with respect to the prevention of infant mortality and the control by vaccines and insecticides of lethal diseases such as malaria and yellow fever. There are still dramatic differences between the infant death rates that prevail in different parts of the world. The average rate of infant deaths per thousand live births in Europe and North America is less than 25, whereas in Cameroon it is 137; in Chile, Ecuador, Peru, and Guatemala it is close to 100 per thousand. All over the developing world the rate is well over fifty per thousand, or more than double, often triple, what it is in the richer countries. Such a statistical comparison emphasizes how much potential there is for added population growth in these world regions, as it is far easier to bring death rates down than it is to alter birth rates. This potential for population growth becomes even larger if one takes account of the age profile in the countries having the highest birth rates. In most of the poorer countries, somewhat over 40 percent of the population is under 15 years of age, resulting in a larger *proportion* of the population being of child-bearing age than at previous periods before the full effects of the falling death rate became evident in the relative age of national population. In the American context a similar phenomenon, on a far smaller scale, has resulted from "the baby boom" in the decade after the end of the Depression and World War II, when there was a sudden spurt of child-bearing. This enlarged birth activity between 1947 and 1957 is just now bringing to child-bearing maturity this bumper postwar baby crop, with the consequence that the population would continue to grow even if the population as a whole achieved a standard of reproduction that equaled replacement.

The relation between *falling* death rates and *static* birth rates illustrates the dramatic and dangerous effects of uneven rates of change. Medical technology, especially epidemiology, has been able to find very effective and rather inexpensive techniques to protect population groups against the ravages of certain kinds of killer diseases. These techniques can be superimposed upon even a primitive social structure at a low, bearable cost, and bring underprivileged societies some of the benefits of modern medicine. Birth control technology is not nearly so easy to disseminate, nor is it nearly so popular. Everyone agrees that it is desirable for a government to do what it can to prevent death. Very few are now prepared to say that it is desirable to prevent birth. In fact, among poorer people everywhere, the experience of child-bearing and child-rearing is one of the few genuine pleasures that attends an otherwise miserable existence. Also, attitudes toward family size are undoubtedly shaped by expectations in earlier generations of much higher infant mortality, often as high as 50 percent, and these attitudes are retained in the collective subconscious of societies and may take many generations to reflect a new situation in which 85 to 99 percent of all infants born survive and have an excellent prospect of reaching adulthood. Some recent data have supported the conclusion that improvement in living conditions, especially survival prospects, reduces the average desired family size in a community. Dr. William Greenough interprets his field experience in East Pakistan as follows:

> The village people I talked with and treated who experienced the loss of one or more of their children as a frequent occurrence seemed to be little inclined to control their fertility. . . . Thus in one particular overpopulated area at a personal level it seemed crystal clear that the regular loss of a very large number of offspring provided an urgent drive toward a high birth rate.

The thrust of this observation is to cut the ground from under neo-Malthusian contentions that to improve the health and survival prospects of an overpopulated, poor community merely defers its day of disaster. Harold Frederiksen's study suggests that after a period of continuing population growth in a poor and backward country, the birth rate begins to fall if the health and welfare of a society are improved. Such an observation has significance for foreign aid and for policy-making, suggesting that the causes of the elimination of poverty and the control of population are compatible, rather than mutually inconsistent.

It is not surprising, therefore, that women in India or Chile, however poor, continue to want on the average at least four children and are not generally even interested in birth control until after the fourth or fifth child. Therefore, cutting down on population growth is not a matter of getting the right kind of birth control pill or intrauterine device (IUD) into their hands. The development of a socially constructive *attitude* toward birth control is as important as the provision of adequate birth control *technology*. It also takes several generations to establish *the point* that large numbers of children are not a free labor source for a family that has moved to the city with a set of rural attitudes. Education, propaganda, and family-planning subsidies can accelerate the learning process, but not in any dramatic fashion under most circumstances. The underlying issue of human choice is far more complex than it is with the avoidance or deferment of death by the introduction of better medical facilities or by the distribution of better drugs. Rumors and horror stories travel fast from mouth to mouth in a backward society. Thus, reports that the pill raises the chances of cancer of the cervix or that the intrauterine loop leads to internal bleeding are likely to discourage use on a widespread basis, leading women to be more reluctant to practice birth control. These reports and rumors are likely to be believed, whether or not true, especially in countries where education is low, superstitions widespread, and many people are generally suspicious about the bad effects of the "modern" world. As of now it is thought that less than 3 percent of child-bearing women in the world practice birth control, a percentage far less than the supply of birth control techniques would permit.

The experience of the Irish famine is chastening as an indication of how a disaster tends to stabilize the population of a poor country, even though it is a Catholic society with strong taboos against birth control. Late in the seventeenth century the potato was introduced into Irish agriculture with great results so far as increasing food supply. Population rose rapidly from two million to eight million by 1845. As Kenneth Boulding has put it, the potato—"a technological improvement"—had as its principal effect "to quadruple the amount of human misery on that unfortunate island." There was more food available leading to a larger population, but no improvement in life quality: such a pattern is a grim reminder that "the green revolution" of the agriculturalists does not necessarily alleviate mass misery, although it may expand the ranks of the miserable. In 1845 the potato crop failed in Ireland; a massive famine ensued. Two million Irish died of starvation and another two million emigrated. Since 1845 Ireland has kept its population quite stable at around four million, mainly by a rather mysterious postponement of marriage until a later average age. The Irish experience, including the disaster and its aftermath, suggests one image for the future. It also suggests the positive results that can flow from such a simple expedient as late marriage: no technology, but a widespread change in attitude presumably reinforced by social pressure in a culturally homogeneous, proud, and self-conscious society. Shifting attitudes toward childbearing and family size is of great importance.

Those countries—Taiwan, South Korea, Japan—that have achieved relatively stable populations in recent years by birth control, abortion, and education, enjoy relatively high standards of living, rates of economic growth, and literacy; evidently there has been spread a sense that further population increase and large families would jeopardize gains already achieved. Such a set of circumstances does not pertain in most of the developing world where the level of mass existence has remained close to or below the poverty line; it is virtually impossible to persuade impoverished people that they should practice birth control to protect their economic position. Also, such a circumstance does not apply to the propagation by the affluent who cannot envision under present circumstances how larger families on their part will imperil their living standards and who do not yet comprehend the extent to which rich families burden the environment of modern society more than poor families. In fact, prevailing beliefs are generally the opposite, undergirded by a eugenic ideology that maintains that the gene pool is shaped by those genetic types who reproduce most. The argument is that the main evolutionary influence, now that modern medicine and hygiene allow the less fit, even the unfit, to survive, arises out of differential rates of reproduction among the various ethnic, religious, economic, and regional groups in a society. The ethical implication caught within the analysis is that those who are socially responsible tend to limit their family size to their means, whereas those who are not do not. Hence, the gene pool tends to become more and more skewed in favor of the least socially responsible elements in the population. This eugenic kind of analysis may be persuasive in its own terms, but it needs to be considered in relation to an ecological analysis which leads in the opposite direction, relating environmental decay directly to higher per capita consumption.

The overall population outlook suggests that the projected totals—which have, in recent decades, always turned out to be underestimated—will contribute to a condition of mass misery in many countries and cause a great deal of harm to planetary prospects for survival. The size of the world population seems likely in the decades ahead to push up against the planet's carrying capacity, just as fruit flies will multiply up to, and beyond, the tolerance capacity of the bottle they are placed within. There does not seem to be much hope of bringing the population below the UN median variant projection of 6.5 billion for the year 2000, and there is a considerable likelihood that it will be 7 billion or beyond by that time and 8 billion by 2005. We do not know yet exactly what this kind of expected population growth will mean, in terms of either human suffering, ecomanagement, or political behavior, but we can be fairly confident that the situation will be worse than it is today, and that it is already very serious indeed. There seem also to be grounds for believing that the size of world population is directly correlated with the level of carbon dioxide in the atmosphere, thereby posing serious, if uncertain, ecological hazards for the world as a whole. These hazards are abetted by an increase in the world level of industrialization, and especially by the gradual shift from electrical to nuclear energy as the principal source of power. It is obvious that population size also complicates greatly the task of all other forms of pollution abatement, including especially the problem of waste and its disposal. There appears to be a Malthusian law of garbage increase: for every arithmetic increase in GNP per capita there is a geometric increase in per capita output of garbage. Precisely the same secondary pressure—with multiplier characteristics—exists in relation to resource use and to the destruction of the natural environment of flora and fauna. Urban crowding in circumstances of poverty also creates a setting in which disease flourishes, and can spawn epidemics and even increase the danger of crippling or lethal pandemics.

In many ways population pressure underlies the entire crisis of planetary organization. Recourse to war has often been motivated in the past by the desire for living space—recall Hitler's call for *Lebensraum*—and by the search for assured sources of raw materials and markets. The Japanese period of "expansionism," culminating in the attack on Pearl Harbor, was in part a search for "a co-prosperity sphere" of Asian countries that provided both the materials and the markets for a rising Japanese industry that was being squeezed out of world trade by Western states. It is simplistic to conclude that population increase *as such* causes war, but it builds the conditions whereby violence is likely to play a larger and larger role in the internal and external affairs of states: (1) Crowding creates pressure for expansion into emptier spaces. (2) Resource demands create incentives to assure resource supplies. (3) Mass misery per square mile, especially as trends toward internal migration involve the expansion of the urban sector, encourage the emergence of repressive systems of domestic government that maintain "law and order" and give rise to extremist ideologies that promise "a new deal" by revolutionary strategies of violence. (4) Widening gaps between per capita GNP in rich and poor countries will tend to intensify "the sense of injustice" concerning economic and political inequality in world society; perceived deprivation has been found to underlie the frequency and magnitude of internal strife, and is likely also to be related to the level and frequency of external strife. (5) Population pressure in rich countries will reinforce existing barriers to migration across boundaries and to the provision of food assistance, and will reduce further the present meager willingness of rich countries to alleviate the situation of the poor ones. (6) The arms market is likely to make available the kind of weapons that a state seeks to possess in order to pursue an aggressive foreign policy.

These conditions will take place in a technological environment that will increasingly induce recourse to the war system as a strategy of change: (1) The improved techniques of mass communication—radio and TV—will make people more aware of inequality and deprivation, and more susceptible to mass mobilization by emotional appeals to irrational feelings of hostility. (2) The spread of nuclear energy facilities will provide almost every significant country with the technological base needed to produce a significant number of nuclear bombs per year, and even if this option is not exercised, almost any government or militant political group will be able discreetly to produce or obtain bacteriological and chemical weapons.

Population pressure continues to be dealt with almost exclusively as a matter of *national policy.* In the language of international law the issue of population policy continues to be a matter of *domestic jurisdiction,* that is, a matter on which governments are entitled to act without subjecting their action to any international right of review. On an international level, facts about population growth are collected and disseminated and expressions of international concern are made. There is a small UN effort to promote birth control by educational and financial help, and there is a concerted effort to help countries raise their food supply, to organize famine relief, and to help prevent the spread of infectious disease. At this time, however, international institutions lack both *the mandate* and *the capabilities* to do anything about standardizing national population policies so as to discover and protect *the world interest.* So far as population policy is concerned, the world operates on a laissez-faire basis, the future appears to depend on the degree to which principal national governments *adopt* and *implement* enlightened national policies that uphold the world interest.

We are faced, then, with a double level of permissiveness in trying to evolve a rational population policy. The world interest depends on the wisdom and conscience of national governments, and national governments depend on the wisdom and conscience of their individual citizens. We have noted already how emphatically President Nixon combined his concern about population pressure with a reaffirmation of an individual right to refrain from family planning and to create a family that accords with his own interests, values, and preferences. The UN Declaration of Human Rights avoids any direct affirmation of such a right, but does endorse individual freedom to found and build a family without governmental interference. The prospect for international regulation, then, illustrates the workings of the paradox of aggregation in a situation where the dangers of repeating the tragedy of the commons seem great indeed. On a subnational basis we find some groups that even contend that a call for birth control is tantamount to preaching "genocide." It is plausible to seek an increased share of political and economic power, especially in a democratic society, by maintaining a higher birth rate than prevails among rival groups in the society. Such a logic applies, of course, both to the many militant minorities that want a larger slice of wealth and power and to the ascendant groups that are eager to retain their advantageous position. In fact, such a strategic view of relative population size generates a competitive dynamic that resembles the arms race. Each actor is stimulated to increase its efforts in response to its perception of the efforts of a rival to improve its relative position.

The essential point appears to be that neither the pattern of human *attitudes* nor the *structures of power* seem turned on to the realities of population pressure. And, in fact, it is not clear that a world structure dominated by sovereign states could ever be expected to collaborate closely enough to protect effectively the world interest in population control. At present, the world interest is vulnerable to irresponsible or ineffectual action on the national level. In this structural setting, the highly centralized societies of the communist world appear to have an advantage, being organized and efficient in the articulation and implementation of national standards of individual behavior, although even these governments have experienced an inability to remold family structure or to abolish discrimination against women. The traditions of Western liberal society are deeply opposed to efforts by government to interfere with behavior within the family unit. American society continues to maintain a constitutional right to bear arms and is especially protective of family sovereignty. The approaches taken by the United States to population control, due to its prominence and power in the world today, are almost certain to exert a broad influence on the approaches adopted in much of the non-communist part of the world. Even intelligent people seem to believe that if population pressure were really dangerous for national and global welfare then our leadership in Washington would know about it and take appropriate steps. Such confidence in the ability of a government to protect a society against downfall and decay has never been historically vindicated. Governments have not been quick to grasp novel dangers to their welfare and survival, and even when such dangers have been appreciated, the capacity to make adjustment has been rarely evident. The Bible is filled with tales about unheeded prophets who accurately warned societies and their leaders about imminent disaster.

The first corrective step that needs to be taken is to recognize that population policy is a matter of world interest. It would then be desirable to arrive at some sense of what constitutes an optimum world population and what needs to be done to approximate its attainment. A second step would be to move national thinking, especially in key countries such as the United States, from a *mood of concern* toward *programs of action* based again on some image of what an optimum population would be under varying conditions of technology, ecomanagement, and social justice. At present, tax and welfare policies subsidize births rather than reward continence. By taking these steps it would appear clear that progress toward fulfilling the world interest depends on overcoming some of the *structures of national separateness* that have grown out of the ideology and history of the sovereign state. Similarly, progress toward fulfilling the national interest depends on breaking down some of *the structures of family autonomy* that have developed out of an earlier laissez-faire view of the relationship between the individual and national government.

3. THE INSUFFICIENCY OF RESOURCES.   The Natural Resources Committee created by the National Academy of Sciences in response to a request by President Ken-

nedy in 1961 defined a natural resource as "any naturally
occurring element, product, or force that can be utilized
by man in his contemporary environment." Such a definition of resources is a characteristic expression of
human egoism, of man's dominion in relation to his environment that is there to be used as the interests of man
dictate. Such a view of resources contrasts with a more
reciprocal image of man as an element of nature, as a
part of the whole, as living in harmony with his environment.

Since the early decades of the industrial revolution
there have been recurrent expressions of pessimism about
the exhaustion of resource stocks upon which the existing pattern of economic development rests. A professor of law, Earl Finbar Murphy, has called attention
to the pessimistic analysis of Stanley Jevons, a noted
English economist, who in 1865 published a short book
called *The Coal Question,* in which he anticipated a
dwindling supply of coal and, therefore, a definite upper
limit upon the capacity of the steel industry, and hence
the economy as a whole, to experience further growth.
Jevons' analysis was ill received in Victorian society,
where an optimistic belief in an ever-improving future
had taken hold partly because a growing industry would
be providing society with more and better goods with each
passing year. In fact, further large reserves of coal were
located, although it is true that it was only as a result of
the timely and quite distinct discovery of the coking
process that the British steel industry was rescued from
the brink of collapse and proved Jevons' pessimism to be
an inaccurate anticipation of the future.

And so also with the other great English pessimists of
the nineteenth century—Malthus, David Ricardo, John
Stuart Mill—all of whom thought that increasing human
per capita demands together with a continuous expansion of human population would inevitably lead to an
exhaustion of the finite, and apparently rather limited,
supply of nonrenewable resources upon which economic

progress rested. How could finite resources satisfy infinite wants over an indefinite period of time?

America, too, spawned its own breed of resource pessimists who foresaw the early exhaustion of even our abundant natural resources if current habits of use were to persist. In the 1890's W. J. McGee and Gifford Pinchot, for instance, warned about the imminent depletion of oil, coal, and minerals within a matter of decades and urged upon the nation rigorous policies of resource conservation. But as with the coking process man's ingenuity has seemed always to foil the pessimistic view of the resource future. Always new economically usable stocks of resources have been discovered that have provided for all of human needs, and always technological innovations have enabled the successful location and extraction of resources beyond wildest hopes and have developed satisfactory substitutes for those resources that have not been found in sufficient natural supply. Despite all the growth in petroleum use over the last few decades, the present estimate of proven reserves leads to more optimistic views than existed at times of less use. By and large the progress of technology has discredited pessimists who have consistently understated the prospects for innovation. What appeared to be problematic has been resolved in time to avoid either resource shortage or societal hardship.

As a result, resource needs are now not often emphasized in discussions of the *effective limits* of human potentiality. More precisely, the projections to the year 2000, or even beyond, that involve higher per capita GNP for a more populous world are not generally thought to be jeopardized by anticipations about the availability of resources. Even so acute and sensitive a student of environmental issues as Professor Murphy writes: "In the latter part of the twentieth century, world reserves of ores and fuel present no serious problems. With the aid of technology, ample quantities of both will be found in deposits of low content or in remote locations, and modern transport will move them cheaply from source to use." As an illustration of this future pattern of resource sufficiency, Professor Murphy calls our attention to the fact that "As early as the mid-1950's American coal, extracted by almost completely automated methods, was selling at the pitheads of the Ruhr for less than the local coal; technology will make this sort of thing common in the future." The gleam in the technologist's eye has grown even brighter now with improving prospects for mining activity on the continental shelf and ocean seabed and for rather cheap sources of energy through nuclear power developments.

Instead of projecting the resource future on the basis of existing knowledge, an assumption is made by analysts that the capacity for continuing payoff from technological innovation will take ever better care of human needs in the future. Thus, in a major study of the resource outlook in the twentieth century the report concludes that "the main escape hatch from scarcity is technological advance across a broad front, and behind this have to be large, varied, effective programs of research and development in science engineering, economics, and management. And to back up these efforts, in turn, there must be a strong system of general education at all levels." Particular stress is placed on the need for major increases in research and development investments "if an ample flow of low cost materials is to be continuously available and if the resource base is to be maintained and utilized effectively." The conclusions are generally optimistic, including the assumption that innovative applications must (and therefore will) occur at a rapid rate to meet the resource requirements of the future: "The U.S. historical data do not point to increasing scarcity in any general sense. Indications as to future technology and supply possibilities, when matched against projected demand for the next four decades, likewise do not indicate a general tendency toward increase in scarcity. A continued rise in the level of living seems assured."

Serious speculation about the future by Daniel Bell, Herman Kahn and Anthony Wiener, and Zbigniew Brzezinski accords no attention whatsoever to problems of resource scarcity in their attempts to provide comprehensive depiction of the problems facing postindustrial society, despite their projections of ever-rising rates of per capita consumption. There is an implicit assumption of resource sufficiency, perhaps a spillover from an overall mood of technological omnipotence, that does not even deign to take note of the sharply rising demand curve for finite stockpiles of nonrenewable resources. In the literature of economic development this mood is so dominant that the issue of resource sufficiency in a world of expanding GNP is rarely even discussed. Almost all the attention of development specialists is given over to economic, social, political, and cultural factors that might distort the flow of resources or hamper the rate of economic growth.

Our informational base appears to be very poor when it comes to assessing the resource stocks of the developing countries. Joseph Fisher and Neal Potter of Resources for the Future, among the most respected and influential specialists on resource policy, observe that "In the less developed areas of the world, the data are extremely thin and projections hazardous," but they "are not persuaded that the next few decades will see any general and marked deterioration of living levels because of increasing scarcity of raw materials." Such a guarded kind of optimism does not seem reassuring, especially when it is accompanied by an acknowledgment of the poorness of the data. Fisher and Potter "do not believe that shortages and inadequacies of natural resources and raw materials are likely to make the more favorable outcomes [in terms of economic growth] impossible of achievement." They also see no impediment to the growth of the Soviet Union or Eastern Europe (China is left out of account) arising out of raw-material scarcity.

There are occasional expressions of concern, but their

influence has tended to be restricted and has not yet been taken into account in the general interpretations of the future. Charles F. Park, Jr., a geology professor at Stanford University, has written a careful book, *Affluence in Jeopardy: Minerals and the Political Economy,* in which he compares the consumption curves of such critical minerals as iron ore, lead, and copper with world population curves to demonstrate that future demand is soon likely to outpace future supplies, with very damaging consequences for the political economy of the nation. Professor Park's perspective is given in the following paragraph:

> If a person were asked to predict what will happen in the future, would not the answer have to be that, if the population growth continues as predicted to the year 2000, greater shortages of all kinds will exist and there will be more and deeper poverty in the world than at present? If greater human misery and grief are to be avoided, can the population growth be permitted to continue at the present rate? More than our affluence may be in jeopardy. It would seem inevitable that shortages will become increasingly critical; even today shortages are developing in silver, mercury, and tin, and the future of these metals is far from assured.

Park's analysis is based on projecting resource requirements, assuming that by the year 2000 the expected world population will be consuming at U.S. standards prevailing in 1960. Such a level of economic growth for the world as a whole is not projected, although the advanced countries will be making demands on resources that are between 200 and 300 percent over 1960 U.S. consumptive standards. Thus, Park's overall appraisal seems reasonably realistic because the excessive per capita growth attributed to poor countries is more or less balanced by the failure to take account of the prospects for per capita growth in the rich countries. Park is skeptical about the prospects for cheap energy in sufficient quantities to make feasible the smelting of low-grade ore that will be needed as fossil fuel supplies begin to disappear. He also does not hold out great overall hope for finding adequate substitutes for many resources that are of basic importance for the operation of a modern economy: "nothing now known can take the place of steel where strength is needed, as in great skyscrapers and dams or in the high-temperature alloys for parts of a jet engine; nothing now known will substitute for cobalt in the manufacture of the strong permanent magnets needed in all modern communications systems; and no other metal will, like mercury, become liquid at ordinary temperatures and therefore be usable in temperature and pressure control equipment." Park is also convincingly doubtful about the contributions to resource supply that can be made by improved reclamation and recycling techniques; for instance, silver is mainly used in photography and electronics, but in such small and dispersed quantities as to make reclamation uneconomical. He also does not hold out much hope of meeting the anticipated scarcity problem by such exotic technologies as deep-sea mining of extractive minerals. In general, then, Professor Park returns us to the pessimistic mood and outlook of Professor Jevons with which we started this discussion.

Partial confirmation, at least, of Professor Park's concern with resource sufficiency is suggested in the 1969 report of the National Academy of Science's Committee on Resources and Man. The Introduction and Recommendations section of the study insists that "Man must also look with equal urgency [to that directed at the need for food] to his nonrenewable resources—to mineral fuels, to metals, to chemicals, and to construction materials. These are the heritage of all mankind. Their overconsumption or waste for the temporary benefit of the few who currently possess the capability to exploit them cannot be tolerated." The text goes on to observe that "true shortages exist or threaten many substances that are considered essential for current industrial society: mercury, tin, tungsten, and helium for example. Known and now-prospective reserves of these substances will be nearly exhausted by the end of this century or early in the next, and new sources or substitutes to satisfy even these relatively near-term needs will have to be found." Although the committee is not alarmist, it does advocate strong government initiatives to develop "appropriate laws or codes restructuring economic incentives" in such a way as to "facilitate conservative recovery, more efficient use, and reuse, thereby appreciably extending now foreseeable commodity lifetimes." And although the tone is non-alarmist, neither is it complacent about the limitations that might arise in the future as a result of resource insufficiency: "It is not certain whether, in the next century or two, further industrial development based on mineral resources will be foreclosed by limitations of supply. The biggest unknowns are population and rates of consumption. If population and demand level off at some reasonable plateau, and if resources are used widely, industrial society can endure for centuries or perhaps millennia." With GNP per capita rising faster than population, many times faster in rich countries, we again observe the greater extent to which demographic and economic trends in the rich countries are of greater relevance to the planetary crisis than are these same phenomena in the poorer countries. The resource consumption of each American is many times as great as that of an average Indian or Ugandan.

Already industry-oriented outlooks have perceived the relationship between the projected trends for a higher per capita GNP in a larger population as creating an urgent *national* need to acquire a large share of mineral resources from the oceans for the United States. In a pamphlet published by the Government, Walter J. Hickel, the Secretary of the Interior, writes in the Introduction that "the strength of this Nation has been founded in

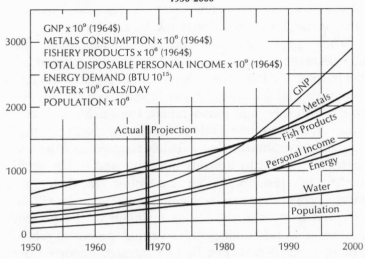

NATIONAL ANNUAL REQUIREMENTS OF NATURAL RESOURCES
1950-2000

great part upon the wealth of the natural resources with which it has been blessed. Indeed, the extent to which a nation possesses or controls natural resources is probably the most important determinant of its world position. We will disregard this principle at our peril." The title of the pamphlet, "Marine Resources Development—A National Opportunity," is built around the *national* need to obtain a large sector of marine resources—both renewable and nonrenewable resources—or find it impossible to accommodate expected growth curves. Lamenting the failure of the United States to match Soviet and Japanese aggressiveness on the high seas, Mr. Hickel adopts an insistent tone: "We must reverse this trend. At stake is not simply our ability to feed our own people. Of this I have little doubt. *The real stake is leadership in a protein-hungry world.*" This kind of competitive view of ocean resources development is also shared, to a large extent, by private industry. The idea is to satisfy national requirements by gaining control over the necessary resources, regardless of the merit of competing claims. A very strident nationalism underlies, for instance, the publication of the National Petroleum Council's elaborate booklet "Petroleum Resources Under the Ocean Floor."

What makes Professor Park's pessimism about the adequacy of resource supplies especially relevant to the purposes of this book is his stress on global consumptive goals and his central belief in the interrelations between affluence and demographic trends. Most discussions of the resource outlook assume a constant international environment that is being adapted to expected patterns of needs by a highly inventive and benign technology. For instance, Fisher and Potter rest their optimistic prognosis on several conditions: " . . . it will be essential to extend, or at least maintain to the present degree, a world trading and investing system in which raw material deficits can be met through imports from other countries with a surplus of these materials." They also warn that "any devastating war would, of course, throw off all the projections undertaken in this paper as well as conclusions drawn from them." What seems clear is that an environment experiencing the degree of disruptive change that is occurring in international society is highly unlikely to provide the sort of stable setting that prevailing optimism is so firmly conditioned upon. Professor Park, whose specific concern is to protect the American economy against the pressures of the future, points out that "the United States, in the shrinking world of the future, will find increasing difficulties in continuing to prosper while our neighbors to the south, upon whom we depend for so much, continue in a state of poverty and hunger and limited education." Here, at least, is an explicit linkage between the issue of social justice on a global scale and resource sufficiency in the richer countries. It is very important to grasp the extent to which the affluent economies of the world need imports to overcome what Fisher and Potter have called "raw material deficits." The United States imports as much as 80 percent of its annual mineral requirements and consumes perhaps as much as half (the calculations vary from 40 to 60 percent) of all the nonreplaceable raw materials produced by the world each year. And as Park more generally observes, "a large part of the raw materials needed by industrialized nations is obtainable only from underdeveloped countries." Why is it that raw-material purchasers enjoy, in general, so much higher standards of living than do raw-material suppliers? By close observation it seems clear that the two kinds of economic activity are integrated into a single controlled system with one part dependent on and subordinate to the prosperity of the other. But the entire structure of the world economy, including emerging patterns of trade and investment, seems designed to perpetuate, if not further expand, existing disparities between rich and poor countries. Poor countries depend on foreign exchange earnings from raw-material sales in order to permit them to import finished products; some reform-minded economists in these countries have started to advocate higher tariff walls on imported goods and greater domestic use of raw materials to build up industry at home. The domestic economies of many poor countries are themselves highly unjust in terms of the distribution of income, and the import pattern in many of these countries is geared to satisfy the demands of a small upper class which alone has the purchasing power to acquire goods beyond subsistence. The dominant economic classes in these societies often have more in common with the rich foreign investor groups from the wealthy countries than they do with their own populations.

A modern economy rests upon its technological base, which in turn rests upon suitable education, high rates of investment in research and development, and a sophisticated manufacturing establishment that is in a position to make use of computers, automation, and other labor-intensive techniques. The social costs of this kind of industrial development may be high where a growing portion of an increasing labor force is unemployed and concentrated in sprawling urban slums. All elements conspire, including "the brain drain" which tends to draw talented members of poorer societies to the richer countries where better job opportunities exist, where the social and political atmosphere may be in less turmoil, and where the facilities and stimulation needed for creative individual development abound. Such patterns of cross-national migration reinforce prospects for continued poverty in Asia and Africa. The extent of the brain drain has been increasing through time.

The exploration of space shows the extent of the gap between the most powerful of the developed countries and the others. Even Western Europe shows signs of being unable to compete with the United States because of its inability to keep pace at the rapidly receding frontiers of hyper-modern technology. The developing, poorer countries cannot even begin to compete in this technological rivalry. As a consequence, the countries that most need to augment their wealth are almost completely cut off from participating fully in the profitable exploitation of the mineral wealth of the sea. Proposals have been made to share this wealth by turning over its administration and exploitation to a world institution for ocean mineral resources, but few observers expect that the forces of private and national economic gain will be seriously subordinated to some kind of planetary regime of control for the benefit of mankind. Ocean mineral resources present a test case for the conflict between the competitive ethos and the world community ethos in a domain in which no sovereign or other *vested interests* have yet had occasion to develop. Existing patterns of power and influence suggest that ocean mineral wealth will make the rich richer, and will neither benefit the poor nor build up the financial independence or functional role of world community institutions.

The United States has also embarked on a costly, but promising, Earth Resources Tracking Satellite (ERTS) program. ERTS provides very sophisticated methods of locating deposits of minerals and fuels under earth or water, by means of the newest kinds of photography and elaborate systems of sensor detection. Once again the principal beneficiary is not expected to be the world community, as such, but the United States. We witness, then, the reinforcement of existing economic disparities due to the extent of control over new technologies by the most affluent and strongest countries. The placement of an American flag on the moon is more than a symbol, it is a manifestation of the sovereignty-shaped application of technology in this era. Such inequality that results is bound to lead to a spiral of upward resentment by those sectors of the world that are excluded from these exploits and their economic benefit. This is a world in which the rather simple basic technology of the war system can be acquired rapidly by even a pre-industrial society such as China, even if the rest of modern technology is beyond its reach. We need to grow sensitive to the degree to which "the new imperialism" seems to depend, on a world level, upon technological superiority. The disadvantaged world can challenge the consequences of this superiority only in the military sphere; it does not possess the possibility of challenging the rich countries through economic competition, especially given the world structure of bargaining and pricing that controls the sale of raw materials and given the success of the multinational corporation in organizing the production of goods and services in the poorer countries.

It should be appreciated that domestic revolutions have arisen out of the perception that human misery is not a necessary aspect of human life, but merely a consequence of the fact that economic and political power is being wielded for the sake of the few, at the expense of the many. Hannah Arendt writes about the extent to which "the social question" underlies modern revolutionary activity:

> The social question began to play a revolutionary role only when, in the modern age and not before, men began to doubt that poverty is inherent in the human condition, to doubt that the distinction between the few, who through circumstances or strength or fraud had succeeded in liberating themselves from the shackles of poverty, and the laboring poverty-stricken multitude was inevitable and eternal. This doubt, or rather the conviction that life on earth might be blessed with abundance instead of being cursed by scarcity, was prerevolutionary and American in origin; it grew directly out of the American colonial experience.

Certainly such a focus on the social question underlay the movements that produced the French and Russian revolutions, and is at the center of the political movements for national liberation presently found throughout the Third World. Militant radicals in the United States—the New Left, the Black Panther Party, Students for a Democratic Society—all base their demands for a new political order on the assumption that misery is not an essential ingredient of human existence, but a matter of the way in which power is organized and wealth distributed. These demands are becoming more and more transnational in character, blacks in America identifying their cause with liberation movements elsewhere and vice versa. The call for structural change is related to national configurations of power, and not to the international structure of stratification based on separate national units of greatly disparate wealth, technological development, and bureaucratic organization. It seems likely that governments and transnational groups from the Third World will increasingly blame their own distress on the structures of stratification that exist in international society, and issue, in effect, a revolutionary call for a radical reordering of international society. In fact, the seeds of such a call are apparent in the principal Afro-Asian forums such as the meetings of UNCTAD and in various conferences of non-aligned states. So long as the rich countries are able, by and large, to maintain disunity and incoherence in the relations among poorer countries, and to work with conservative and traditional elites in the Third World, the challenge will be ineffectual; these elites benefit from international stratification, work with foreign interests, and would be displaced, perhaps brutally, by social revolutions within their own societies. There is then a link between the political development within the impoverished countries and the stability of a highly stratified international economy.

The link between "the social question" and world resources arises from the evident relationships between mass poverty, the dominance of technology, the economics of international trade in raw materials, and the effectiveness of the multinational corporation as a form of economic organization that protects the privileged position of social classes in investor and host countries. It seems as if organizational changes on the international level could lead to higher earnings for the poor countries. Even in the area of food resources, the realities of trade show that underneath the pretension of charity the rich countries are exploiting the resources of the poor ones. The food specialist Georg Borgstrom has pointed out, for instance, that the agricultural surpluses of the West are a consequence of high-protein fertilizers and fish meal imported from Latin America, Asia, and Africa to fatten poultry and grazing stock: "The 2.5 million tons of grain protein delivered annually by the rich and well-fed are counterbalanced by a flow to the Western world of no less than 3.5 million tons of other proteins of superior rating in the form of soybeans, oil-seed cakes, and fish meal. We in the Western world are actually making what amounts to an almost treacherous exchange." Borgstrom sees the situation as one in which "the world now faces the alternative of either a global class struggle between nations or the farsighted creation of a mechanism of universal solidarity." A very firm conclusion takes shape here: the affluence of the advanced countries is interdependent with the poverty of the poor ones.

Orthodox economics does not give us much insight into the gap between the rich and poor countries, because it fails to take any account of power relationships. Instead, we are shown GNP accounts and the entire emphasis is put upon devising economic policies that will raise the rate of economic growth by a point or two. Such an approach fails to appreciate that the whole structure is organized on behalf of the rich economies. A well-known Marxist economist in the United States, Paul M. Sweezy, has outlined an analysis of these problems in a brief, provocative passage:

(a) From the beginning, the development of today's advanced capitalist countries has been based on subjugation and exploitation of Third World countries. The latter's pre-existing societies were largely destroyed, and they were then reorganized to serve the purposes of the conquerors. The wealth transferred to the advanced countries was one of their chief sources of capital accumulation.

(b) The relations established between the two groups of countries—trade, investment, and more recently so-called aid—have been such as to promote development in the one and underdevelopment in the other.

(c) There is therefore nothing at all mysterious about either the gap or its widening. Both are the inevitable consequence of the global structure of the capitalist system.

(d) It follows that the situation can be changed and real development take place in the Third World only if the existing pattern of relations is decisively broken. The countries of the Third World must secure control over their own economic surplus and use it not for the enrichment of others but for their own development and their own people. This means thoroughgoing revolution to overthrow imperialism and its local allies and agents.

Obviously Sweezy's formulation is polemical and written within an anticapitalist ideological tradition. Nevertheless, it clearly states a way of looking at the world economy that bypasses the operation of the market and the presentation of GNP accounts and puts the focus where it belongs—on the structures that maintain the present system, on the system's inadequate fulfillment of human needs, on its injustice, and on the required drastic changes in the social and political structures of the dependent countries if adjustments are to be made. Such an analysis challenges the whole marginalist view of Western economics, which limits reform to minor modifications at the margin of the social, political, and economic system.

The West has so far been successful in spreading the belief that modernization or economic development, whether on a free enterprise or Marxist basis, provides the cure to Third World problems. Park has shown that the global mineral resource stock of the world could not be expected to sustain a growing world population at anything approaching 1960 consumptive standards in the United States. Borgstrom, on the basis of an even more convincing demonstration, concludes that "It can be safely asserted that our particular form of civilization with regard to the use of energy, water, forests, and soils cannot be copied on a global scale." Given existing world economic structures, even without taking into account demographic prospects, there is little hope for the poor countries to escape their fate as proletarian societies delivering their wealth to sustain the living standards of the advanced ones (and of a tiny national elite).

In this light we must reconsider the earlier discussions of the sufficiency of resources to meet the expected demands during the decades ahead. Such an idea of sufficiency is based on *demand schedules* that do not meet *human needs*. What does it mean to have sufficient resources? It is a supply/demand relationship that moves goods as market conditions dictate. But the United States has discovered during the Vietnam War that even its great wealth does not permit guns *and* butter, without severe dislocations in the form of inflationary pressures, balance of payments deficits, and cutbacks in welfare and education spending. Does the United States have sufficient resources to transform its cities from *urban ghettos* into *decent human communities?* Increasingly, a question of this sort is treated as raising an issue of priorities within domestic societies that calls for reallocation of resources from the defense sector to the welfare sector. In an editorial opposing the ABM, the *New York Times* wrote: "There are other valid arguments against Safeguard deployment. The vast sums that would go into a futile new arms race are desperately needed to solve urgent domestic problems." The implication is one of *choice* rather than *sufficiency*. Resources exist to permit a sustained rate of economic growth with relatively full employment—so-called "pockets of poverty" in the United States are not regarded as structural problems, nor are the extraordinary levels of luxury-spending.

It might be contended that the issues of social justice raised are ones that call for a better distribution of the fruits of economic effort, that is, a new domestic and world order system that establishes minimum consumptive levels for all people and diminishes the extent of inequality and the volume of wasteful spending. But here the finite limits set by available resources suggest that no more just or efficient distribution system could hope to establish high global standards. Greater distributive justice would mean that the entire world population would be drawn down toward subsistence, or below. Even under present conditions it is ominous to note that "equal distribution of all available food would only make hunger universal and shared by everyone." Such a circumstance exists in a world with 3.5 billion. What can we hope for in a world of 6 to 8 billion? We should immediately come to terms with the stark actuality that we live in an environment in which most critical resources are *scarce*, not *abundant*, and that we rest our destiny with forms of social organization that would make sense only if resources were *abundant*. Furthermore, the nature of man is such that sufficiency of resources needs to be measured both in relation to *minimum thresholds* (are *needs* being met?) and in relation to *disparities* (how unequal are the benefits and burdens?). On both these social scores, the world system conceived as a whole is doing a dreadful job in overcoming mass poverty and gross disparity, considering the total level of production that is taking place. The latent social facts are being brought into the realm of political consciousness by increasing degrees of awareness on the part of those who are suffering most in the world as it is now constituted. The sense of deprivation once linked to a structural explanation creates the setting in which a revolutionary challenge is likely to emerge.

The resource literature does not, as we have seen, measure sufficiency in accordance with distributive criteria by taking account of *human needs* and the *disparities* in disposable income available between groups in a society or between regions of the world. If one can postulate a constant social and political context for future decades—a wildly utopian assumption—then resource stocks appear generally sufficient for minerals, although, as we have noted, there is some disagreement among resource experts. These experts agree that rising populations who enjoy improving living standards will confront a variety of problems in relation to the principal renewable resources: land, forest, soil, air, water. Many of these problems will be considered in the next section in the course of discussing environmental deterioration as a result of the many varieties of pollution. Earl Finbar Murphy has contrasted the situation between renewable and nonrenewable resources as follows:

> Everything is very vague, and there is the chance that this present generation may be as wrong about the renewable resources as Jevons was in 1865 and as the American conservationists were any time before 1920 about the stock resources. In any event, stock resources are not the problem. Renewable resources, on the other hand, are a matter of pressing concern for the present generation. The problem may be induced to go away, or it may be something for planning to eliminate. But it *is* here. This is not another Coal Question, nor will any royal commissions exorcise it as a myth. Man, in relation to his environment, may be the only reality by 2000.

Both the reassurance and the concern of this passage express the mood of the time, confidence about natural resources, great anxiety about the deterioration of renewable resources through misuse of the environment. The extent of the earth's stock of resources, always exceeding man's present inventory, has induced an unwarranted sense of resource sufficiency, as if new reserves will always turn up, or even if not, then man as alchemist can devise a suitable, plentiful substitute to satisfy comfortably the demand. We hear now about new processes that will turn a ton of garbage into 40 barrels of oil, with almost no residue; perhaps such a prospect is sheer fantasy, but certainly it functions as part of the abiding mood of complacency. Such attitudes prevent any appreciation of the underlying reality of the earth as a limited and limiting island in space which at some point in time will not provide man with the raw materials needed to sustain further industrial growth.

And should there be a major war, then it may not be possible to rebuild a modern industrial society because the technology needed to mine low-grade ores or deposits in remote places may not be available and the easily accessible ore deposits needed to start over no longer exist, having been substantially used up.

Finally, the sufficiency of resources, even given optimistic assessments, tends to rely on a very short time series—a few decades at most. As man begins to accept a place in the wider cosmos of his existence, such a limited time frame seems to betray a disregard of the planet's future. The growth of the human spirit in a manner that was more closely governed by ecological norms would almost certainly start to measure resource sufficiency in terms of recycling processes that could ensure the indefinite maintenance of the material basis of human survival. With such a perspective—one that now seems totally absent from the relevant specialized literature—there would be serious concern, even if not of the emergency kind applicable to the crisis of environmental deterioration that imperils our stock of renewable resources, for the resource consumptive patterns of the modern world. Can we move toward a restored harmony with the world around us that does not need the threat of disaster to prompt constructive action? As with the struggle of conservationists to preserve the wilderness, so with our stock of natural resources, it is necessary that we preserve what we have rather than relegate the future to the will and whim of "the developer," "the engineer," and "the GNP-minded" public official. In an ultimate sense, our hopes for a positive response rest on what Nancy Newhall has aptly called "the crucial resource": "Of all resources, the most crucial is Man's spirit. Not dulled, nor lulled, supine, secure, replete does Man create, but out of stern challenge, in sharp excitement, with a burning joy. Man is the hunter still, though his quarry be a hope, a mystery, a dream."

4. ENVIRONMENTAL OVERLOAD.   There presently exists a curious mixture of agitated concern, ignorance, and nonchalance about the deterioration of the environment. Of course, *happenings* such as Earth Day, a smog attack, the struggle to ban the use of DDT, or a report that a new jetport is about to be located on some open land arouse interest in environmental issues. But these are only the first glimmerings of a *coherent awareness* of the environmental crisis that is sending out its warning emanations with ever-increasing frequency. There is still a tendency to regard *Torrey Canyon* incidents as isolated disasters that are an inevitable cost of economic progress, and to regard the slow accumulations of poison in the air or ocean as merely an inconvenience that is of no greater consequence than the disappearance of an exotic variety of bird or fish. Among men generally there abides an overriding sense of security about the capacity of the environment to sustain human society in its present patterns of collective existence. This confidence is most shaken in the United States where the evidence of an overburdened environment is becoming overwhelming. Senator Gaylord Nelson of Wisconsin, an early crusader for environmental issues, put the current state of affairs this way: "Progress—American-style—adds up each year to 200 million tons of smoke and fumes, 7 million junked cars, 20 million tons of paper, 48 billion cans and 28 billion bottles."

As of now, mainstream writing about the future of mankind continues to overlook the uncertain status of the environment in its most basic dimensions. Kahn and Wiener's *The Year 2000,* published in 1967, has not a single reference to environmental and ecological issues in their avowedly synoptic account of the future. Many pages are devoted to the problems posed by increasing leisure and alienation in the post-industrial society, but literally not a word about poisons, garbage, the loss of wilderness, and the declining purity of water, land, and air. Such an omission is especially startling because the authors use a shopping-list approach in their effort to anticipate even implausible future eventualities. The rapid rise of concern about environmental quality in the United States would make such an oversight impossible today.

The rise of environmental issues to prominence has constituted the one really new issue near the top of the governmental agenda. But whether it is high on the agenda only in a nominal sense still remains a well-grounded suspicion. Despite many anxious words, the Nixon Administration has yet to assign staff or resources proportionate to the significance it avowedly attaches to these matters. There has been some speculation that the sudden embrace of environmental issues by politicians represents an effort to distract attention from the unsolved problems of the blacks and the ghettos and to shift the center of public interest to a set of grievances shared by the dominant middle-class constituency in American politics. As yet, it is too soon to probe the motivation or seriousness of concern about environmental quality, except to note that its prominence represents the most abrupt and sweeping shift in concerns that has occurred in the United States in recent years, and has built the base for a genuine political movement based on environmental defense and rehabilitation.

Discussions of the international setting are particularly devoid of emphasis on problems of environmental quality, and yet it is not at all clear that the risks of poisoning the oceans or heating up the atmosphere are not as great as that of nuclear war. But specialists in foreign policy and world affairs still seem oblivious to the relevance of environmental hazard to their professional concern with the management of national power. Henry Kissinger, Hans Morgenthau, or George Ball will write at great length about alliance patterns, the balance of power, bipolar relations with the Soviet Union, the strategic arms race, but will not devote a single word to the ecological fragility of the world.

How are we to explain this curious process of psychological denial—the sorting out of facts so as to deny certain actualities? Part of the explanation is simply unawareness. We live in an age of specialists who are expert on a particular cluster of issues. It is difficult for a new group of experts—the ecologists, in this case—to break into political consciousness sufficiently to have their presentations affect the way the agenda of public issues is constituted. It is almost equally difficult for established experts to take account of new issues and subject matters. Buckminster Fuller has pointed out that biologists studying the extinction of *animal species* and anthropologists studying the extinction of *human tribes* have interestingly enough come up with a common explanation: "Extinction in both cases was the consequence of overspecialization." There is a tendency to cut out from the traditional field of awareness anything that falls outside the traditional definition of "the problem" or the traditional focus of "the subject." Academicians with their "disciplines" and their "departments" seem especially slow to revise their sense of a subject to take account of an altered situation.

But why would Herman Kahn—a self-proclaimed anti-academic, anti-pigeon-hole kind of "think-tank" thinker who specializes, although with a somewhat morbid penchant, in the "unthinkable"—let us down so completely on these environmental issues? Here the explanation is more complex, and would seem primarily to be a by-product of a highly corrosive mood of technological confidence. The Kahn-Wiener outlook endows man with a capacity to govern everything except possibly the aggressiveness of others and the alienation within himself. Man is the only unconquerable foe, and human groups must be wary of one another or face the prospect of conquest and subjugation. Nature is the backdrop over which man exercises his dominion to achieve ever-expanding economic growth. The critical index of "progress" relates to GNP and GNP per capita, and the relevance of population trends is limited primarily to their impact upon the various economic outlooks for diverse national societies around the world. To the extent that the future holds disaster in store for mankind, it has to do with the occurrence of major warfare or the spread of totalitarian government. It is a wholly unreal future derived by extrapolation from the prominent attributes of the past.

Of course, until very recently environmental issues seemed to be mainly the preserve of eccentric bird-watchers, butterfly-chasers, and overstrenuous hikers. There surely was something quaint about the Audubon Society and the Conservation Foundation, undertakings that appealed to the super-rich as benign, very much akin in spirit to some genteel civic group concerned with raising funds for a worthy cause by staging an opulent charity ball. In other words, with some slight local exceptions, such as where a fight among well-entrenched interest groups arose, the province of the modern conservationist was thought to fall mainly within the area of high-minded, socially restricted philanthropy. It did not touch upon the grand issues of the day and it did not in any sense belong on the agenda of the nation or of the world. If anything, the conservationist was a nuisance to the society, always raising some rather fuzzy objection to the projects that promised to bring more profits, more jobs, and more goods to the bulk of the community.

To break these barriers down has not been easy. In the advanced countries generalists are distrusted and discounted, prophets of doom are discredited along with other forms of apocalyptic and utopian thinking. The presentation to the public of the planetary crisis inevitably has an alarmist quality that arouses fear more credibly than it gives hope. As a result, public immobility of a sort results from the magnitude of the pessimism that is engendered. Increasingly, the more enlightened members of the public are vaguely aware of the cumulative dangers to their environment that arise from the exploits of modern technological living, but by and large the situation is supposed be less bad than the most prominent of the alarmists, such as C. P. Snow, Lewis Mumford, Buckminster Fuller, David Brower, and Kenneth Boulding, have made it out to be. Of course, such men of eminence are always easier to discount than to read, and any serious exposure to their ideas and arguments makes it that much more difficult to remain complacent about environmental issues. Certainly, the presentation of their case has influenced statesmen of conscience and intelligence, especially in the United States, to take greater account of these problems in their principal statements to the public. Stewart Udall and Robert McNamara have both become eloquent and cogent spokesmen for a new and growing endangered-planet constituency, not yet potent enough to redefine the agenda of the nation and the world, but beginning to receive a public hearing.

Sverker Åstrom, the Swedish Ambassador at the United Nations, has taken the lead in organizing a world conference for 1972 under UN auspices to focus concern upon environmental problems. Early in his Administration, President Nixon set up an Environmental Quality Council at cabinet level with himself as Chairman and his Science Adviser, Lee DuBridge, as Executive Secretary. This Council is serviced by a Citizens' Advisory Committee on Environmental Quality with Laurance Rockefeller as its Chairman. Such institutional action naturally prompts (and to some extent is prompted by) the efforts of mass media. *Life* magazine, for instance, has been publishing a series of articles to demonstrate the extent to which the environment of man is running down. Their introductory editorial on August 1, 1969, put their own relationship to the environmental situation this way:

"We will report the threats and attempt to show how they can be opposed and, if possible, defeated. We believe that the threats are serious, that the potential losses are critical, and that the need for action is urgent." By the time of the State of the Union Address in 1970 the initial task of apparent persuasion had been clearly won; concern for environmental quality dominated President Nixon's presentation of the problems facing the nation. But note that the definition of the problem continues to emphasize its domestic character and its sharp separation from concerns about population pressure, resource depletion, and the war system. Nothing in the recent flourishes of public concern expresses the realization that we need to revamp our entire concept of "national security" and "economic growth" if we are to solve the problems of environmental decay. And it is more than new concepts that are needed; we need implementing allocations of energy, imagination, and resources. And to receive these new allocations we may need an organized movement that competes for political power with entrenched groups.

Putting a man on the moon has also, somewhat paradoxically, helped focus attention on the condition of planet earth. Seeing the earth suspended as a lonely sphere in space helps us to grasp its wholeness, and to overcome these man-made separations that have produced national boundaries and rival centers of conflicting political authority. The prime symbol of the earth rotating through space and man sending out spaceships farther and farther beyond his atmosphere has given potency to the metaphor of the earth as itself a spaceship. Buckminster Fuller has carried this metaphor especially far, deriving all sorts of imperatives about the need for austere living and integrative thought and action from the operation of a spaceship that are directly applicable to issues of resource conservation and the substitutions of a needs-oriented productive system for a market-based one. Also the dangers of space journey are a solemn reminder that the earth's own space travel is now endangered, that coordinated intelligence, wise central guidance, and immense organization are needed to avoid a disastrous collision. Can one imagine an Apollo Mission being administered by rival control centers that coexist in a condition of distrust, tension, and hostility? Perhaps the Apollo Mission can provide us with a model of unified coordination that can be put to use on the perilous survival journey of the human race over the next few decades.

The travel to the moon also has given a first taste of prominence to the integrative discipline of ecology. For many years the ideas of ecology have permeated the study of plants and animals in relation to their natural habitat. But it is only recently that human ecology has started to make an impact, because only recently have we come to realize that we are living in such a way as to destroy the quality of our environment. We need to appreciate environmental issues from the totality of concern that is the enterprise of ecology. It has begun to be evident that we need to study seriously and systematically the intricate network of interconnections that join man to his total environment. As Paul Shepard has put it: "Man is in the world and his ecology is the nature of the *inness*. He is in the world as in a room, and in transience, as in the belly of a tiger or in love." When human ecologists go to work, long-suppressed interconnections between conservation issues, population growth, pollution, and urban sprawl begin to grow evident, as does the global scope of the processes that underlie both a deteriorating and a viable relationship between man and nature. Boundaries between academic disciplines seem as artificial as boundaries between sovereign states. Shepard and Daniel McKinley, editors of an excellent introduction into the character and ramifications of human ecology, have shown their awareness of these implications by entitling their collection *The Subversive Science: Essays Toward an Ecology of Man.* Shepard has put the case in very explicit terms:

The ideological status of ecology is that of a resistance movement. Its Rachel Carsons and Aldo Leopolds are subversive. . . . They challenge the public or private right to pollute the environment, to systematically destroy predatory animals, to spread chemical pesticides indiscriminately, to meddle chemically with food and water, to appropriate without hindrance space and surface for technological and military ends; they oppose the uninhibited growth of human populations, some forms of "aid" to "underdeveloped" peoples, the needless additions of radioactivity to the landscape, the extinction of species of plants and animals, the domestication of all wild places, large-scale manipulation of the atmosphere or the sea, and most other purely engineering solutions to problems of and intrusions into the organic world.

Such a posture of concern and position makes of human ecology a kind of ethics of survival. It is a science that relies on careful procedures of inquiry, data collection, and detailed observation as the basis of inference, explanation, and prediction. But it also involves a moral commitment to survival and to the enhancement of the natural habitat of man. As such it involves opposition to the formidable pressures of encroachment that have been aggregated in modern industrial society. As such, human ecology moves in the direction of politics and action and away from the Weberian model of the social scientist and expert as a detached observer who rigidly maintains an attitude of neutrality toward value considerations. Such an ecological consciousness accords with the rhythm of our times, with the demands of students for a relevant kind of educational experience. Human ecology would seem to offer an academic and social response that combines the search for knowledge with an active social concern that entails political action.

But ecology is, in essence, a natural science. As such,

it has not yet intermeshed the phenomena of man and nature with the political structures of power and authority that are operative at various levels of human organization. Part of this book's purpose is to make some effort to bridge the gap between the ecological outlook and the work of political science, especially as applied to the affairs of mankind as a whole. The objective is one of achieving cross-benefits: to make ecology more aware of the political setting of human activity and to make social scientists more ecology-minded.

Despite the more adequate attention that is now being given to environmental issues there remains the great task of mobilizing the relevant national communities to take effective action. Paul Ehrlich, the crusading biologist, has effectively stated one difficulty: "The trouble with almost all environmental problems is that by the time we have enough evidence to convince people, you're dead." Most societies remain committed to the primacy of economic growth as a condition of social health and are insensitive to the underlying contradiction between GNP-mindedness and the ecological outlook. The early discussions of the location of a new jetport to serve southern Florida after the capacity of the Miami National Airport will be exhausted in the late 1970's concentrated upon the selection of a *site,* not upon placing *ceilings* on growth so that another jet facility would not be needed in the area. Officials of Dade County, civil servants with the Federal Aviation Agency, conservationists, and various vested-interest political groups all take it as given that *the health* of the Miami area depends on continuing to expand its population base and to maintain, or even to improve, its rate of economic growth. "Planning" involves accommodating these expansions without suffering any unnecessary burdens and this calls for thinking ahead, sometimes as much as several decades. Planning rarely involves the establishment of an optimum environment to sustain a harmonious relationship between man and nature. There is some special sensitivity to the preservation of a good environment in Florida because the prosperity of the area depends on what are "the amenities," that is pure air and water. A preliminary decision as to site is reached on the basis of intergroup bargaining among various interest groups, governmental actors at a county, state, and national level, and a vague kind of deference to what "the public" would prefer, or better, tolerate. One principal set of considerations involves profits, jobs, growth, whereas another principal set involves the alligators, rare birds, drainage basins, and so on that constitute the basis for the healthy maintenance of the adjacent Everglades National Park. In such settings the former group of interests has tended to prevail, being more precise, more persuasive to more people, and more directly associated with the configurations of power. Sometimes the debate takes the form of alleged "trade-offs" between the natural and the human environment, as if there were a real distinction; such a debate compares the costs of jet noise close to an urban population with that of inconveniencing varieties of marsh life on some "open land" being "wasted" on the edge of cities. Few of us even now realize that such wasted acres may be producing as much as ten times the protein of even highly fertile cropland in the form of organic soils, oysters, clams, and fish. The point is that if the issue is framed in this kind of either-or way there has been a strong disposition in the past to favor the human community at the expense of the natural environment. The cumulative impact of these many separate decisions is to produce a steady deterioration of the natural environment upon which we consciously and unconsciously depend for the richness of our spiritual, and even of our material, existences. It is, perhaps, indicative of the new mood of concern that President Nixon intervened at the last hour and decided to refuse permission for the development of the Everglades Jetport. It is hard to imagine that conservation-minded groups would have been able to gain such effective access to an American President even a year earlier. The new order of ecological priorities, at least on major questions, has started to emerge.

A further inhibition on the growth of coherent awareness is a tendency to consider specific reports of environmental disaster as *isolated instances.* There is, so far, an insufficiently clear and cogent presentation of the underlying societal situation that makes jetports invade the wilderness, that produces dangerous "accidents" in atomic facilities, and that leads huge oil tankers to collide or run aground. The first consideration is the absence of a planned and preventive attitude toward potential risks; we tend too often to *react,* rather than to *prevent,* and given the scale of our newer undertakings the consequences of error are increasing at a fast pace.

Second, the space needs of a rising population and an expanding industrial base naturally collide with the dedication of land to other uses. This process is made even more significant as a consequence of the increase in the urban sector and by the increasing per capita space requirements of each resident. In the United States alone over 20,000 strip mines are cut out of the land surface at the rate of 153,000 acres every year. By 1980 more than 5,000,000 acres will have been defaced by surface mining alone. David Ehrenfeld, a biologist with an ecological outlook, reports "that between 1965 and 2000 the average town-dweller in France will have doubled his space consumption. Because the population will also double, the total urban area of France will quadruple in one generation." Ehrenfeld also reports: "Urban land area in the United States is expected to increase from 21 million acres in 1960 to 45 million acres by the end of the century."

Third, the kind of technology that is coming into use— huge tankers with carrying capacities of more than 500,-000 tons of crude oil, large nuclear power plants, highly toxic stockpiles of nerve gas and biological agents—make accidents likely to occur on an unprecedented scale with possibly far-reaching ecological consequences.

Fourth, cumulative buildup of environmental threats may go on for years without any politically significant notice being taken. It required Rachel Carson's *Silent Spring* to move the danger caused by insecticides from the pages of obscure scholarly journals into the field of public awareness. It has required a previously unknown professor of physics at the University of Pittsburgh, Dr. Eugene Sternglass, to make some of us aware that a fantastic, unexplained correlation may exist between radioactive fallout at so-called "permissible" levels and the extent of infant mortality. His findings, which would have probably otherwise remained in the closed circuits of academic journals, received a flurry of attention because they were published in *Esquire* magazine during the summer of 1969. The preliminary analysis of Dr. Sternglass suggests that as many as 400,000 have lost their chance of life as a consequence of the radioactive clouds carried across the United States after the first atomic test explosion of 1945 in New Mexico. Clement Attlee, who was a member of the British War Cabinet, acknowledged that the decision to use atomic weapons was made without any awareness by the officials involved that genetic damage to future generations might result:

> We knew nothing whatever at that time about the genetic effects of an atomic explosion. I knew nothing at all about fall-out and all the rest that emerged after Hiroshima. As far as I know, President Truman and Winston Churchill knew nothing of these things either, nor did Sir John Anderson who co-ordinated research on our side. Whether the scientists directly concerned knew, or guessed, I do not know. But if they did, then so far as I am aware they said nothing of it to those who had to make the decision.

Throughout the nuclear era the administering agencies have failed to assess or have underestimated the environmental risks attendant upon nuclear programs of weapons testing and power development. It was the name and resources of magnate Howard Hughes that finally led the general public to appreciate that the geologic substructure of the earth might be seriously altered by underground nuclear weapons testing carried on in accordance with AEC guidelines in the Nevada desert. It also became clear that radioactive contamination of underground water supplies and induced earthquake activity in distant lands were also unstated risks of the weapons tests. We have also grown familiar with "accidents" involving nerve gases that have not behaved as expected and have caused, or nearly caused, great damage to animals or people. The point is that we are playing with the atom for stakes we do not and cannot adequately comprehend. From the perspective of the AEC all inhibitions on the use of nuclear energy, whether in the form of arms control agreements or safety standards, are in fact resisted as dangerous to the national interest and interferences with the pursuit of military capability, the development of civilian technology, and the advance of scientific inquiry. The outlook of the AEC shows the extent to which the development ethos has come to control the behavior of governmental centers of policy and action.

Fifth, the priority of interests and values means that developers usually, but not always, prevail or virtually prevail over those groups seeking to protect the environment against the forces of "progress." Then there are various institutional groups within and without government that grow up to facilitate the material goals that a society adopts. In the United States we have grown familiar with the pressure of the Army Corps of Engineers to build spectacular public works, dams, bridges, and highways so as to justify their professional importance, to increase their appropriations and staff, and to carry out their ideas of how a society best grows. We take for granted eight-lane highways through lightly traveled landscapes. We also find rather doubtful coalitions between rich ranchers and farmers and public water conservancy programs, which often involve making the taxpayers finance the irrigation of croplands or grazing acreage belonging to wealthy farmers under the guise of "conservation." Disguised economic self-interest—in the form of the profit motive—turns out to be everywhere evident when the activities of government, on any level, touch upon the prospects for economic gain. We have grown accustomed to the role of industry in the defense sector, but we are less aware that in every aspect of our society there is an effort by private entrepreneurs to gain cheap access to valued ends: mining companies seeking permits to build access roads into national forests and to engage in exploratory drilling, oil companies seeking offshore drilling leases, real estate developers destroying the marine life of a bay by dredging and filling to create valuable high-rise apartment complexes, gasoline companies refusing to eliminate leaded gasoline despite the clear links between health and the dosages of atmospheric lead. The expectations of government-industry cooperation are clearly set forth in the publication *Marine Resources Development—A National Opportunity:* the government will chart the resources, help with the costly exploratory work, and then leave the field free to private industry once the operations look profitable. In the italicized language of the Report, *"government activity should cease as soon as industry is able to take over."* Such encouragement naturally leads industrial firms to be very nationalistic and argue in favor of sovereign appropriation rather than shared use of new sources of wealth and potency.

Only by constant vigilance can occasional victories be won against the special interest groups bent on high profits and maximum development. The eight-year movement to save San Francisco Bay finally, in August 1969, after much damage had already been done, scored a legislative victory by enacting a law placing stringent control on shoreline development. The uncoordinated landfill and dumping activities had already halved the water area and done great damage to the ecology of the Bay. Opponents of the conservation legislation included various commercial interests, most prominently the Westbay Community Association in which David Rockefeller, Chairman of the Board of the Chase Manhattan Bank, was the leading member. Another major opponent was the Leslie Salt Company, which owned 50,000 acres of shoreline land that it was already in the process of transforming into housing sites.

Sixth, as John Kenneth Galbraith has made clear, our kind of governing process is not very successful in its provision of *public goods,* i.e., something of benefit to the community as a whole rather than to a segment of it. The paradox of aggregation operates to lead particular individuals, groups, companies, or nations to shift the cost elsewhere in the community whenever possible. There is no locus of responsibility. Defense spending is a public good of which the cost can be justified out of fear and greed; but environmental control—especially if related to such intangibles as beauty or the protection of rare flora and fauna—seems dissociated from any immediate need of the community. Also, until environmental concerns build their own bureaucracy it is impossible to achieve the kind of momentum needed to support the values at stake. Our domestic political process works through trade-offs of various special interest groups, by marginal shifts in resource allocation patterns so as not to disturb the bureaucracy, and by aroused responses to various "states of emergency" that emerge in the wake of a disaster that is widely attributed to an underlying social ill. The protection of the environment depends on producing a coherent awareness that begins to appreciate the extent of danger and the degree of sacrifice and adjustment needed to make curative words into effective deeds. As we have said, the rapid buildup of recent concern has been remarkable and the important task now is to accelerate the process of understanding so that appropriate policies can be put into operation before it is too late. Paul Ehrlich has given one estimate on the time scale: "We must realize that unless we are extremely lucky, everybody will disappear in a cloud of blue smoke in 20 years." He added that "If we had 100 years to solve" pollution problems "I'd be optimistic." And the point is that we do not know when it becomes too late. We are beginning to realize that it may be too late if a major nuclear exchange occurs. We are now beginning to have evidence that we are damaging, perhaps fatally, the thin envelope of life-sustaining atmosphere—the biosphere— by the persisting character of large-scale pollution. We seem now to be getting some sense that the long-life chemicals that we dump into the sea may be poisoning many varieties of marine life and may eventually hamper the oxygen-producing processes of photosynthesis upon which our earthly existence depends.

Barry Commoner, a biologist who has played a major role in arousing public interest in environmental issues, described what is happening to us as a result of our abusive treatment of the earth: "A new generation is being raised—with DDT in their fat, carbon monoxide in their systems and lead in their bones. That is technological man." Of course, such dramatic indictments tend to be ignored by civilizations inclined toward downfall. Cassandra, although correct, can never make herself really heard, or heeded. We prefer not to have Cassandras around, or if around, then as with Commoner, we habituate ourselves to their presence through a cover story in *Time* magazine. Even the prophet of doom becomes an ornament of the age, an expressive reminder of our lush diversity and tolerance. To take seriously the counsel of Commoner, Ehrlich, and others would shatter the mood of human complacency and suggest the urgent need to alter our style in basic respects. We resist pleas for change out of sheer sloth, itself a deadly sin. Nietzsche devised a relevant maxim: "Even the bravest of us rarely has the courage for what he really *knows*. . . ."

We are concerned with several sets of issues: (1) the deterioration of the basic renewable resources of air, water, and land that support life on earth; (2) the conservation of the natural splendors of wilderness, the diversities of flora and fauna, and the conditions of nature that enable men and other animals to have fulfilled existences. In particular, we are eager to underscore the international setting of these problems. A prevalent misconception persists that national efforts, if sufficient, will guard the environment. The misconception arises from a failure to appreciate the actualities of *interdependence:* what we do needs to be coordinated with what others do if beneficial results are to be achieved.

Such interdependence generally produces bad effects because *contradictory priorities* exist among principal national groups: the concern for food production or for the prevention of disease may outweigh and cancel the concern for preserving the quality of the oceans. Also, there exist *transnational diffusion* patterns that arise from the circumstance prevailing in foreign societies: unsanitary urban crowding may help induce a wave of infectious disease that spreads across the globe.

We have grown aware of the extent to which we are vulnerable to pandemics by our helplessness in relation to the so-called Asian flu that sweeps across the earth every few years. Such a virus could just as easily be a killer-strain that leaves millions of dead in its wake. The stability of our lives depends on the good fortune that no Andromeda strain has thus far been at the base of an epidemic. The recent isolation by scientists of the deadly Lassa virus from Nigeria lends plausibility to this set of concerns.

If China tests nuclear weapons in the atmosphere and

the wind blows the fallout to the rest of the world, then the irrelevance of boundaries becomes apparent and a decision to respect the sanctity of behavior performed within national territory amounts to an invitation to self-destruction. The danger of "irresponsible" disposal of radioactive wastes from nuclear energy plants is probably more of a threat to the security of other states than is the danger of war and conquest. A nuclear exchange in the atmosphere could easily saturate the earth's atmosphere with lethal fallout such that "a neutral state" would suffer more than it would if attacked directly by the most cruel and rapacious of military conquerors.

If the movement to ban DDT in the United States succeeds, it may still not make much difference. On July 9, 1969, Dr. M. G. Candau, the Director General of the World Health Organization (WHO) and a Brazilian, said that DDT should not be outlawed until economically feasible substitutes are found for malaria eradication: "Outlawing DDT would cause grave health problems in many of the developing countries." In addition, at present there is no cheap insecticide that protects agriculture nearly so well as DDT. Just a week later on July 16, René Dubos, also speaking at the same WHO 22nd Assembly, said, "The world will get into trouble with DDT. The insecticide is stored in man's cells like a time-bomb and under stress its side effects become dangerous. . . . I am well aware of the immense usefulness of DDT, but my suspicion is that we will come to regret it." LaMont Cole, a famous ecologist, has reported that DDT seems to inhibit photosynthesis in the oceans by poisoning marine diatoms. It is important to realize that photosynthesis in the oceans produces 70 percent of the oxygen present in the earth's atmosphere.

Of course, caught up in the whole concern with environmental control are the dynamics of world population growth. As if pursued by ancient furies, the outlook of more and more earthlings nullifies almost every other good intention. If we reach the point in population projections (anywhere above 5.8 billion, according to one very conjectural set of projections) when there is not enough oxygen to go around—even assuming present purity levels can be maintained—we can appreciate the need for a planetary perspective. It matters not whether most of the *too many* breathers stand on one side or the other of various national frontiers.

Perhaps an even clearer case is created by pollution that either heats or cools the earth's atmosphere and might cause either "the greenhouse effect" or a solar shielding effect. If the quantum of breathing, or the discharges of the effluence of industrial society, raises the earth's temperature even as little as 4 to 5 degrees then the polar ice caps would melt, as some specialists contend is now happening. The consequence could be to raise sea level as much as 40 feet in a few decades, as much as 300 feet in a century or so, with the result that many of the world's great cities would be submerged beneath many feet of water. "The greenhouse effect" arises from the great amount of carbon dioxide put into the atmosphere (Glenn T. Seaborg estimates six billion tons a year) that makes the earth warmer under "the nonconductive blanket formed by carbon dioxide." In effect, we may create a situation such as exists in a greenhouse where the heat gets in but it cannot get back out. As a result the planet could reach a saturation point where it begins to roast its populations to death.

Some scientists foresee the opposite peril: that a new Ice Age is on its way, sending glaciers across temperate land masses. The reasoning is as follows: the cloud cover over the earth will continue to thicken as additional quantities of dust, fumes, and water vapor are spewed into the air from car and truck exhaust, factory smokestacks, and incinerating garbage dumps. This thickening blanket of cloud will screen out the sun's heat and the planet will cool, the water vapor will descend to earth and freeze, and the chill reality of a new Ice Age will result. Whether the greenhouse effect of screening heat in the earth's atmosphere or the blanketing effect of screening heat out is correct, we do not know, and, needless to add, it is highly improbable that the two effects will neatly balance each other off. Specialists disagree and acknowledge that much more information is needed before questions of this kind can be resolved. The crucial point for our purposes, however, is the extent to which human activity is threatening the fragile ecological balance upon which our life depends. By persisting in these dangerous ways we invite some kind of ecological collapse, even if its exact character cannot be anticipated in advance.

These dramatic possibilities—the precise risk of which we are unable to assess—emphasize the *fragility* of planetary conditions of human existence. Even if we deny ourselves the benefit of prudence, we may want to avoid these perils for our children and their children. As soon as we lengthen the time scale beyond our lives we begin to grasp the incredible risks that we are taking by living as we do. To bring these risks under some degree of human control requires awareness, action, and new modes of social and political organization at all levels of human life. Even under the best of circumstances the planet is vulnerable to events beyond its control: a one percent shift in the sun's temperature would either burn or freeze us into extinction, as might some kind of shift in galactic fields of gravity.

On all sides, the situation that already seems serious promises to become worse. The prospects for smog control in large cities can serve as an example. Already breathing for a day in a polluted city equals, according to the U.S. Public Health Service, smoking two packs of cigarettes, with all attendant health consequences. As early as 1964 the single city of Chicago each day put 25,000 tons of pollutants in its air. Much of this air pollution consists of carbon monoxide from cars (more than half the total), although there are also large quantities of sulphur dioxide, particulate matter—soot, fly ash, hydrocarbons, and nitrogen oxides fed into our urban air. Each car, unless equipped with an anti-pollution device, dumps about seven pounds of waste into the atmosphere each day, from the action of its gas tank, carburetor, crankcase vent, and tail pipe. Since a large proportion of driving is within cities—Jaro Mayda, a law professor in Puerto Rico, reports "ninety-three percent of all auto trips in the mid-1960's were made entirely or partially within cities"—given the prospects for rapid rates of global urbanization and of large increases in the number of vehicles, the problem of urban air pollution is likely to reach highly dangerous proportions within a decade or so. One fairly reliable estimate shows that in the United States alone annual auto production is likely to rise from 6.7 million in 1960 to 12.6 million in 1980 to 25.9 million in 2000. Such projected increases call for heavy roadbuilding activity which places additional strains upon resource supplies and will tend to withdraw rural land for urban use. The same process as reported in the United States is going on elsewhere in the world. Just to cite one example, Spain reports a booming auto industry growing at the rate of 10 to 12 percent per year, and is now giving it the capacity to produce 400,000 vehicular units a year.

Even now if smog is trapped in the city for several days by "an inversion," severe health hazards arise, including large increases in deaths reported for lung disease. As a result there is the possibility, not at all remote, that urban populations will experience critical shortages of breathable air so that a plan to erect a plastic dome over the center city or a municipal ordinance requiring that gas masks be worn to work are practical possibilities rather than weird fantasies. Already one sees improvised face masks worn in heavy smog; Tokyo's traffic police wear masks at present; some models of future cities show a translucent dome that lets light in but keeps the air out.

The rapid total and per capita increase in garbage is another critical dimension of the urban pollution problem. New kinds of plastic containers have become very important for market distribution and add all sorts of chemical gases to the air when incinerated, including phosgene, which was actually used in World War I as a weapon of gas warfare. The figure of per capita combustible garbage in the United States has risen from 500 tons in 1930 to 1,100 tons in 1980 and the totals keep rising. Elaborate proposals are being put forward to ship garbage to disposal dumps as far away from main cities as 375 miles or to send it off into outer space. As matters now stand, garbage and sewage volumes are growing beyond the capacities of treatment facilities and are making urban water as well as urban air increasingly impure. Urban jetports add further to the toxicity of the atmosphere around the cities of the world and establish highly polluted corridors of air in the upper atmosphere.

The intricate problems of noise pollution are growing steadily worse, damaging our health and possibly imperiling our sanity; noise levels in cities from the increasing vehicular population, omnipresent construction activity, and the growing chorus of in-home appliances already frequently exceed the danger threshold of the 85 decible level, above which the U.S. Air Force recommends wearing "ear defenders." Communities living near jetports are subjected to such troublesome noise that many residents cannot remain in the area, or if they do, are bothered by severe symptoms of disorder such as insomnia, nausea, and acute nervousness. It is estimated that $200 million worth of lawsuits for this kind of damage are now pending in the courts. Everywhere one looks, the bad situation is likely to grow worse as a consequence of growth of numbers, expected rises in living standards, and expansion of public and industrial facilities. Countermeasures are being taken, but not such as to question the premises of permissive economic growth, the underlying disorder. Proposals to eliminate the internal combustion engine or to find a substitute for it are occasionally discussed, but little attention is given to banning cars from the inner city. It is even difficult to make mandatory relatively inexpensive anti-pollution devices for cars in such an affluent, car-conscious society as the United States. We are confronted with double obstacles to adaptation: the paradox of aggregation inhibits *private adjustment* whereas the absence of distinct interest groups with much at stake discourages adequate public expenditure. Behind these general obstacles lies an uneven pattern of benefits and burdens in the event that stiffer anti-pollution standards are adopted. Those profit-making entities that are confronted by new costs will resist the new regulatory claims, or argue that their own efforts at socially responsible action are sufficient. Even the National Petroleum Council—the lobby of the oil and gas industry—supposedly offered to send speakers out at its expense in support of the Environmental Teach-Ins scheduled for April 22, 1970. Presidents of power companies and airlines have jumped aboard the ecological bandwagon proclaiming themselves "fellow-evnironmentalists."

Any effort to impose regulations on a powerful sector of the economy generates a very prolonged and difficult struggle. The outcome, if the evidence favoring regulation is overwhelming and the political organization pushing for new standards of regulation is well financed and sophisticated, is likely to be a compromise stretched out over years, impairing the profit picture as little as possible and softening the defense of the public interest so as to secure the voluntary participation of the corporate interests at stake. The efforts by Government to regulate media advertisements by the cigarette industry provide a typical case study. In the end the cigarette companies have agreed to stop all television advertising by the end of 1971. To implement the voluntary plan of the cigarette companies requires broadcast companies to release cigarette manufacturers from their long-term contracts. The regulatory compact was itself a compromise, allowing the end of TV advertising to be voluntary, not mandatory, and assuring cigarette companies that efforts to advertise in newspapers and magazines would not be subject to prohibition. But why should the broadcast networks acquiesce in such a scheme carried out mainly at their expense? The ABC network has provisionally refused to release cigarette companies from their contracts, CBS has reluctantly acquiesced. But why, if health hazards exist, allow only certain media to advertise the noxious products? We observe here the undertow of contradictory pressures that pull even the most elemental pursuit of the public interest badly off course. The same undertow of pressures and interests exists in relation to environmental issues. It is even more serious with respect to the evolution of a planetary policy as regulatory authority is fragmented among the main national governments. The reality of national sovereignty signifies the absence of a unified government in world affairs and makes difficult the negotiation of compromise policies in situations of diverse interests. The present *organization* of international life does not facilitate the growth of *coherent awareness* or the proclamation of *a state of planetary emergency.*

The oceans used to seem so vast that no amount of waste disposal could cause a pollution problem. But even the oceans have their tolerance limits, and can absorb wastes only up to a point. Thor Heyerdahl crossing the Atlantic by papyrus boat in 1969 found "plastic bottles, oil blobs and other detritus of civilization adrift on huge patches of ocean far from the nearest ship or shore." Ocean pollution is today one of the most serious areas of environmental danger. The oceans are governed by classical ideas about "the freedom of the seas." As such, they represent an international setting wherein Garrett Hardin's analysis of "the tragedy of the commons" fits in most illuminating fashion. All governments are free, in theory, to fish and to make other profitable uses of the ocean. There exists no effective common regime able to enforce standards in the interests of the world community. This is a governmental deficiency of far-reaching consequences: there is no reason not to anticipate a tragedy of the oceans being enacted in the years ahead. The facts vis-à-vis pollution are alarming, and, as elsewhere, get worse with each increase in the world industrial base and population. Statistics are not the whole story, but they do convey some sense of magnitude and provide an informational base upon which to rest proposals for cure. LaMont Cole, the ecologist from Cornell University, estimates that 500,000 chemical compounds per year are put into the air, and most of these find their way into the oceans. Airborne lead from anti-knock gasoline falls into the sea at the rate of 250,000 tons a year. Lead poisoning causes kidney failure, brain damage, death. According to one estimate "the United States society uses annually 7.5 tons of fuel and 5 tons of minerals, food and forest products for each person. Nearly all of these materials used by man in social, agricultural, and industrial activities are not retained by him, but are dispersed, often in degraded forms, about the surface of the earth. The oceans by intent of man, by their very expanse or by their chemical and physical characteristics, receive a major portion of this discharge." Edward D. Goldberg, a foremost oceanographer, summarizes some of the basic sources of oceanic pollution: mercury used in agriculture, where it functions as a very effective fungicide, has entered the oceans at the rate of 4,000 to 5,000 tons per year. In Sweden mercury contents of birds has increased 10 to 20 times in the last decade. Mercury is a poison for man if consumed in any significant quantity. A mercury compound had been discharged into

Minamata Bay in southern Kyushu Island of Japan and is being concentrated in fish and shellfish. Mercury poisoning caused severe illness and a considerable number of deaths beginning in 1960; Minamata disease reached "near epidemic proportions" as a result of mercury poisoning. The Minamata phenomenon is an accident of the sort of the *Torrey Canyon* oil spill—that is, it is an accident that is bound to occur, given the technological modes of exploiting the ocean. Only the time, place, and extent of disaster are uncertain.

A great deal of oil enters the sea as a consequence of its transport from one place to another. Current estimates of the losses due to leakage and spillage are calculated at .1 percent of the total transported volume, which adds up to 1,000,000 tons per year. *Quantity becomes quality* in relation to environmental decay: increasing *per capita* and *aggregate* demands will contribute a proportionate increase to the quantum of ocean oil pollution. Local disasters such as those caused by a tanker breakup or the leakage from offshore oil wells bring immediate ruin to nearby beaches and bird life.

Industrial chemicals are being discharged into the oceans in increasing quantities and at an increasing rate. Large quantities of acetone, butyraldehyde, and methy-(-ethy)ketone have been found in surface waters in the Florida Straits and Amazon estuaries. A group of synthetic chemicals used in plastic, rubber, and paint manufacture, called polychlorinated biphenyls (PCB), are found in the bodies of sea birds of the Pacific in an amount that is double that of DDT: even heavier concentrations of PCB are beginning to be found in waters close to industrial areas. PCB induces sterility in birds, and acts upon them in a manner similar to that of pesticides.

The burning of fossil fuels releases into the atmosphere great quantities of carbon dioxide. Since the beginning of the industrial revolution there has been an increase in the percentage content of carbon dioxide in the atmosphere above the ocean. The process of photosynthesis is able to absorb by reconversion just so much carbon dioxide, and the consequences of further increases of CO_2 over a term of years are difficult at present to assess, but potentially dangerous for the purity of air and the earth's climate.

Pesticide residues, especially DDT, but also dieldrin and endrin, have been carried from land use into the oceans. "At the base of the marine food chain are the single-celled algae whose photosynthetic activity is decreased by only a few parts per billion of DDT." Various sea birds, widely dispersed, have exhibited growing concentrations of DDT: skuas, peregrine falcons, shearwaters, black petrels, pelicans. The result has been to decimate the populations and to endanger the survival of several bird species.

Nuclear testing and discharges from nuclear reactor facilities have also introduced significant quantities of radioactive material into the oceans. Among the radioactive elements have been Sr^{90} and Cs^{137}, as well as the radioactive species C^{14} and H^3. It is as yet uncertain what effect these substances have had or will have on marine ecology.

Finally, there is an enormous quantity of domestic sewage deposited in the ocean each year. Several million tons of solid sewage are contributed each year to the world ocean. Heavy sewage discharges apparently cause a reduction in "the number of species of organisms" although the total population of organisms may remain the same if adjacent and less affected areas are taken into account.

It is evident that the worsening pollution of the ocean is a consequence of the scale of industrial development, the life-style of affluent societies, and the continuing increase in world population. Some steps have been taken to abate pollution, in general, and specific prohibitions often follow upon demonstrations of dangerous results. The ongoing fight against DDT and the prohibition of mercury fungicides in Sweden are illustrative of these efforts. In the United States the National Wildlife Federation has concluded that clean water in the United States could be produced by a $26.3 billion expenditure spread over five years, whereas present expenditure levels are at $3.7 billion per year. One is impressed by the projected expense, by the existing effort at pollution abatement, and by the margin between the two spending levels. Even in 1969 President Nixon's budget request for water pollution control was $214 million and when Congress, aroused by the concern of the electorate, appropriated $886 million, Nixon refused to spend most of these additional funds. Within the international setting the problems associated with the paradox of aggregation are present. One country has no strong incentive to safeguard the oceans at great expense unless all countries are adhering to similar standards. But for countries confronted by mass misery and semi-industrialized economy, even the most dire consequences attributed to ocean pollution seem remote by comparison. Voluntary action by national governments is very unlikely to preserve the overall quality of the oceans, although the leadership of rich societies can make a tremendous difference in setting the tone for the whole world. At the present time the U.S. Government tends to be the spokesman for the private interests that are responsible for most of the harm; the Department of Interior is explicitly concerned with encouraging by subsidy, if necessary, greater private investment by American industry in ocean development, in relation to both food and mineral resources. In other words, the public interest is conceived of in terms of national competition, and the role of government is to increase the relative share of wealth and income enjoyed by its national subjects. Such a conception is at odds with any pursuit of planetary welfare through cooperation among governments. National governments have mortgaged their discretion over policy to such an extent as to defeat any hope for promoting an enlightened version of the national good, let alone work for the benefit of some wider community of mankind than the sovereign state.

We do not yet know the full danger arising from various forms of ocean pollution. We do know that there are many preliminary warnings that the ocean as a source of food supply and oxygen may be endangered by "the chemical invasion of the oceans," that the fight has already been lost to save certain species of plants, fish, and birds from extinction by human predation, and that the future will bring about increased hazards as a consequence of proposed mineral exploitation, larger oil tankers, and a consistent pattern of increase throughout the world in the discharge of wastes of all kinds.

The destruction of the environment seems to center around one basic kind of human activity: disposal. How can we get rid of waste without poisoning the life-sustaining elements of the environment and without burying ourselves in a garbage dump? Such a problem is an inevitable by-product of what physicists call "the conservation of matter." One can shift the form and state of matter by various strategies of *dispersal,* but there is apparently no way of getting rid of dangerous ingredients that do damage to natural ecosystems when introduced in certain quantities, although some exotic solutions by way of hypercompression composting have been recently proposed. The character of modern technology, the extent of contemporary population, and the life-style of the relatively affluent have given waste disposal its urgent character. We are dealing with more and more poisonous substances in larger and larger quantities.

A "small" incident suggests what may lie in store for us: in 1945 the British Government ordered Germany to get rid of 20,000 tons of chemical weapons material by dumping it into the Baltic Sea. Twenty-four years later it is believed that various of these chemicals, including a lethal nerve gas called Tabun, started escaping from the rusting containers. An incident in the summer of 1969 involving six injured fishermen reportedly "brought panic" to vacation resorts in southern Sweden and in the Danish island of Bornholm. Thousands of tons of fish were thought to be contaminated after a Danish trawler arrived in Bornholm with various degrees of mustard gas burn. Two members of the crew were in such serious condition after handling contaminated nets that they had to be flown to a Copenhagen hospital for skin transplants. The Bornholm affair is really a modest event, given the character of the times. At present, larger quantities of lethal gases and various radioactive substances are being buried in the earth or thrown into the ocean in various kinds of containers of uncertain reliability. We are all hostages of the prudence and foresight of whatever organization is entrusted with the particular dumping operation. Even exceptional care may be of no avail, given the unknown character of some risks over time.

George R. Stewart points out that there really are only two basic disposal strategies: dumping or recycling. Dumping is appropriate to a more sparsely settled primitive environment in which man merely gets rid of waste products in the cheapest and most convenient way. Often dumping into rivers or oceans carries the waste off or disperses it in a harmless way so that whatever is noxious can be gradually reconstituted.

But the air and water can no longer absorb in a satisfactory way the wastes of a chemically oriented agriculture (fertilizers, fungicides, insecticides) or of a car-oriented mass society (with its exhaust fumes). Part of the problem involves resource allocation: many standard forms of pollution could be rendered harmless by spending enough money on pollution abatement technology. It is a matter of appreciating the need and organizing the concern so as to overcome entrenched interest groups. Pollution abatement essentially involves the recycling of waste so that the process of dispersion is managed in such a way as to convert garbage and waste back into a productive relationship with the environment without too much intervening damage.

The point is that we need a new set of social attitudes: something needs to be done and it can be done. That something, as has been pointed out in many ways, is mostly of a social and political nature, but it also involves serious economic and technological adjustments. An ecologist, Frank E. Egler, emphasizes the political underpinning of pollution control: "The problem of pesticides in the human environment is 95 percent a problem—not in the scientific knowledge of pesticides, not in the scientific knowledge of the environment—but in the knowledge of human behavior. The problem has come into existence because of a revolt of an intelligent minority against the pseudo-scientific technology of our age. . . ." We know what to do, but not how to get it done. This poses a problem of politics and morality, not of technology.

In such a situation we note the archaic character of regulatory authority pertaining to the oceans. So far as the oceans themselves are concerned, there is "freedom" by states to act in any way that does not inflict harm on others. This freedom has been argued by prominent international lawyers to include even the right to test nuclear weapons on the high seas. Land-based pollution—crop insecticides and fungicides—is subject to the *exclusive* control of national governments. Inland states, upriver states, non-marine states have different priorities and interests from coastal, fishing, or maritime states. The absence of regimes that are more embracing than these kinds of exercises of sporadic and fragmented control seems to be a very serious deficiency at this point. The case for community institutions to protect the environment of ocean and air seems overwhelming, as does the insufficiency of national institutions of control. As O. M. Solandt, Chairman of the Science Council of Canada, has suggested, "mankind is steering a precarious course" between the Scylla of "uncontrolled exploitation of new technology" and the Charybdis of "rigid control." To reach the middle course, Mr. Solandt writes, "Obviously action must begin in individual nations, but it should quickly become international in scope because so many of the potential problems are worldwide." The question beneath such a sensible view is whether the structure of international society would be able to undertake large-scale guidance of important subject matter like the quality of the oceans or the defense of the world environment, given disparities of ideology, affluence, and stage of development existing in various national societies. Will nations be able to achieve common understandings, given their very different national circumstances, backgrounds, and priorities? We find out how hard it is to resolve issues of social control within the context of national society where highly developed and effective traditions of government and authority exist. How much more difficult it will be to iron out differences of interest and outlook in a global setting where such traditions are absent and where prevailing modes of negotiation rely so heavily on threats, self-help, boundaries, and a false sense of national honor. Compromise and coherent control are very hard to establish under political conditions beset by conflict and self-assertion. For this reason it seems crucial to examine the international dimension of environmental problems in relation to the prevailing world-order system. When this examination is made it becomes clear that we are continuing to handle novel problems of great magnitude with the outmoded attitudes and obsolete political structure that served us in the past.

5. SOME CONCLUSIONS. This survey of endangered-planet dimensions has sought to emphasize certain features of the crisis: (1) its urgency; (2) its planetary scope; (3) the interconnected character of the four main dimensions of the crisis; (4) the role of *archaic attitudes* in inhibiting adequate responses; (5) the role of *archaic organizational forms*—especially in the relations among states—in preventing a clear definition of the problems and the design of solutions; (6) the role of modern technology, the expansion of human numbers, the rise of consumptive affluence, and the consequences of dumping strategies of waste disposal as central causative agents of the crisis; (7) the role of cumulative processes of isolated, trivial acts in building up "quiet crises," noiseless time bombs ticking away until their abrupt detonation; (8) the potentiality for adaptive change provided effective action is taken immediately on an emergency basis in the main centers of political authority throughout the world; (9) the overriding importance of putting these issues high on the political agenda of our time for *all* ideologies and countries. The immediate first step is to awaken human consciousness as to the magnitude of the danger and the necessity for drastic social and political change at all levels of human existence.

Underlying the entire discussion is the validity of a certain amount of *exponential thinking*—that is, worrying about the future because of trends and rates of change in the present. A more conventional and convenient view of the future arises from assumptions of *automatic adjustment processes.* Something will turn up, a technological solution can be found, action will be taken when the situation becomes really dangerous. We all have the feeling that the government would be doing more if the situation was really as serious as the evidence suggests; somehow the evidence must be presented in a one-sided or exaggerated form. None of us can really take seriously a set of facts that require us to make some basic changes, including the sacrifice of some valued ends.

Neither *exponential-mindedness* nor the prospect of *automatic adjustment* seems to locate the situation quite accurately in *political space-time.* Every recorded civilization has been blind to the threats directed at its ascendancy. Who could have believed in the fall of Rome when the Empire was at its height? Or in the collapse of the great civilizations of Egypt, China, Greece, Latin America when they had gained ascendancy? The blindness of man to the causes of his own downfall is a persistent strain of historical experience. Because of the scope of the crisis and the technology of warfare there appears to be more likelihood that the downfall will imperil the entire world, not just a civilization, and that the earth processes of recovery and continuation will be more difficult than anything so far experienced in the history of the planet.

V. World Order Today: The Quest for Stability

The idea of world order is a very elusive one. It is often used in a way that confuses what *exists* with what *ought to be*. There is a great deal of wishful thinking and utopianism present in world-order literature. My intention is to be descriptive. The use of the term "world order" intends only to describe the *characteristic forms of behavior* by which security and change are pursued by states and other international actors. As such, the study of world order is concerned with the *structures of authority, types of conflict, role of violence,* and *methods of settlement* that are relied upon by international actors in pursuit of their goals.

A system of world order can be appraised in relation to its capacity to satisfy the goals of these actors at acceptable costs and in an atmosphere of tolerable risks. Thus, it will be necessary to examine whether nuclear deterrence works as a principal way to uphold national security, whether it costs too much, and whether the risks and consequences of failure are too high.

In later chapters I shall consider other ways of organizing international society. These other ways amount to alternative systems of world order.

We are dealing with a world political system that is dominated by the sovereign state, although a variety of other political and economic entities play major roles. Earlier chapters have argued that this pattern of organization, and the attitudes and traditions that have grown up with it, are the principal obstacles to *adaptive change*. Governments of most sovereign states continue to be dedicated to a political agenda that gives scant recognition to the problems of an endangered planet. The dominant world ideologies are national (and transnational) in their orientation, and continue to view the domain of politics as coterminous with the problems of *man-in-society*, emphasizing patterns of conflict and collaboration among societies.

States are very different in size and strength. Nevertheless, the formal acquisition of statehood in international affairs shapes the basic character of international life in countless ways. The actuality of sovereign equality of states is a significant political fact.

Relative power, by and large, does not determine the routine relationships that dominate life across boundaries. States are reluctant to resolve most international disputes by threatening or using force, even when the disparity is great between the aggrieved state and the provoking state. The United States has been reluctant to rely upon any coercive solution for its dispute with Ecuador, Peru, or Chile arising from the decision of these countries to extend their territorial waters 200 miles from shore so as to exclude unlicensed foreign fishing vessels. Numerous American fishing vessels have been seized and their owners fined as a punishment for engaging in activity that the United States Government continues to uphold as legal. The Soviet Union has refrained from threatening China with nuclear attack during its long and bitter dispute over title to almost 1,000,000 square miles of territory now occupied by Soviet troops.

Black African countries of tiny capability refuse to grant South African commercial aircraft permission to overfly their territory. South Africa complies with these regulations.

In most situations an ambassador from a small country enjoys the same status and diplomatic privileges as an ambassador from a principal state. Similarly, a passport validly issued by a minor country, even by mini-states such as Liechtenstein or Luxembourg, gives an individual a right to international travel and to claim alien residence, rights that can be obtained by "a stateless person" only with the greatest difficulty.

When the Iraqi Government in 1969 ordered its 15,000 nationals living in Lebanon to leave the country in twenty-four hours or face penalties, over 8,000 left within the day's time. Almost all Iraqis are reported to have complied with the extension of this edict to 48 hours despite the inconvenience and hardship involved in cutting long-standing business ties so quickly under pressure; the order was motivated by an anti-Lebanese campaign of uncertain character in Iraq.

These rather commonplace illustrations are intended to show the extent to which individuals continue to shape their world view by reference to the primacy of their national affiliation and by the degree to which national governments even of unequal size and capability uphold in practice the prerogatives of state sovereignty. The state represents the central political entity in world affairs, the basic distribution of authority that gives to international relations their special character. It is no accident that standard world maps are drawn to emphasize the critical importance of national boundaries rather than ethnic, linguistic, or topographical groupings, nor is it surprising that so many of the most intense international disputes involve the location of political boundaries. It continues to matter greatly whether particular areas of land or groups of people are subject to one or the other of two disputing national governments. In fact, such an issue continues to be a matter of life and death in many parts of the world. The location or relocation of a boundary remains a cause worth dying for, almost everywhere.

As such, the ideology of sovereignty remains very much alive in the modern world. In fact, the hard-won independence of many formerly colonial states has given a new world dynamism to nationalism as a political cause. The strength of state sovereignty and the differences in priorities and outlooks among states explain the high degree of incoherence and fragmentation in international society. The United States would not share its weapons secrets with its closest allies even during the confrontation stages of the Cold War. Arab countries, although formally joined in common cause against Israel, remain incapable of aggregating their separate capabilities in a single joint effort. Similarly, black African countries have shown little ability to cooperate against their common enemy, South Africa, except in a variety of rhetorical respects, such as votes of censure in the Organization of African Unity (OAU) or in organs of the United Nations. Even when agreement on goals and means is reached it is difficult to overcome the separateness of sovereign states except in situations of shared emergency, as during a common war effort such as occurred during World Wars I and II. War efforts represent the most radical experiments in international cooperation that we know about.

The only other kind of coherent action arises out of situations of domination where the dependent member is not in a position to assert its autonomy. Explicit and covert empires may be considered as a form of involuntary supranationalism.

In the present historical period several sharp alterations in the structure of international society have occurred that bear upon its character and capacity for change:

1. The disintegration of colonialism and movements of national self-determination have produced a vast increase in the number of sovereign states participating in world affairs. As a consequence, the network of relationships has grown more complex, but the operative structure of international society has remained virtually unchanged.

2. The technology of war, including especially nuclear weapons and the means for their accurate delivery at great distances, has meant that every state is vulnerable to rapid destruction at the will of either nuclear super-power. The competition for an edge in nuclear weapons posture has generated an expensive arms competition that only the United States and the Soviet Union can afford. The rapid rate of technological innovation in weapons design has also led to a flourishing arms trade and to the existence of stockpiles of slightly outdated, but surplus, conventional weaponry. As a result, states that have gained their independence since 1945, especially if engaged in regional conflicts, have devoted vast resources (proportionate to their overall budgets) to arms buildups to maintain internal order and to cope with foreign policy issues.

3. Among the most powerful states, indicators of relative power are increasingly related to the level of technology, the size of GNP, and the scope and extent of indirect political influence. There is less inclination on the part of expansive states to seek the conquest of territory, although there may be sharp struggles arising over boundary disputes. External power of an imperial variety is now likely to be imposed as in Eastern Europe and Latin America by indirect and informal means, often disguised beneath the mantle of "regional action." States are more reluctant now to intervene over various economic defaults or to restore civil order in a foreign society. Eco-

nomic power, also, tends more and more to assume a shape other than direct foreign investment, and is likely to be connected with the operations of a multinational corporation or with foreign ownership of a nationally incorporated enterprise.

4. The postcolonial world of Asia and Africa is beset by political turmoil, especially pitting those who seek to maintain the traditional social order against those who wish to give political control to a group of radical innovators. Everywhere, regardless of governmental form, the goals of political leaders are to build as rapidly as possible a modern society, which means above all else a society that is experiencing rapid industrial growth as expressed in GNP accounts. This growth rests on the application of knowledge to the problems of society, especially through the buildup of specialists, attitudes, and techniques suitable for the growth of large-scale industry. In terms relevant for an endangered-planet analysis Dankwart Rustow defines modernization as "rapidly widening control over nature through closer cooperation among men." It remains the prevailing assumption throughout the world that the prime political task is to organize domestic society so as to facilitate "control over nature," and that the measure of good government is correlated closely with the extent of this control, although other measures involving domestic stability and the absence of repression may also be taken into account. From the perspective of almost every national government, the summary goals of policy are ever-increasing control over nature in an atmosphere of political stability. The search for a new basis of world order or for cooperative action to protect the deterioration of environment is not on the operative political agenda, although it may occasionally receive some lip service. The immediate priorities of nations induce maximum exploitation of nature at minimum economic cost, and the collective world interests in maintain-

ing the environment for the benefit of all is given little attention.

5. The world system has experienced since 1945 a sharp ideological cleavage between the communist and non-communist spheres of influence, and more recently, a sharp cleavage within the communist sphere in the form of the Sino-Soviet clash. The collapse of communist cohesion has been accompanied by some weakening of bonds within the Atlantic Alliance and by certain efforts at cooperation between principal ideological rivals, the Soviet Union and the United States. Both of these states have engaged in highly unpopular uses of military power within areas defined as falling *within their political zone of dominance,* although *outside their national territory.*

6. A variety of actors other than the state have been playing increasingly important and autonomous roles in international society. The multinational corporation has emerged in recent decades as a very significant determinant of economic policy on an international level, in many settings, a more powerful center of decision and influence than governments. Through its command of large capital resources—including budgets larger than many states—the directorates of multinational corporations determine which of several alternative national societies shall be selected as the site for further industrial growth. If the corporate structure is dominated by nationals of a single country, then the multinational corporation can both achieve a measure of freedom from national regulation and at the same time operate as a means of penetrating foreign societies on behalf of a national government. Jean Servan-Schreiber's analysis of the rise of American economic imperialism in Europe rests upon the control in key industries maintained by American entrepreneurial leadership and aggressive investment policies.

Functional agencies dealing with a variety of matters ranging from health to money are becoming increasingly active on the international level. Transnational groups often manage to create their own patterns of affiliation and develop ways to elude the control of the national government.

Such denationalizing developments are not taking place uniformly throughout international society. They are especially prominent in the West and throughout the Third World. In highly centralized societies such as the Soviet Union or China all foreign contact is filtered through the state apparatus and tends to reflect a unified set of national policies.

Erosion of state sovereignty, the emergence of an ecological consciousness, and a realistic program of transformation constitute the world-order needs of our time. Drastic readjustments in the behavior and outlooks of governing groups and their populations need to take place. An adequate appreciation of the current situation by those in control of national government is also needed. We outline in this chapter some salient features of world order today.

State and Nation. There are profound differences between state boundaries and the configuration of national, ethnic, and linguistic groups. The great states of the world are all multinational, with important cleavages among the main elements of their population. The fiction of the nation-state as the basic unit of world affairs is a false and misleading distortion. In most political entities that qualify as states, the authority of government is vested in the hands of a dominant national or ethnic group and subservient groups are subject to varying degrees of discrimination and suppression, and experience some degree of disaffection. There are vast differences in situation, but almost without exception, no major state defines the reach of its sovereignty in a way to correspond with the ethnic or national boundaries of its ruling group. As a consequence, there are varying degrees of instability introduced into the domestic life of all state systems that

generate intimidation and violence to maintain and stimulate various forms of opposition to challenge the existing
pattern of control. Even in Europe, where the modern
state first arose, there are movements for subnational
autonomy or drives for equality in almost every heterogeneous state. The international significance of this
domestic circumstance has not been often noticed in an
appropriate way: it means that violence is endemic to the
process of government and that even governments with
democratic traditions rely upon violence to thwart the
counterviolence of those who are excluded from the full
fruits of dignity, wealth, and power within the political
community. Such a reliance on violence for internal control also reinforces a resistance to structural change on an
international level. The primacy of domestic politics
means that a shift to a more volitional system of world
order, as seems implicit in a political solution for an endangered planet, would appear to imperil domestic
structures of order, as well as to weaken the capacity of
states to maintain control over external spheres of influence. The first principal point, then, is to notice the
discontinuity between political stability and the kind of
strongly centralized apparatus of government that has
emerged in relation to the rise of the modern industrial
state.

It may be useful to distinguish between nation-states
and state-nations, depending upon whether the sense of
national consciousness preceded the formation of a state
apparatus or followed from it. In Europe, for instance,
"a degree of national and cultural consciousness preceded
the formation of the state in Germany . . . whereas in
France the situation was reversed and the monarchial
state preceded national consciousness." Most of the
political entities that have emerged in the non-Western
world since 1945 are state-nations in the sense that they
have established the formal structure of governance and
have acquired sovereign status in internal and external

affairs. But the psychosocial basis of membership—the sense of affiliation that generates pride in nationality and citizenship—is often lacking; accordingly, individuals identify much more closely with subnational categories (religion, tribe, caste) than with the nation. In this sense, we should appreciate the emphasis on nation-building in the developing world. Above all, leaders in developing countries seek to arouse a sense of nationhood on the part of the population governed by the state. They seek to consolidate sentiment with structure in such a way as to forge the kind of durable political unit that we associate with the modern state.

But the bonds of affiliation, especially in more heterogeneous societies, are often under severe strain. In many parts of the world—in old states and new ones—ethnic, religious, or regional groups are seeking either to establish greater subnational autonomy or the full benefits of economic and political opportunity enjoyed by other groups. In some cases the effort by a subnational group is to break away altogether to form a separate state in which the bonds of affiliation will be more natural and more satisfying, and less a consequence of coercion. The Kurds in Iran seek subnational autonomy, the Quebeckers in Canada, the Turks in Greece, the Catholics in North Ireland, whereas the Ibos had sought to secede from Nigeria to form the state of Biafra, as the Katangese had sought to break away from the Congo to establish their national independence.

From the perspective of world order it is essential to take account of a wide variety of national circumstances. Many governments of sovereign states rely upon a domestic war system to maintain control over their own population. There is a curious kind of paradox that arises from the dominance of the state form in international affairs and the inability of most states to achieve a voluntary internal association of peoples that transcends differences of race, religion, and subnationality. In many international and national settings, group identity is more important as a basis of solidarity than is the ideological orientation of a government. Thus Arab governments of left and right are united in their struggle against Israel, just as black African governments of liberal, moderate, and radical persuasion are united in their struggle against racism and colonialism in the southern part of the African continent. These deep internal cleavages suggest both *the strength of transnational bonds* (that is, between race, ideology, religion, class, and national character) and *the limits of international cooperation* among states that are not able to secure domestic tranquillity except, if at all, by threats and uses of violence. These structures of domination encourage very limited ventures in international collaboration, except in circumstances involving preparation for struggle against a common enemy; since "the enemy" provides the rationale for cooperation, the scope of cooperation is always less than global, and represents one part of the world organizing to oppose or fight another part. A vital connection exists between a peaceful world order and the quality of domestic order.

In the setting of an endangered planet these issues seem even more important, as the governors of most states are so absorbed by the tasks of maintaining their own control that they have little excess capacity to absorb the significance of more fundamental, but also more remote and very recently perceived, threats arising from the cumulative effects of technology upon the environment. Political energies of rulers are directed toward traditional and day-to-day concerns of power and order, and these concerns do not encourage an appreciation of ecological factors. Larger states tend to be particularly absorbed with the maintenance of power in internal and external relationships—partly because of the artificial "community" beneath their sway—and, hence, are resistant to changes in the traditional agenda of politics.

Sovereignty and Independence. The formal requisites

of sovereignty have to do with the identity of a state, including access to international institutions and participation in diplomatic intercourse. A state is usually defined as a delimited area of territory administered by a government in control of people and resources. Sovereign states in this *formal* sense may or may not enjoy *political independence*. Many formally sovereign states are subject to various forms of domination, penetration, and dependence; intervention in internal affairs of sovereign states is a regular part of international society. If the intervention occurs within an acknowledged or effective sphere of influence, then even rival states will tend to confine their response to verbal protests. If there are indistinct spheres of influence or contradictory claims to exert control, then a highly dangerous situation is present that can produce very serious kinds of warfare. The colonial system represented a formalization of spheres of influence, a denial of formal sovereignty to the colony, and an extreme assertion of prerogatives of control and exploitation by stronger states. In the present postcolonial period (remnants of the colonial system survive, principally in southern Africa), the attributes of formal sovereignty are normally bestowed even upon states whose independence is severely infringed by a foreign government that may have put and kept in power a governing group without a firm base of political support inside the country in question. The Soviet role in Eastern Europe since 1945 and the American role since 1954 in Southeast Asia illustrate this kind of relationship. The rhetoric of sovereignty influences certain kinds of behavior, restricting for instance the freedom of action that the dominant government may have to induce "its puppet" to behave in a certain way. The United States Government has seemed especially susceptible to arguments based on its own formal insistence that it is respecting the independence of client states. As a result, the powerful government may find itself entrapped within "a commitment" that it made to a dependent government. The relationship between the United States and Formosa illustrates this inversion of control, by which the weaker member manipulates the stronger one.

The result of this discontinuity between *sovereignty* and *independence* is to create a gap between *behavior* and *language* in world politics. The *actualities* of inequality in power and variation in ideology and ambition overcome the *formalities* associated with the primacy and equality of the sovereign state as an actor on the world scene. The effect of this gap is to inhibit structural change because dominant states depend on their freedom of action, including the threat and use of military power, to sustain their relative position. Anything that compromises this capacity to dominate will appear to impair their current advantages, including access to markets, investment outlets, and assured sources of cheaper raw materials at favorable terms of trade.

Size, Function, and Authority. International society continues to remain organized around the existence of sovereign states, and is dominated by the activities and policies of several principal states. The nuclear superpowers are of ultimate significance, holding within their top echelons of government the capability to initiate (perhaps without or even contrary to intentions) nuclear war. The state became a dominant form of political organization in the late Middle Ages at a time when it was possible to provide for the common defense of the realm against external enemies and to organize a domestic society into a relatively self-sufficient economic unit. Because of interdependence today, many poorer countries can be brought to a situation of ruin by a mere shift in the buying habits of a richer country. For instance, the economy of Ghana would be severely damaged by the development of synthetic cocoa in Western Europe or North America.

Increasingly, the rationale of state sovereignty has been undermined in several, almost contradictory, ways. First of all, the scale of industry and the character of modern technology create strong incentives to increase the size of the basic political unit to enlarge the market, avoid expensive duplication, pool resources for the newer technologies, and take advantage of industrial specialization. Such reasoning—primarily economic in its significance—although bearing on relative power, induces larger-sized units and creates pressure for economic and political integration. These forces have stimulated a regional movement that has enjoyed a certain limited success in all parts of international society except Asia. Given the resource requirements of modern industry and defense there is an increasing sense that the sovereign state is an inadequate unit in all but the few instances when its size is of continental magnitude (U.S., U.S.S.R., China, India, Brazil, Canada, Australia). But large state size causes its own problems of governance, especially if there is a large diverse population.

Secondly, the management of certain common interests in world affairs creates an increasing quest for *unitary* forms of organization—whether the issue is the hijacking of commercial aircraft, the protection of the blue whale against further depletion, or the response of the inadequacy of the *fragmented* system of controls embodied in the state system. *The scope of functional concern vastly exceeds the scope of political authority of even the most powerful state.* Even if its government is enlightened and benign it has no way of safeguarding its own interests or of upholding the world interest, given the organizational forms that now exist and their necessarily limited jurisdiction. Every state short of a world state is too small to manage the countless problems being caused by present-day pressures on the world ecosystem. And the future promises to underscore further the inability of governments to reach out on their own and control circumstances and behavior that will impinge upon their most vital concerns of health, welfare, and even survival.

At the same time, the very shortcomings of a state as a

functional center of authority are undermining its present base of domestic support, especially in modern societies. The rise of demands throughout the world for subnational autonomy reflects a growing dissatisfaction with the state as the basis for political community. The artificial concentration of power in the center of a large society seems vindicated only if the functional payoff is high in terms of wealth, power, and prestige. But fewer and fewer modern states are able to achieve any degree of functional autonomy, and thus grow dependent on the defense establishment and economy of a superpower, thereby eroding the legitimacy of their claim to rule.

In sum, then, we can say that from a functional perspective, there is growing pressure to increase the *size* of medium states beyond their present scope in the direction of continental scale to achieve a competitive position vis-à-vis the superpowers in the areas of science, industry, trade, and defense. There is also emerging pressure on *all* states to acquire unitary control over processes that imperil their individual welfare in a variety of ways. And, conversely, there is pressure to break down traditional aggregations of sovereign authority to enable more authentic political communities to emerge as distinct entities with real control.

These diverse pressures underlie many of the instabilities evident in international society that are both abetted and obscured by the more obvious struggles for control being waged by the forces of the left and of the right in various parts of the world. My assertion is that there exists an *organizational* level of conflict that gives substance and direction to the *ideological* level of struggle. And beyond this, if the outcome of ideological struggles is only to shift the orientation of elites toward the problems of man-in-society, then these struggles are of little consequence to the struggle to save an endangered planet, except as sources of distraction and delay in the effort to grasp the new set of political realities.

Charter and Westphalia Logic. Much of the normative

confusion in international society arises out of the affirmation of contradictory logics of world order—that derived from the Peace of Westphalia in 1648 and that derived from the Charter of the United Nations. The logic of Westphalia established the state system as the basis of world order. National governments enjoy exclusive control over internal affairs and are the exclusive formal actors on an international level. Sovereign status is the essential ingredient of formal participation in international society. Out of this basic Westphalian premise arose many fundamental doctrines of international law—domestic jurisdiction, the sovereign equality of states, diplomatic and sovereign immunity, the doctrine of non-intervention, and the doctrine of recognition of new states and governments. These doctrines continue to have a prominent ordering role in international life. The logic of Westphalia stressed the formal attributes of territorial sovereignty and the resulting equality of states. Such a conception collided with *inequality* in political and economic fact and the interrelatedness of international life, inducing a variety of *de facto* systems of intervention and domination. Some of these inequalities were introduced into formal doctrine. For instance, imposed treaties of peace were legally valid, dependent sub-sovereign status —for instance, colonies and dependencies—was created by contractual arrangement, and a variety of rules authorized intervention to protect foreign investment, to collect public debts, or to protect endangered nationals. The primacy of inequality rested, above all else, on the discretionary right of government to wage war as an instrument of foreign policy. Such a right confirmed the option of the strong to impose their will upon the weak and induced the formation of alliances to discourage a stronger state from conquering, or otherwise infringing upon the liberty and territory of, a weaker state.

World Wars I and II produced a widespread reaction against the consequences of such a decentralized and uncoordinated system of world order. The great waste and suffering caused by these wars—and the prospect of even worse wars in the future—led to a general demand for the reform of international relations. The goal of reform was to reduce the prospect of major war and to shift from the national level the authority to embark upon war. Formal instruments were drawn up to prohibit the use of force in international affairs except in the exercise of self-defense. The League of Nations incorporated into an organic document—the Covenant—a general community responsibility for the maintenance of peace and security. But the prohibition of aggressive war and the creation of world community machinery and responsibility had little visible effect on ingrained attitudes and patterns of behavior. States with grievances and ambitions still relied upon military means to achieve them. Germany, Italy, and Japan, with major claims for the revision of international political and economic structure, embarked on a systematic program of expansion through military conquest. World War II ended with atomic explosions at Hiroshima and Nagasaki and a renewed demand for a truly effective implementation of the League idea. The leaders of Germany and Japan were prosecuted and punished as war criminals and a new enterprise—the United Nations—was launched to safeguard the welfare of mankind.

The United Nations was created as a new organization to symbolize the determination of statesmen to succeed where the League had failed. Only "enemy" countries were excluded, at first, from participation and the idea was to evolve a truly universal organization capable of protecting the peace and security of mankind. The Charter of the United Nations was the document that embodied this new conception of a community-centered basis of world order. But as with the League, the shift in *rhetoric* and *sentiment* was not accompanied by a corresponding shift in *capabilities* or *goals*. Despite participation in a *cooperative* framework that was supposed to represent the *world community* as a whole, the basic pattern of international relations remained *competitive* and shaped by diverse calculations of national advantage.

At the same time a new language of politics has resulted from a certain deference to the ideals and norms of the Charter. There is a genuine renunciation of war by governments as an explicit instrument of explicit policy and an enormous effort by all governments to reconcile their behavior with Charter standards. Governments are not, however, prepared to accept adverse judgments by the United Nations, and criticize the Organization rather than accord respect to its decisions and recommendations or take advantage of its procedures. South Africa, Israel, and the Soviet Union have each demonstrated their defiance of overwhelming condemnations by the United Nations, and the United States has been able, so far, to maneuver in such a way as to avoid such a confrontation, although its role in Vietnam has been repeatedly attacked by the Secretary General of the Organization, Mr. U Thant.

The resulting interplay between Westphalia and Charter logics produces a contradiction between the persisting *autonomy* of the state as *actor* and the acceptance by states of certain *constraints* on their *freedom of action.* In particular, the international status of violence is confused by the affirmation of these two conflicting normative traditions. The logic of Westphalia emphasizes the idea that state governments enjoy paramount authority within their national domain and that there is no higher external authority that can pass judgment upon a claim by a rational government to wage war. Higher norms may exist—including the renunciation of war as an instrument of national policy except in self-defense—but the national government is the *exclusive* and *final* interpreter of whether its behavior is compatible with its legal duties. Self-interpretation tends to be self-serving.

Therefore, an official justification of war by the war-making government ends any kind of legal scrutiny. In most respects states continue to adhere to the logic of Westphalia, retaining the capabilities to implement national claims to use force and to prevent any kind of serious community review in the United Nations or any other supranational forum.

In contrast, the logic of the United Nations Charter implies restrictions on state discretion to have recourse to violence in an international dispute. Uses of force are made subject to review by the UN even when a government claims to be acting in self-defense. As Article 51 of the Charter puts it: "Measures taken by Members in the exercise of this right of self-defense shall be immediately reported to the Security Council and shall not in any way affect the authority and responsibility of the Security Council under the present Charter to take at any time action as it deems necessary in order to maintain or restore international peace and security." Article 2 (4) obligates "All Members" to "refrain in their international relations from the threat or use of force against the territorial integrity or political independence of any state" and Article 33 (1) requires a state that is a party to a dispute to seek settlement by use of all peaceful means. These Charter provisions are part of *a state of mind* embodied in the Preamble of the treaty instrument as creating the conditions needed "to save succeeding generations from the scourge of war." The entire language of the Preamble and the operative provisions of the Charter suggest a commitment to a new type of world order in which a community of states would be joined together to serve the common interests of mankind in peace and security. A community judgment is supposed to take precedence over the determination reached at the national level. The ideology of the Charter is somewhat compromised by reaffirmations of sovereign prerogative and by the veto power given to the five Permanent

Members of the Security Council (although the veto has itself been revised by a wavering tendency in practice to shift authority to the General Assembly—subject to a two-thirds voting majority—in the event that the Council is blocked).

On a normative level, then, we have a very confused system of world order, especially as it pertains to the critical issue of the use of violence by states to pursue external ends of policy. Westphalian logic locates discretion on a national level and acknowledges the fragmentary authority pattern that corresponds to the *organizational structure* of the *state system.* Charter logic centralizes authority, without repudiating the Westphalian outlook, and more significantly, without providing the kind of organizational structure and capabilities that are essential to the performance of Charter functions in the area of peace and security.

One of the unexpected results of UN activity has been to "revise" in practice the Charter definitions of prohibitions on the use of violence. In the Charter only *defensive* force seems permissible, aside from *peace-keeping activities* of the UN itself or of *regional actors* carrying out UN mandates. But in UN practice two developments have occurred: (1) The strong Afro-Asian group has generated a consensus in support of *legislative uses* of force against South Africa and the other regimes under colonial and racist control in southern Africa. This consensus alters the normative status of government conduct, including acts taken to support the ends of liberation groups. It moves the UN system from one based on *substantive standards* (self-defense) to one based on *political majorities* (authorization of violence against South Africa). (2) The ideological split in the UN, together with certain cleavages within the spheres of influence subject to U.S. and Soviet control, has encouraged the detachment of *regional processes of authorization* from UN control. Regional groupings such as the Organiza-

tion of American States and the Socialist Common-
wealth provide a kind of supranational appearance of
legitimacy for international uses of violence that would
not be generally endorsed at the global level.

In the present world system recourse to war by governments is always accompanied by some kind of plea in self-
justification. War, conquest, and empire are no longer
accepted as legitimate explicit goals of states. The rhetorical terms of reference have been revised both by the UN
Charter and by the two developments in practice that
have altered the Charter. In addition to rhetoric, the
dangers of general war are more evident to principal
governments, inducing greater caution in provocative
situations and an apparent trend away from testing intentions by a confrontation of opposed wills.

There is, however, a great degree of frustration and
confusion caused by the tension between the logics of
Westphalia and of the Charter. Partly, the rhetoric and
the organizational capacities are not coordinated to permit either the implementation of Charter norms or the
abandonment of Charter rhetoric. Even where the international community is apparently unified, as in relation
to the southern African issues, it seems impossible to
mobilize the capabilities needed to implement the sense
of community; such failures of implementation deepen
the sense of impotence. At the same time, leaders are
socialized by their diplomatic style, and human nature
appears to require that controversial action be based on
some kind of credible justification. During the period
when criticism of American involvement in the Vietnam
War grew severe, President Johnson is supposed to have
carried a copy of the Gulf of Tonkin Resolution around
with him so as to provide any opponent of the war with
ready evidence of the authority he possessed to extend the
war to the territory of North Vietnam. The need for a
sense of legitimacy is great and widespread, but it seems
easily satisfied by a self-serving and formalistic reliance on

documents of doubtful validity or of authorizations of questionable relevance. Therefore, in the final analysis, national governments still appear to base their attitude toward war on a combination of egocentric factors, including varying attitudes of rulers toward risk-taking and prudence.

It has been traditional for international lawyers to divide their subject into two principal parts: "war" and "peace." This discussion of the tension between the Westphalia and Charter conceptions primarily relates world order to warfare and the status of violence in international society. In the next section of this chapter we examine the basic forms of order that underlie the assertion of international rights and duties in times of peace. Naturally, the interplay is intricate between these two basic settings of world affairs, and conditions of peaceful competition often contain an element of threatened violence and active coercion.

Fundamental Ordering Principles. Recalling the discussion in Chapter III of the tragedy of the commons, the disaster to English country grazing land arose as a consequence of combining private ownership of animals with common ownership of the grazing lands under *conditions of scarcity. Public ownership* of the grazing land might have been able to fix stable limits to the carrying capacity of the acreage and achieve some kind of enforceable quota system to limit the total animal population. Alternatively, shared or equal profits would have tended to produce a stable animal population within the carrying limits of the grazing lands. Without such limits the contradictory logics of private and common ownership tend to produce overpopulation, despite its bad consequences for all concerned.

In international society the two fundamental ordering principles parallel the ideas of private ownership and common use that we have just been considering. The land area of the earth's surface is subject to sovereign appropriation, establishing a regime of exclusive authority that is comparable to private property within the state. The water area of the earth's surface, with the exception of a small band of water contiguous to land—the territorial sea—is treated as a commons, as open for the use and exploration of all states in accordance with the overriding principle of "the freedom of the seas." Airspace has been treated as subject to the regime governing the surface area below, hence subject to national control if above land and to common use if above the high seas. These general ordering principles have grown subject to numerous qualifying exceptions over the years, but their fundamental relevance to world order remains clear.

Two developments of increasing importance are, however, undermining these principles. First, the idea of exclusive authority (lawyers call it "jurisdiction") over territory presupposed the general correspondence between events and their territorial impact. Characteristic forms of behavior within national boundaries were of direct significance only to that society. The impacts of behavior were, in general, *local* in this sense. Of course, exceptions always existed, as when currency in one country was counterfeited in another or foreign investment was allowed and then later confiscated by the territorial sovereign. Basic norms of behavior that reflected common interest—avoiding counterfeit currency, assuring the flow of mail, protecting foreign diplomats—have been sustained by the logic of reciprocity. Namely, country X upholds a world community norm to encourage countries Y, Z, etc., to do the same. Where reciprocity does *not* apply, as often with foreign investment, where the site of investment often lacks investments of its own to protect, then the community norms depend for their stability either upon the perceived *self-interest* of the territorial sovereign (in attracting investment or avoiding other repercussions) or on the capacity of the investor state to *coerce* compliance by a variety of methods. In the nineteenth and early twentieth centuries it was common for investor countries to protect the investments of their nationals by military intervention, if necessary. In more recent decades, however, the use of force for such purposes has fallen into increasing disrepute. As a consequence, the protection of foreign investment tries to rely either on reciprocity (through the creation of mutual benefits of approximate equality) or on indirect forms of coercion (for instance, through the withholding of economic or other favors from the infringing government). The Congress in the United States, for instance, through the so-called "Hickenlooper Amendment," requires the President to suspend foreign aid to any government that does not provide full compensation for any expropriation of American-owned property within six months.

The growing visibility of national events to world scrutiny also breaks down the indulgence accorded national governments to do whatever they might wish to their own population. The rise of the concern with human rights expresses a growing, although as yet feeble, insistence by world public opinion that governments be held accountable for certain minimum standards of decency even in their domestic behavior. No barriers of sovereignty should permit a government the discretion to engage in genocide against a portion of its own population. The focus of world concern has been upon racial discrimination, especially as practiced by the white minority governments of southern Africa against their black populations. The struggle against *apartheid* has disclosed both the wide *rhetorical* consensus in favor of racial equality and the very weak disposition to take effective economic or military action to implement the will of the international community. In effect, the *formal* prerogatives of state sovereignty are being steadily eroded in the area of human rights, especially by the ever more far-reaching claims of interference in South Africa's internal affairs, but national governments maintain classical zones of *effective* and *unencumbered authority.* As a consequence, there is a sense of resentment on the part of conservative forces which oppose any encroachment on the ideals of national sovereignty, and of disillusionment on the part of progressive forces which are keenly sensitive about their inability to translate verbal imperatives into appropriate action.

In general, however, a national government retains exclusive rights to exert control over such matters as population, economic, and environmental policy. These matters are deemed to fall within "domestic jurisdiction," and are not subject to any international norms. The strain on world order arises from a growing awareness of the interrelatedness of national societies, the global impact of domestic policy, the impulse to gain extensive control over ocean fisheries and subsurface minerals, and the general inability to correlate sensibly the assertion of authority with a given unit of space. It is beginning to become evident that national policies generate damaging pollution on a world level. More elusive, yet clearly emergent, is the global impact of increasing population upon the oxygen-carrying capacity of the earth, upon its climate, and upon its susceptibility to pandemic patterns of disease. As we grasp the significance of this interrelatedness we will become convinced that the premise underlying separate zones of national sovereignty is highly irrational and intolerable unless qualified by the effective enforcement of *common minimum standards.*

The zones of "freedom" are also subject to growing strain. As the marine resources of the seas grow scarce in relation to growing demand and an increasing capacity for exploitation, there arise both a drive toward conservation of the resources at the highest sustainable yields and rivalry for a maximum share of the total withdrawn supply. States have different priorities and capabilities, and so press for inconsistent regimes of control. In particular, technologically advanced countries tend to support free competition because their comparative advantage will lead to a bigger share of the market, whereas technologically backward countries prefer proprietary regimes that acknowledge special prerogatives for contiguous land areas. The previously mentioned claims of Chile, Ecuador, and Peru to exercise sovereign control over ocean area within 200 miles of their coasts illustrate an effort by poorer countries to secure for themselves valuable fishery resources and to exclude from competition the fishing fleets of more advanced countries. The extension of the idea of private ownership onto the high seas can also be vindicated by an appeal to conservation considerations. One way to avoid the tragedy of the commons is for one farmer to take over the entire acreage of grazing land. The unification of ownership provides the owner with a real incentive to maintain a resource at the highest sustainable level of productivity, provided he knows how to do this at a tolerable cost and that present yields produce acceptable rates of return on the investment. Of course, perspectives on time will vary, and a private owner may be concerned only with maximizing profits during his lifetime (or that of his children) and hence tend to overgraze. By and large governments are somewhat less shortsighted, and take a longer view of conservation, although their time scale will vary depending on domestic pressure and on the orientation of a government.

With mineral prospecting taking place along the continental shelf, and eventually on the ocean bed, there arises an additional incentive to extend ideas of sovereignty beyond land areas. The United States, already in 1945, claimed sovereign rights over the continental shelf, the width of which is still a matter of controversy, but at a minimum may involve several hundreds of miles depending on the contours of the extension of continental land masses into the ocean. Other states have followed this precedent, especially those that border on continental shelves. The incentive here is to preserve a potentially valuable area for exclusive national gain and to encourage entrepreneurial capital by assuring protection of high initial investments in exploration and development. Without a unified system of ownership or leasing, latecomers can avoid most of the early expenses of developing a mining opportunity. Therefore, both *development* and *conservation* interests tend to undermine original ideas in favor of freedom in the oceans.

Military operations also tend more and more to infringe upon the freedom principle. Nuclear testing or radioactive disposal in the oceans imperils both the use of the seas by others and the welfare of life on land. Vulnerability to attack from the sea leads to tracking operations and the seaward extension of air identification zones that depart from earlier ideas of "freedom." The rationale of a regime of freedom implies both a condition of *abundance*— there is enough of whatever is valued to satisfy the demands of all users—and of *compatible* use—the use of the sea for one purpose does not damage its use for other purposes and the use of the land does not interfere with the use of the sea. Conditions of abundance *and* compatibility are less and less present in relation to the oceans. The situation cannot be successfully dealt with by working out a series of cooperative arrangements, but establishes clearly the need for unified control over oceanic use. Even such limited cooperative efforts at conservation as now exist are likely to fail because international cooperation continues to rest so largely upon the ethos of voluntary compliance, and volunteerism appears unable to organize a successful defense of an overused commons. Nuclear testing is a prime illustration of *incompatible use,* precluding maritime passage in the contaminated area for a considerable period of years, arousing anxiety among consumers of fish products, and destroying marine life over a wide and indefinite area (depending on blast and fallout characteristics). Of course, the test site can be located in such a way as to minimize its incompatibility, but the perception of incompatibility varies greatly with the priority associated with narrow uses of the sea. This perception also varies with the assessment made of the genetic and medical damage that may be done by various densities of strontium-90 and other components of fallout. As with other aspects of technology, we know now that original hazards were understated and that detrimental effects are increasingly spread out in both time and space. Also the possibility of exceeding the saturation capacities of water and air creates some danger of irreversible change and ecological disaster. These various elements of the situation suggest that the earlier idea of "freedom" as an organizational principle is coming under increasing pressure with each passing year.

Outer space is the new domain of "freedom." Partly this freedom exists because there is no capacity to exert more stable forms of control and partly because to date patterns of use appear compatible with one another. Also, of course, the vastness of space is the clearest expression in human experience of a setting of such abundance that every actor can do all that he wants without "crowding" the resource. Cooperation may still be sensible (to avoid duplicating expensive exploits) or, in limited settings, almost imperative (to avoid accidents or bankrupting arms races in space). Such cooperation is compatible with freedom, and can occur whenever states have clearly perceived common interests, as generally existed on the oceans for instance, in relation to piracy, and may now exist in the airways in relation to the prevention and punishment of skyjacking of commercial aircraft. The reality of reciprocal common interests (or the striking of bargains to overcome nonreciprocal relationships) establishes the outer limit of international cooperation in the present world. Air piracy illustrates the vulnerability of modern life to interference, and also, the character and limits of an ordering process that rests upon a combination of *sovereign prerogative* and *reci-*

procity. So long as some countries benefit from skyjacking and other countries respect the sovereignty of the beneficiary countries, it is virtually impossible to establish effective methods of control. The return of skyjackers through extradition arrangements would lead to criminal convictions and would probably diminish greatly incentives to commit air piracy. But extradition presupposes the establishment of mutual interests, and that probably requires some kind of tacit trade-off between commercial air states and the principal havens of skyjackers. Perhaps Cuba would agree to extradition in tacit exchange for a reduction of economic pressure by the United States.

If common interests are absent and cannot be brought into being, then the maintenance of order tends to be shaped by imperial considerations in the event that the dispute involves an issue that is deemed important by the parties, and the will of the stronger prevails either through some variety of intervention or because of its threat. War and violence is the outside case of an imperial contest, and is the final limiting condition in a political setting that lacks institutions of government capable of keeping the peace, settling disputes by impartial means, and instituting changes in existing rights and duties. Some disputes, such as the Arab-Israeli conflict, where the underlying demands are contradictory, defy any kind of peaceful settlement satisfactory for the opposing parties.

The present world system, despite an increasing reliance by statesmen upon rhetoric of global solidarity, continues to be dominated by competitive patterns of behavior. We live in *a period of transition* in which *the language* of politics looks toward a future world-order system based on greater human *solidarity* and more fundamental *cooperation* among states, whereas *political and economic behavior* continues to reflect earlier values, entrenched interests, and competitive forms. Human reason can adapt language to new realities more readily than behavior.

World Order Systems: Functions, Authority, and Capability. The *scope* of modern world-order tasks—the agenda of an endangered planet—creates more and more *functions* that require *global management.* The state system can neither provide a functional solution through the enlightened assertion of its traditional *authority,* nor often forge cooperative links with other states that can enlarge the *scale of authority* to correspond with *the scale of the problems.* In certain situations of simplicity and urgency as with conservation efforts to save an endangered animal species such as the blue whale, it becomes possible for relevant governments to agree upon a common regime of management, including standards and central authority. Even in such relatively simple circumstances the absence of international enforcement machinery impairs, if it does not altogether negate, the effort at cooperative behavior. Nationals of one or more states are likely to continue to withdraw members of the protected species so long as profitable because they will not be detected, because their catch is so small in relation to the endangered population as to make no perceptible difference to the conservation effort, or because since others are not complying, or thought not to be, there is no reason to forgo profits out of respect for a vague and hypothetical community interest that is, in any event, being infringed upon.

The United Nations Charter is based upon a centralization of authority that is premature. It was not accompanied either by the grant of resources or by expectations of enforcement such as would be needed to make realistic its claims to exercise authority over governments in the field of peace and security. As a consequence, the experience of the United Nations produces contradictory reactions of *fear* and *hostility* (its authority claims collide with traditional ideas of national sovereignty) and of *disenchantment* (its failure to keep the peace, end *apartheid,* or establish an acceptable code of international behavior). Both the fear and hostility of world-order conservatives and the disenchantment of world-order idealists are by and large unwarranted. The United Nations cannot be expected to act beyond its capabilities, and the assertion of far greater authority in its Charter and its formal acts is bound to be ineffectual, unless it happens to coincide in a particular instance with the preferences of major governments. The capabilities for enforcement are possessed by a few principal states that are in a highly competitive relationship with one another in matters of wealth, influence, and ideology and are, therefore, rarely able to unite on a common course of action carried out under UN auspices.

A more promising, if unexpected, role of the United Nations is to focus attention on the world-order requirements of the day. Its public presence can serve educational and promotional needs, especially in relation to setting forth evidence that the threats to human welfare are not being met by existing organizational forms. As we have already argued, a better awareness, especially if widely shared by divergent ideologies and cultures, is an essential precondition for developing a system of world order based upon more solid cooperative patterns of national behavior. Whether the UN can do much to facilitate an awareness of this situation is unclear, especially because the cumulative impact of these threats does not fall equally or at the same time upon states and because some states are under far greater immediate pressures than others. The experience of the past suggests that disaster and breakdowns are needed to crystallize an understanding of a common threat. Even the menace of a potential enemy is often not enough to prompt preventive action. Officials of the United States Government ignored a variety of advance signals conveying the probability of a surprise attack by Japan on Pearl Harbor; the process of denial works strongly to prevent effective adaptation to common threats, especially when basic changes in behavior are called for. Once Pearl Harbor came, the vividness of the event enabled a rapid mobilization of *common effort* in support of the United States war effort. Do we require a planetary Pearl Harbor in order to respond adequately to the threats being posed in relation to the existing world order system?

In this chapter I have outlined the main features of the present system of world order. Attention was given to both the persistence of the sovereign state as the principal center of power in the world and to its declining capacity to promote the welfare and survival of its own population. I have also discussed new organizational forms—especially the United Nations—that have been created because of dissatisfaction with the propensities of states to engage in costly wars with one another. But the wish to do away with large-scale war cannot be accomplished by merely endowing some central institutional structure with a mission to keep world peace. It is necessary, as well, to build up the capabilities of such an organization and to generate the attitudes that will support its difficult work. The United Nations, as now constituted, represents, at best, a potentiality with respect to problems of war and peace.

In more recent years subtler problems have undermined the capacities of the state system. The interrelatedness of world economic and technological behavior, emerging environments of danger and scarcity, alarming trends in population growth and movement, have pointed up the limitations of national governments to act alone or through improvised efforts at cooperation to meet the challenges of the modern world. The United Nations is beginning to foster a global awareness of these limitations by a variety of endeavors that serve to document mounting dangers to planetary welfare and survival. The UN Conference on the Environment, planned for Stockholm in 1972, is a major attempt to put the agenda of an endangered planet before the whole of mankind in a responsible fashion.

But awareness is not enough. We need to *organize* to meet the new situation of danger. This organizational challenge requires new attitudes and new structures. The next three chapters propose a positive course of response. The essence of the proposal is the development of a political model of a *cooperative system* of world order and a set of *transition strategies* to bring it about. During the period of transition the principal question is whether we can develop effective and rapid *world-order learning experiences* that do not depend on long-term training and disaster to achieve an impact on political imagination. The sooner the process of adjustment begins to take place the more likely it is that extreme varieties of coercion can be avoided. If more centralized structures of authority can be coaxed into being ahead of catastrophe, then there is an improved chance of making orderly and humane shifts of control. If transition is stimulated mainly by a series of catastrophes, then the sense of urgency and desperation are likely to induce highly unstable political responses. Now is the time to begin working for adaptive change and to declare the existence of "a state of planetary emergency." The cost of delay is to place our basic liberties in jeopardy, and invite the solutions of a police state.

A series of factors contributes to the world order difficulties of our age:

1. The emerging *scarcity* of vital resources that have been formerly held *in common,* such as oceans, space, and life-support atmosphere. The basic world-order doctrine of "freedom of the seas" embodies the notion of the commons, making available what is *abundant* to all actors that might want to put the oceans to good use. Only under conditions of scarcity do we confront the prospect of "the tragedy of the commons." When the

quantity and character of use threaten the carrying capacity of the oceans, then adherence to the earlier idea of freedom becomes destructive for the entire community. This destruction can occur because unregulated use grows dangerous, as for instance through oil spillage, and the need for effective regulation becomes evident. It can also occur if the volume of a particular use reduces or even eliminates the prospect for future use, as when over-fishing occurs, thereby creating the need for an agreed-upon and enforced system of sharing *limited* rights of use. The same analysis extends to the limited capacity of air, soil, and water to absorb or recycle wastes at the rates and in the quantities at which they are now being released, or of the oxygen supply to remain sufficient to accommodate world population above certain levels. The point is that the emerging situation of scarcity almost necessitates a shift from "freedom" as the basis of *unlimited* rights to "control" and "regulation" as the basis of *limited* rights in several key dimensions of international life.

2. The increasing degree of *extraterritoriality of critical events.* Traditional world order rests upon "freedom" plus "sovereignty." The idea of sovereignty—property on a national scale vis-a-vis other states—has been evolved to reinforce claims by national governments to exercise exclusive claims over events located within national territory. But, increasingly, events that bear centrally on the fundamental welfare of one's own society take place beyond the reach of territorial sovereignty. Therefore, national societies are shaped by, and in turn, shape foreign societies. There is a loss of the *vital connection* between *the locus of physical occurrence* and *the locus of practical significance.* As such, one of the principal justifications for national sovereignty is undermined. A territorial sovereign, however enlightened its governors, is unable to protect its society against these impacts from abroad. Satellite broadcasting, urban centers of disease

propagation, and national policy toward nuclear energy, mercury fungicides, and hard pesticides are further illustrations of the non-territoriality of "events," and of the tendency of non-territoriality to make the tradition of sovereign prerogative peculiarly ill-adapted to overall planetary needs.

3. The continuously *diminishing margin of decisional error*. The political rudder has become more delicate, the shoals more treacherous, the currents swifter and more uncertain. Steering the collective destinies of mankind has grown more hazardous and demanding. This vulnerability to accident is deepened by the absence of a center of navigational control in human affairs. Separate sovereignties are not coordinated to facilitate joint steerage in response to common menaces. Deployed ABM systems will provide a chief executive no more than fifteen minutes to decide whether to launch nuclear intercept missiles and risk a nuclear explosion. Time has been contracted as a result of the acceleration of mechanical operations by modern technology. Such an effect is manifest with regard to a discrete dramatic decision, such as whether to initiate or respond to a nuclear attack. The consequence of such a decision may not be readily reversible, especially should the nuclear exchange saturate the atmosphere with fallout and destroy the main centers of industry. There is even some question as to whether mankind could ever restore pre-nuclear conditions of existence either because reproductive capability was destroyed by fallout or because there were no easily accessible high-grade mineral resources with which to re-create the high technology needed for access to lower-grade resources.

The other kind of diminishing margin relates to the quiet buildup of cumulative decay, whether in the oceans, air, or land. We do not know how to identify thresholds of irreversibility except in retrospect. We do not know at what point the pollution of the oceans initiates a death process similar to that which has killed so many fresh-water lakes in the world. Those with interests favoring existing use patterns have been effective in confusing the public and its leaders as to the need for decisive protective efforts. There are always ambiguities in the evidence of alleged harm that can be relied upon by vested interests to oppose public regulation of profit-making activities or products. The buffering effect may disguise the deterioration of an ecosystem until the moment of abrupt collapse. When it becomes too late to save an environment, it hardly matters whether further pollution occurs. After the relevant ecosystem has been destroyed there is no longer a persuasive reason to establish strong regulations. At such a point of public outrage there is often a call to lock the barn despite the fact that the horses have all escaped or been stolen. Such a situation of cumulative effect arising out of a myriad of low-visibility acts—the grains of sand that add up to a beach or the snowflakes that together compose a blizzard—applies to all of nature's renewable resources: air, earth, water.

Therefore, the narrowing margin of human decision can imperil human welfare, and even survival, either by the sudden unexpected act or by the disguised cumulative change. In either case the result may be irreversible deterioration for some major life-support system, or even, possibly, the entire planet. It is no longer a single state or even an entire civilization that is in a condition of jeopardy, but the planet as a whole. "I'll tell you what we worry about most," says David Gates, an eminent ecologist who runs the Missouri Botanical Garden. "An irreversible catastrophe. A number of pesticide spills, for example, in those areas of the ocean where large colonies of marine organisms produce much of the world's oxygen. If you plot out the frequency of this kind of event, they're getting closer and closer."

4. The absence of knowledge about the maintenance of

the planetary ecosystem. Gates describes the state of ecological knowledge: "We're still in the Stone Age."

Closely related to the narrowing margin of error is the failure to achieve any effective consensus as to the probable or assured consequences of present trends or patterns of behavior. We are still experiencing a confusion of tongues, a contest of experts; world order as an edifice resembles the Tower of Babel.

In such circumstances of uncertain knowledge there is a tendency to persist in existing folkways, however dangerous. The prospects for world-order change would be greatly improved by the emergence of a clear statement of ecological consensus, even if it were a minimum statement to the effect that we are risking permanent imbalance and consequent destruction of the fragile ecosystems that sustain life on the planet. As matters now stand, the protection of the planetary ecosystem is virtually excluded from the province of world politics, and is given little attention by most governments whether they are acting in isolation or in concert.

5. The persisting tendency of governments to resolve vital disputes among states by violence and the threat of violence. The pervasiveness and persistence of collective violence distract energy and resources from the other dimensions of world order. The character of collective violence is made more dangerous by the access of statesmen to the latest kinds of technological innovation and by the failure to centralize control over the permissible quantity and quality of weapons possessed by national governments. The present system of world order allows national governments to have as many weapons as they want and of any variety. There are a few partial and not very effective limits on use (e.g., poison gas).

Even national governments have failed, by and large, to establish peace systems within their boundaries. When basic policy is in dispute the prospect of collective violence remains as strong in domestic as it does in international affairs.

The problem of world order needs to be linked up with the limitations of political order in general. The issue has planetary status, once again, because of the apparently indissoluble marriage between the violent propensities of men and the dazzling accomplishments of technology.

6. The dynamic character of military technology and the strategic doctrine animating its role in international diplomacy constantly undermine international stability. The competitive quality of international relations leads to a continuous search for exploitable weaknesses of the adversary and a correlative effort to overcome one's own vulnerability. The high velocity of technological change requires action in anticipation, a constant spiraling of actions and reactions that produces frantic interludes whenever one side thinks it is falling behind or is on the verge of a breakthrough to a position of superiority. The arms race and the mentality of secrecy and suspicion which accompanies it generate tensions and place situations of apparent equilibrium in frequent jeopardy. There is no stable balance of terror; at best, it is indeed, to borrow the language of one celebrated war thinker, "a delicate balance of terror." Deterrence offers no genuine prospect of international serenity, much less peace; it provides a makeshift standoff between rivals that is always in jeopardy and is susceptible to collapse as a result of miscalculation, misinformation, adventurism, accident, or anxiety. To rely on mutual deterrence as anything other than a desperate interim security posture, tolerated during a period of transition, is to become reconciled to a catastrophic destiny for mankind.

7. The existence of external constraints upon voluntary patterns of coordination among sovereign states. The competitive basis of world society leads to efforts at ostracism and postures of alienation on the part of national governments. China and South Africa stand out as con-

temporary examples of important governments that are effectively excluded and self-excluded from the voluntary patterns of international cooperation. These exclusions set dramatic limits to the ability of governments to act in concert on behalf of a threatened world interest. Such limits are not necessarily a permanent part of international society, but they have frequently existed whenever a state seeks a new economic or political order in world affairs (China) or whenever a state pursues domestic policies abhorrent to the world community as a whole (South Africa). Other actors, such as Israel and the Arab countries, or India and Pakistan, are so deeply locked in regional conflict as to orient virtually their entire foreign policy around immediate issues of prewar, low-visibility war, or deterrent preparations. As such, these governments are unable to give attention to wider issues of planetary concern. The result is that the process of cooperation among national governments, even if regarded in a positive light, is severely handicapped by the limited participation in world society of a large number of critical states.

Additional constraints on the potential for adaptive change through voluntary action of states arise from the central dynamics of economic and ideological rivalry. Even among states that are closely tied by military alliances, such as the NATO countries, there occurs very sharp competition. Weaker countries may perceive both the loss of economic autonomy and the weakness of their security position arising out of the selfish pursuits of the strongest alliance member. The failure of European economic integration to overcome so far the resistance of France to "supranationalism" is a further illustration of the limits of cooperative effort. The limits imposed by ideological rivalry and hostility are so prominent as to require little illustration. Soviet-American cooperation in relation to strategic arms limitations has been consistently delayed or prevented by the primacy of competi-

tive factors. Even in relation to such clear common interests among the two superpowers as inhibiting nuclear
proliferation, effective cooperation has been badly re-
tarded by their ideological and geopolitical rivalry which
has led the respective governments to take contradictory
positions in relation to nearly every emotionally charged
international conflict.

8. The existence of internal constraints on the pursuit
of international cooperation serving the planetary interest. In all principal societies there are a variety of economic, ideological, and bureaucratic constraints, *internal*
to the state, that make cooperative action and policies
difficult or impossible for a government to pursue. These
constraints arise from vested interests in competitive
patterns held by various parts of the government, such as
the military sector (or even within the military sector as a
result of interservice rivalry) and by various private and
semiprivate groups. Vested interests are also served by
a very strong inertia encountered in almost all large-
scale organizations (especially governments) and by
recruitment of elites based on the *acceptance* of rigid
belief-systems embodying competitive and stereotyped
images of world order. The "military-industrial complex" that grows up within and around national governments enjoys high status, obtains easy access to the
political leaders, possesses well-organized and well-
financed pressure groups, and commands widespread
grass-roots support, especially from patrioteering groups.
Such a value/power base imposes severe constraints
on a government's freedom of action, even if its leaders
are oriented toward more enlightened and cooperative
initiatives in foreign policy. In this sense, it is relevant
to recall our earlier discussion in Chapter IV of "entrapped elites," that is, elites that are effectively constrained from acting beyond the boundaries and expectations set by domestic public opinion and the inward-
oriented power/value structure of national society. Also

special interest groups can be very effective in narrow points of particular concern, such as the distortion of the world economy by protectionist tariffs.

9. The time lag between change and adjustment to institutional outlooks and behavior. Governments do not seem to have the capacity, except in a time of perceived and acknowledged emergency such as war or depression, to adapt to significant changes in the political setting. Samuel Huntington, a political scientist, has described "the lag in the development of political institutions behind social and economic change" as "the primary problem of politics." The learning process of government officials is impaired by the pressure of daily routine, by positive reinforcement on the part of junior bureaucrats, as well as by the general tendency of officialdom to avoid the risks associated with controversial and innovative attitudes and positions. Especially in rigid governmental structures where the elite does not easily circulate, it appears especially difficult to introduce adaptive changes in official policies and practices.

These nine considerations summarize the difficulty of adapting the world-order system to the urgencies of an endangered planet. In the next chapter we shall explore three principal strategies of adaptive change that have been proposed to provide the basis for a more adequate system of world order.

VI. Beyond Deterrence: The Quest for World Peace

At least since the state system was formalized by the Peace of Westphalia in 1648, national security has been based upon threats to defend national territory against any potential attacker in military combat. At various points in international history the balance of threats among the main states produced an equilibrium condition that resulted in a relatively peaceful world. At such time, "the balance of power" was praised by observers as the best basis for peace in a world of sovereign states. The dynamics of military technology and strategy, together with the emergence of expansionist leadership in critical states, have disrupted the central equilibrium of world society over and over again, most recently in 1914 and 1939. The contemporary confidence in "the balance of terror" created by the possession of huge stockpiles of deliverable nuclear weapons by rival governments is only the latest form of this centuries-old reliance on an exchange of threats to diminish or defer the prospect of large-scale warfare.

Such a conception of security establishes "stability" as its proclaimed goal. We have heard much from our leaders over the last decade about upholding the stability of mutual deterrence. Arms policies manifest other, quite inconsistent goals. Governments talk of "negotiating from strength" and of achieving "sufficiency" or even "superiority." New weapons systems are constantly being researched and developed, and not only to protect the deterrent. The dynamics of a competitive world-order system necessarily lead to an ambiguous attitude toward international "stability." A competitive edge—or superiority—is needed for one nation to grow economically and politically at the expense of others, and the pursuit of such superiority combined with the drive to avoid a situation of inferiority leads to continuous pressure on any apparent circumstance of equilibrium in international affairs. This pressure is intensified by the limited character of *space* and *resources* in the world, which means that growth and expansion tend to take place *at the expense* of others. Such a process of international competition for a limited stock of rewards means that no security system can promise much for long in the way of world peace.

In a world system of this sort filled with incredibly destructive nuclear weapons it is only natural that a few voices, soft murmurings usually overheard only in arenas far from the centers of power, call for the establishment of a *genuine peace system* in place of a *spurious security system*. It is difficult to determine, however, developments that would produce a lasting peace in the world of today. In this chapter I propose to discuss some of the main obstacles to the achievement of world peace and some of the principal suggestions on how to overcome these obstacles. I also want to demonstrate that the mounting fourfold threat to human survival alters the conditions of world peace and gives a new urgency to the quest.

We are now living in the dawn of *the ecological age.*
The ecological dimension is rapidly assuming critical
relevance for the future of mankind. It has been fashion-
able, and quite correct, to suggest that a *prenuclear* system
of world order cannot expect to cope with the problems
of *the nuclear age.* One of the distinguishing characteristics of the nuclear age has been the production in
the United States of a generation of cool-blooded civilian war-thinkers the best of whom set forth with clarity, ingenuity, and uncompromising rigor the strategic
implications of nuclear weapons, and proclaim on their
own behalf the virtues of "thinking about the unthinkable." One notices that these nuclear strategists seem to
attack their subject matter with such relish that it seems
fair to suspect the presence of a rather unusually large
appetite for exploring these particular regions of unthink-
ability that fill their more squeamish colleagues with dis-
gust. But no matter; a fact of life in the nuclear age is that
our government takes seriously, pays well, and encourages scholars in our universities and centers of learning to
think coolly about when, how, and whether to use nuclear
weapons, and expects them not to blink at the prospect.
Morton Halperin, who exhibits the characteristic style of
a nuclear strategist but a greater than normal concern
about the dangers, identifies the two main leaps forward
in destructive power in the nuclear age as, first, the so-
called "atomic" or "fission" bomb (of the type used at
Hiroshima and Nagasaki, which achieved 1,000 times
more destructive power per pound than TNT), and
second, the so-called "hydrogen" or "fusion" bomb (the
standard long-range nuclear warhead deployed in the
missile force, which achieved 1,000 times more destructive power per pound than had the atomic bomb). As Halperin indicates, $1,000 \times 1,000$ is $1,000,000$, and it is this
increased capability to destroy that summarizes the transition from prenuclear to nuclear age world politics.
Halperin points out: "Another way to indicate this

changed magnitude is to say that one American bomber now carries more than the total destructive power of all the weapons dropped in all the wars in human history." Given human fallibility, the persistence of hostility, secrecy, and distrust, and the occurrence of tests of will and credibility, there arises an inescapable apocalyptic dimension in the nuclear age. Even experts at the defense-minded RAND Corporation analyze the prospects of a nuclear war in terms that should make sane men tremble: a poll of experts conducted at RAND in 1964 as part of a project to study the future concluded that there was a 10 percent probability of major war within 10 years, a 25 percent probability in 25 years. We have grown resigned to living literally minutes away from the precipice of doom, and it is this proximity to catastrophe that gives the nuclear age its special historical quality. But now, this unassimilated nuclear assault upon the fundamental ideas and structures of our world has been aggravated by the new awareness that the fragile equilibrium of our biosphere is being degraded and threatened by the range and scale of human society. We are confronted by this second more fundamental assault on our ideas and institutions that results from the emerging ecological crisis. To meet the added challenges of the ecological age we need to develop, then, ideas and an entire cosmology that go beyond even the realities of the nuclear age. It is itself unprecedented that the transition from the nuclear age to the ecological age has come about in a matter of decades, in such a short interval of time; the political leadership of the world acts largely in accordance with precepts, values, and institutions that are two major historical transformations behind the actualities of the present situation.

It seems appropriate at this stage to offer some positive recommendations for thought and action. We need an ample vision of the future to guide the steerage mechanisms that operate within the structures of government. Without a credible image of the future there is no solid ground for hope. Without a solid ground for hope, the basis of action itself is drawn into question. Immobility results, and the existing patterns persist with only minor adjustment until some breakdown arrests the process and shatters the mood of complacent despair. Disaster-learning is costly and inefficient, and tends to be superficial and contrived, producing responses to the disaster itself rather than to its root causes. As a consequence, the disaster-producing pattern is kept intact, although the shape of subsequent disaster may vary as a result of a variety of environmental changes. The relentless history of warfare discloses both the persistence of the disaster pattern and its constantly evolving form.

Well-intentioned efforts to eliminate war from the center of international life display the futility of reforms that concentrate upon symptoms rather than causes. First, major efforts at the reorganization of international society have been, invariably, hastily contrived postwar initiatives in reaction to a specific disaster-learning experience. Second, these efforts have proceeded on the basis of modifications of *form*—the formal status of war—without any corresponding shift in war-making capabilities or change in the attitudes surrounding the exercise of discretion by government officials acting on behalf of a sovereign state. Furthermore, the impulse to eliminate war has never been clearly associated with the removal of the social, economic, and political *causes* of war. At a minimum it seems hopelessly naive, and dangerous, to eliminate collective violence without providing procedures to promote and implement *social change*. Only a highly effective totalitarian system could hope to reduce recourse to violence in the event that nonviolent prospects for change were nonexistent. In a world of rapid value change and constantly changing life circumstances it seems self-deceptive and sentimental to seek the elimination of *war* without an accompanying assurance that *real substitutes* for war will be established and maintained. Experience with the United Nations suggests that only modest achievements can result from a world organization that lacks police capabilities, independent financing, an assured right to pass judgment on governmental conduct, and the mandate to implement the will of the community of states with respect to minimum conditions of human dignity and social equity. A world-order system is shaped by structures of power and by supporting belief-systems that determine the occasions, forms, and uses of such power. In this central respect the Westphalia system persists, national governments remain predominant, and rivalry among highly unequal sovereign states continues to be the principal preoccupation of statesmen. Proposals for world-order change, to be at all realistic, must proceed from this kind of acknowledgment.

Background Perspectives. In this section we shall clarify the various considerations of choice that bear on the search for a world-order solution that responds to the crisis of an endangered planet and is concerned about the politics of world-order transformation. We need both a *model* and a *strategy* to provide a *sense of direction.* The *model* is a provisional statement of *goals,* not a *blueprint* of an *ideal* world-order system. Progress toward the model will necessarily and properly be accompanied by its constant *reinterpretation* and *respecification.* Unlike *utopian plans* of the past, the solution I advocate is not a *static image,* but a set of *preliminary signposts* to guide a journey into the future. In the course of the journey there will be many discoveries made that bear on the direction and character of a proper destination. These discoveries will require changes of direction, and a constant willingness to erect new signposts. The process of transformation is itself a *learning experience* that is in constant interaction with itself.

To begin with, it is desirable to .clarify a point of departure in the search for a new world-order system by setting forth some preferences and giving some indication of what I think is possible.

Gradualism and Radicalism in World-Order Postures. Among government officials there exists an almost automatic assumption that the *gradual reform* of the world-order system is the only *practical* and *realistic* way to proceed and that proposals for *abrupt modification,* whether by way of global confederation or drastic disarmament, are *idealistic* and *unrealistic.* This prevailing bias in favor of gradualism seems unfounded *in fact* and dangerous *in theory.* Given the rate of deterioration taking place in the world, there is not enough *time* available to entrust *adaptive change* to the processes of gradual modification. Perhaps more fundamental, the present structure of world order sets rather *rigid boundaries* on the prospects for transformation by gradual change. These boundaries exist in relation to the defining characteristics of the present system of world order, especially with respect to control over the use and development of military capabilities, traditional prerogatives of sovereign states, and the limited authority of international institutions. The groups favoring retention of the old system can be expected to mobilize in defense of these boundaries. As these boundaries are approached, then, either the flow of gradual change is stopped or the advocates of gradualism are led to insist upon abrupt modification.

Another way to describe this issue is to consider the limits of *reform,* given the character of the world-order system. If the *goals* of reform *exceed* these limits, then it is necessary either to abandon the goals or to relinquish a reformist posture and adopt a more revolutionary outlook. The world-order demands and urgencies—that is, the minimum required depth and rate of transformation—of the ecological age lead me to adopt a revolutionary outlook that works for an *abrupt* (although obviously not immediate) and drastic *modification* of the world-order system.

Gradual adjustments may be compatible with the pursuit of radical change, either by setting the stage for abrupt and drastic modification (e.g., certain kinds of education and experiments in regional order) or by providing some safeguards against the breakdown of the present system (e.g., certain types of arms control agreements, such as the Limited Test Ban or the Hot Line). An adequate approach to world-order change calls for a combined strategy—gradualism to keep the present system going and to make it more susceptible to abrupt modification, and radicalism to build up the centers of value, action, and control needed to achieve a coherent process of adaptive change. I am advocating the adoption of a creative combination of gradualist and radical approaches toward world-order change, not one *or* the other.

Persuasive and Coercive Strategies of Transformation. Another central issue confronting advocates of a new system of world order is whether it is necessary to use violence to build the political base needed for change. It has been traditional for authors of peace plans to develop arguments that appeal to reason; these appeals are apolitical and fall upon deaf ears. Those who hold and benefit from power do not give it up except under coercive pressure. Those who benefit from the state system, or suppose that they do, will struggle against serious efforts to transform it.

Efforts at transformation become serious when they begin to make inroads upon the prevailing power/value structure or set up a credible alternative to it. Such inroads depend upon the organization of domestic and transnational movements that advocate drastic changes in world order as part of their platform. The means by which the existing structures are penetrated or toppled varies with the character of the political system. The process will be very different in a highly controlled central state and in a diffuse, liberal democracy. On occasion, political leaders are already beginning to accept the need for abrupt modification of the world-order system, and may be able to make the existing structures more responsive to pressure for change.

Given the spread of nuclear weapons to principal states and given the unlikelihood of devising a reliable defense against their use, it would seem clear that a technological limit has been imposed on the rationality of recourse to *violent means* to promote world-order change. The processes of transformation must rest on an underlying consensus, although it may assume a tacit form, among principal governments. The prevailing international strategy, even supposing a radical posture on world-order change, would have to rest on *persuasive* and largely nonviolent instruments of change. Within national political systems, however, it is far less evident that recourse to illegal, and even violent, means may not be feasible and necessary to achieve a reorientation or replacement of the national governmental elite. The sooner such a reorientation takes place, the more humane and well-conceived the process of transformation is likely to be. It is of critical importance, however, that governing groups in principal societies become aware of the danger to the planet and select *convergent* strategies of response. Clarification of issues, options, and goals may encourage such a convergence and the rise of compatible interpretations of the crisis situation in the different centers of world political activity.

Traumatic and Nontraumatic Transformation. Must the abrupt modification of the world-order system be preceded by the shattering of the old system? Is a catastrophe the necessary stage setting for adaptive change? The experience of international society suggests a high degree of interwar rigidity. The period since the end of

World War II, despite the impending hazard of nuclear war, has proved to be no exception. Only isolated voices of doom have counseled abrupt modification of the world-order system, and these voices have not been influential.

Perhaps, however, the destiny of an endangered planet will intensify pressures toward international cooperation to serve the common good of mankind. Perhaps, also, we can improve the interpretation of preliminary fissures, document disintegrating tendencies, and organize proper responses during the period of crisis and tension that precedes total ecological collapse of life on earth. We do not know. We have unprecedented abilities to transform values and belief-systems through the coordinated use of global media of communication. If an educational movement can be built around a world-order consensus, then perhaps the entire climate of world opinion can be quickly made to support the energies of adaptive change.

To await patiently a world-order catastrophe in the present situation is to invite an end to human history. The potential scale of nuclear warfare and the fragility of the global ecosystem make the danger of *irreversible change* so serious; it may be too late to rectify the situation if the reorientation takes place in the wake of planetary trauma.

The occurrence of war has stimulated proposals for new systems of world order throughout human history. Most of these proposals have been concerned with the creation of a peace system on a global level that corresponded to the ideal of a well-administered domestic society. In all ages and cultures the persistence of warfare has troubled men of conscience. Visions of the future have almost always included the elimination of war from human affairs. As this book makes plain, the present concern with world order has been broadened considerably beyond the search for a war-prevention system; the danger of ecological catastrophe arises, strangely enough, from activities that have been regarded for centuries as beneficial—having children, burning wastes, the disposal of garbage, increasing the productivity of industry and croplands, and bringing greater longevity and higher standards of living to mankind. This *new* vulnerability of life to human progress is not yet well appreciated as a world-order phenomenon. The analysis of world order still proceeds mainly along the well-worn path of war prevention (which more recently has often added concerns about the reduction of world poverty and the creation of legislative alternatives to war), but continues to be almost oblivious to the urgent need to restore harmony between man and nature within the planetary ecosystem. In fact, the liberal vision of a world-order utopia includes programs to bring American-style affluence to the rest of the world, a utopia that would cause a human disaster worse than what it is designed to overcome.

To encourage the development of an ecological view of world order will require us to examine past intellectual efforts to eliminate war from or moderate war in human society. There are three main kinds of analysis, often overlapping to a degree, that have developed in world-order literature:

1. Calls for the modification of international society.
2. Calls for the modification of national society.
3. Calls for the modification of human nature.

Most peace proposals look toward the creation of world government, transferring security functions to a world police agency and disarming national governments. Dante's vision in *De Monarchia* (circa 1315) of an ideal world order in the form of a universal Roman Empire represented an early belief in the capacity of government to maintain domestic peace even on a world scale. More recent proposals, such as are associated with World Federalism, favor a voluntary transfer of military capabilities and police functions to global institutions through a constitution-building process. By and large, these proposals have relied on *argument,* not *politics,* and have had only a slight impact on thought and behavior. The pressures of competitive activity have made it unthinkable that governments, especially dominant and rich ones, would negotiate away their present capabilities and prerogatives. After World War I, and again after World War II, widespread popular dissatisfaction with the failures of governments to keep the peace by balance-of-power diplomacy led to the international creation of new forms (the League of Nations and the United Nations) and new norms (prohibition upon aggression). But the transformation was a matter more of language than of power. No governmental buildup took place on an international level, competitive arms races resumed, ambitious governments mobilized themselves for expansion, and the pervasiveness and persistence of conflict and war returned to the center of international life. The pace of technological innovation as well as the spread of modern weapons on the world arms market has in many ways made the scale, impact, and costs of warfare worse now than ever before.

The incidence of civil strife in the modern world casts deep doubt on a governmental approach to world order. Even in highly developed national societies, recourse to violence is common in the event that deeply felt group grievances are not dealt with in a successful fashion. The outbreak of collective violence during the 1960's in American cities, in Northern Ireland, in Quebec, and in Belgium bear witness to the inability of even the most modern governmental structures to impose a peace system successfully on domestic society. The situation in Latin America, Africa, and Asia is even more dramatic; the presence of government gives no assurance of domestic disarmament or internal peace. Any projected world government would be confronted with a world society containing deep cleavages arising out of grievances and ambitions that could not be contained by police methods. It is not even clear that the disarming of domestic populations reduces the magnitude or duration of violent encounter, although it may under certain circumstances.

Underlying the persistence of war *within* governmental societies are issues of *change* and *justice*. Procedures for change rarely are able to accommodate fundamental justice demands, probably because the apparatus of government is almost always controlled by those groups that benefit from the alleged injustice. Slight concessions may be made by enlightened or frightened governing groups, but such concessions are usually too small and too slow to satisfy aroused and powerful claimants. Impatience and hostility toward government result. Extralegal persuasion ensues, and is often followed by sporadic recourse to violence. The Civil Rights movement has followed this course in the United States. After a certain point is reached, the alternative becomes insurrection for the claimants or oppression by their rulers. Both alternatives rely on violence, and amount to war strategies. The only consistently effective national governments appear to be those that rule small homogeneous societies with no fundamental disputes (Scandinavia) or those that impose, by a huge preponderance of force, totalitarian rule on a subject population (South Africa).

Domestic states were unified in part to establish a unit that could resist external attack. A world state would have no external enemy, and might therefore not be able to maintain its coherence in the face of factionalism.

Several conclusions emerge: (1) war persists even in societies that enjoy effective government; (2) even domestic societies have not been able to provide adequate substitutes for collective violence in the face of fundamental disagreement on vital matters; (3) those who exercise power do not often relinquish it by voluntary means; (4) effective world government in a divided world situation may entail totalitarian methods.

Other advocates of world-order change have emphasized the central importance of the composition and character of the national state. The great German philosopher, Immanuel Kant, for instance, stressed in his essay, *Perpetual Peace,* which was published in 1795, the correlation between liberal democratic national polities and world peace. Kant did not think that it was necessary to build up a world government, but only to vest control in the people over whether wars were waged and revenues used for war-making capabilities. In Kant's view the *exploitative* character of monarchy underlay the war system, since the king could shift the burdens of war to the general population and secure its benefits for the glory of the crown. It now seems evident that Kant underestimated the aggressiveness of the general population, its vulnerability to patriotic arguments and other forms of propaganda, and the ability of vested-interest groups to build up the capabilities and attitudes that assure the central role of war. The modern defense industry is spread across the face of the nation. *U.S. News and World Report,* commenting on the $5.2 billion reduction of the 1970 U.S. military budget, writes that "workers are being laid off by the thousands. Many small cities and towns that depend heavily on military spending are in trouble. . . . Nearly every state will be affected one way or another." Such economic gloom arises despite the fact that U.S. military spending remained in 1970 at $74.4 billion (compared to $79.7 billion for 1969 and compared to a 1970 Soviet military budget of $39.8 billion). The roots of the war system have taken such deep hold on our society that it becomes plausible to regard the whole operation, in the vivid phrase of Richard Barnet, as "the economy of death."

Another variant on the Kantian thesis is the Marxist-Leninist view that the danger of war will disappear with the collapse of the last capitalist society. In this view, the triumph of socialism on a national level is all that is needed by way of world-order reform. The Sino-Soviet boundary disputes, as well as the Soviet interventions to prevent liberalization of East European regimes, indicate that the world success of socialism would offer little hope of the elimination of war. National imperatives of control and domination appear to operate as fully within the socialist group of states as in other sectors of world society.

There are some thinkers who have emphasized that the danger to world peace arises principally from a reliance on coercion by government in domestic society. The solution to the war problem would be to open societies and make sure that domestic government rests on consent rather than terror and intimidation. But again we find little evidence to encourage this view: Haiti, Spain, South Africa are highly coercive toward their own populations and quite peaceful in external relations, whereas the United States is relatively permissive in its domestic sphere and very coercive in foreign affairs.

Radical groups sometimes think new elites or new structures of interests could reorient a government toward peace and cooperation in its world role. Disarmament is itself thought to deemphasize the role of the military and thereby diminish the use of threats and violence in foreign policy planning. But there is little evidence that such approaches are feasible at all unless they are system-wide; otherwise, the wider competitive logic virtually requires all important actors to plan their security in reaction to what their rivals do or might be tempted to do. Among governments we find no significant dropouts from the war system in the world today.

There are several points that emerge: (1) the war system does not seem associated exclusively with any particular political system or cultural tradition; (2) the achievement of ideological, cultural, or ethnic homogeneity among nations does not lead to the elimination of war; (3) the wider logic of international competition makes it almost impossible for any state to opt out of the war system, except possibly to the extent that alliances create a temporary sense of protection or to the extent that the national patrimony is so small as to be untempting to potential aggressors.

The third set of explanations relates the persistence of war and collective violence to the aggressiveness of human nature. Changes in political structure within or between states can have only marginal effects because it is man himself, not his institutions or his social values, who is driven toward aggressive conduct. Besides, these institutions merely reflect human nature, and to propose changing institutions without altering human nature is a hopeless endeavor. Sigmund Freud in *Civilization and Its Discontents* associated the aggressiveness of a community with the repressed sexual energies that arise in the course of civilization. More recent ethological researches by Konrad Lorenz, Desmond Morris, and Robert Ardrey, argue that aggressiveness is characteristic human conduct when man is studied as an animal.

Aggressiveness as a human trait does not seem to be expressed uniformly throughout world society. There are peace communities in which there is little prospect of intergroup violence in the event of a dispute (e.g., the relations between Canada and the United States or among the Scandinavian countries).

There is evidence that human aggressiveness exists and that the predisposition toward violent settlement of basic disputes, especially involving territorial claims, is very characteristic of intergroup conflict. It is, of course, the linkage between *technological progress* and the *persistence of war* that has caused the special modern crisis of world order. If this link could be broken, then many of the dangers associated with modern war would disappear. If war were converted into a symbolic test of resolve or strength, rather than as a mobilization of total destructive capabilities, it would not constitute a threat to human survival, nor even a blight on human existence. Many primates establish hierarchy and dominance within their group by symbolic encounters, grimaces, and belligerent postures in which the weaker contestant normally gives way without violence, bloodshed, or death. Medieval notions of chivalry emphasized reliance on symbolic encounters between knights at ritual tournaments, although there were field battles, too, that caused a number of deaths. War as it has developed in the modern world is an extraordinarily expensive, inefficient, and self-destructive method by which to establish relations of hierarchy and dominance among sovereign states.

In examining these various proposals for the modification of international society, national society (or certain national societies), and human nature, it seems that war is an irrational and unnecessary institution, and therefore reflects *organizational* or *personality* deficiencies. But there are reasons to accord a certain objective status to conflict arising from genuine disagreement about the makeup of a political system, from competition for scarce resources, prestige, and territory, and from contradictory perceptions of facts and rights arising from different outlooks. *Most proposals to organize a peaceful world society have badly underestimated the objective grounds of conflict and the dynamic process of a competitive world order.* These peace plans have relied on the force of reason to induce a change of heart and mind on the part of political leaders and their domestic constituencies. Arguments stressing human solidarity, the desperate need to avoid nuclear holocaust, or the universal benefit of a peaceful world are dismissed by men of power and influence as unrealistic, or worse, as deceptive, or are converted into propaganda slogans used to berate the other side.

Earlier ideas about the reorganization of international society have been devoted almost exclusively to the avoidance of large-scale warfare. This pursuit of peace in the *prenuclear age* was dominated by ethical strivings, the

positive pursuit of a world community and the negative
pursuit of finding less barbarous ways than war to resolve
international disputes. In 1625 Hugo Grotius, the Dutch
jurist who laid the foundation of modern international law,
explained his motivation in an introductory passage to
his great treatise on the law of war: "Throughout the
Christian world I observed a lack of restraint in relation
to war, such as even barbarous races should be ashamed
of; I observed that men rush to arms for slight causes, or
no cause at all, and that when arms have once been taken
up there is no longer any respect for law, divine or
human; it is as if, in accordance with a general decree,
frenzy had openly been let loose for the committing of
all crimes." With the initiation of *the nuclear age* at
Hiroshima, the identification of world order with war
prevention became complete, and the alternatives of the
future seemed to many observers to involve either an
eventual nuclear catastrophe or effective and complete
nuclear disarmament. World-order thought concentrated on the preconditions of effective nuclear disarmament—safeguards, police, incentives for non-nuclear
states. Before examining proposals suitable for *the
ecological age,* we will briefly examine the main approaches to world-order change that have been prompted
by the war-prevention imperative of *the nuclear age.*

THREE APPROACHES TO ADAPTIVE CHANGE.

1. *The Disparity Approach.* Barbara Ward has argued
that disparities in *wealth, ideology, and power* block transition to a more survival-oriented system of world order.
Therefore, the processes of transformation must erode
these disparities so as to make possible the buildup of those
institutions and norms needed to sustain a world community. So long as these disparities exist, *competitive*
pressures and anxieties tend to restrict the potentialities
for international *cooperation;* distinct national governments view one another primarily as rivals pursuing ad-

verse interests rather than as partners pursuing joint and converging interests. (Of course, it is conceivable that if disparities *did not exist* men would wage war to produce them.)

Andrei D. Sakharov's proposals reflect strongly the disparity approach by their dual emphasis upon a fifteen-year tax of 20 percent levied on the GNP of the rich countries for the benefit of the poor countries, and upon an ideological converging of Soviet and American political systems. Another great Soviet physicist, Petr L. Kapitsa, suggested that only by the *convergence* of Western and Communist governments would it be possible to avoid "a fatal clash" between the two superpowers. During this process of convergence Communist states would become more democratic and permissive in political spirit, and Western governments would become gradually more socialist in outlook through the further growth of state planning and welfare economics. As convergence becomes more manifest to leaders in both societies, the hope and expectation is that distrust and conflict will gradually disappear and the potentialities of joining together to serve a common world interest will greatly expand.

Inequalities in power would grow less significant in the event of drastic disarmament, of representative participation by all major societies in the central institutions of planetary administration, and of rapid progress toward minimum economic thresholds for all societies. Within a federal structure, the inequality between California and Rhode Island is not often of great consequence to the satisfaction of human needs, although the inequality in treatment between the state of New York and the city of New York, or between the South as a region and the Northeast as a region, may be an intense political issue. Therefore, the reduction of power disparities needs to be understood in relation to the lessening importance of national and state power as a determinant of international behavior. Subnational movements to achieve internal autonomy for ethnic, tribal, linguistic, or religious minorities tend also to make less significant the aggregated strength of large states and, therefore, disparities among them. The rise of the multinational corporation as an international actor with power and authority of its own also diminishes the zone of effective sovereignty, as do transnational movements of labor, youth, and professional groups and ecumenical movements arising out of religious affiliation. The city as an exploited internal political unit has been identified as creating the basis for a transnational movement. While Lin Piao has called upon the countryside of the world (Africa, Asia, and Latin America) to unite against the cities of the world (Europe and North America), the economist Kenneth Boulding has called upon cities to rise up in arms against their rural oppressors—"cities of the world, unite, you have nothing to lose but your exploitation."

The disparity approach, then, rests on the positive relationship between *adaptive world-order change* and the reduction of principal economic, ideological, and power disparities. Such reductions would seem helpful both in the spirit of *stage-setting* and as a major ingredient of *abrupt modification.*

2. *The Unification Approach.* Almost all proposals for the abrupt modification of world order accept the need for increasing the capabilities and the role of central (or global) institutions. This need for centralization partly expresses the ethical imperative implicit in the ideal of the unity of mankind. This ideal, active in all major traditions of human thought, has persisted despite the prevalence of xenophobia and ethnocentric values and statecraft. A future vision of world order seems to draw inspiration from the expectation of creating a political community that encompasses the entire family of man. Since the end of World War I there has also existed a political imperative relating to the rising costs

of war as a core institution of conflict, change, and security in world affairs. Only through some degree of political unification could a planetary peace system acquire enough credibility to overcome the habits of mind and action that have evolved through centuries of *self-help* in world affairs. Since the atomic explosions of August 1945 there has been a growing expression of concern by governments that warfare is not only costly but tends toward being cataclysmic in its modern aspect. When the Soviet Union tested a hydrogen bomb in 1953 it became apparent that the principal rivals were defenseless against each other's might, that the traditional distinctions between victor and loser or between neutral and belligerent no longer held firm, and that the deterrent system was based upon an exchange of threats to annihilate essentially innocent populations held hostage. Such a security system involving the maintenance of huge defense budgets to assure the continuing vulnerability of the rival society contradicts the ethical imperative to respect human solidarity in a very profound way, creating a very deep split between ethical ideals and political actualities in world affairs. The cost and danger of war and the unreliability of mere promises of renunciation seem, however, to incline any world-order thinker in the direction of an *enforced* and *policed* system of world disarmament.

The political and ethical imperatives have been more recently complemented by the ecological imperative. As I have had frequent occasion to argue throughout this book, the scale and nature of modern industry, life-style, and technology endanger the equilibrium of the planet's biosphere in a variety of ways. To deal with these dangers it is essential to coordinate action in such a way that common standards of behavior are created and maintained, and that no government is free to pursue an ecologically hazardous course just because the locus of its acts are confined to national territory, outer space, or the high seas. Potential global impacts give rise to the need for institutions and authority to protect the safety of the entire world. Fragmentary and partial political systems of control cannot do the job.

These three imperatives join to undergird the argument for centralization in world affairs. Centralization of control does not necessarily or appropriately mean an increase in the bureaucratic control over human affairs. Hopefully, the creation of a planetary world-order system would diminish the extent of governmental interference in the lives of individuals and groups. The partial centralization of authority in Western Europe has lessened to some extent the intrusion of bureaucratic regulation upon travel by individuals within the region. As matters have developed, individuals can travel in Western Europe virtually without reliance on a passport and are rarely encumbered by border-crossing check-points. As the boundaries between sovereign states diminish in critical significance, there is every reason to expect that life could proceed with less, not more, governmental presence. Hence, and quite strikingly, the advocacy of a centralist solution for world-order dangers is quite compatible with a commitment to a more libertarian, even anarchist, political order for domestic society. "Law and order" for world society may be indeed conducive to a less structured and less managed control over less inclusive human communities.

A subordinate argument favors the protection of maximum national diversity, both as an intrinsic value enriching the texture of planetary existence and as some check against the demonic exercise of central power as a means to exploit and enslave the entire human race on behalf of some dominant elite.

3. *The Community Approach.* In this approach to world order the essential element is the creation of *a sense of community* in relation to emerging political forms. Such a sense of community would involve the growth of

loyalty to planetary interests and institutions. Throughout the world there is a widespread sense of individual and collective alienation from the present structures of power and authority that control human lives. A new system of world order would need to increase the participation of all societies in decisions of global importance, and especially in decisions that bear on the particular concerns of a given society. Under present circumstances the governments in Third World countries are not able to play a serious role in shaping the principal policies and practices that determine international economic and military affairs. Alienation results. There is a sense of impotence that induces hostility and indifference.

From our experience within national society it becomes evident that effective participation in decisions bearing on group welfare is necessary if a sense of community is ever to be born. The "black power" argument about community control is a vivid example of the connection between participation by all peoples at all stages. As we have already suggested, the first step in the direction of world-order change is the concern and awareness caused by a comprehension of the endangered-planet argument. Such awareness must itself emerge from *within* the main centers of human activity throughout the world; it cannot be *imposed from without*. A politically potent global consensus on world-order change can result only as the outcome of a collaborative process initiated within each major society and gradually linked together. This process of building will be itself a testing-ground of participation by the various ideological, ethnic, and interest groups active in world affairs.

One educational venture that is building in this direction is the World Order Models Project. This project is organized around groups of scholars operating as independent units of inquiry in the following areas: Europe, Latin America, India, Japan, Africa, the Soviet Union, and North America. The assignment of each group is to work out a conception of a *desirable* and *attainable* model of world order that can be brought into being by the end of the century, the decade 1990. The separate teams are directed by a scholar who serves as a Research Director and works in conjunction with a group of scholars and with an advisory committee of leaders in different phases of the society. The research directors meet together twice a year in different parts of the world to discuss their progress on the project. By the end of 1970 the goal is to have seven distinct models of preferred world-order systems. These models can be expected to exhibit both the diversity of perspective that arises from distinct national, cultural, and ideological predispositions and the convergencies that result from a shared appreciation of the increasing inadequacy and danger embodied in the present system of world order. The World Order Models Project itself illustrates the kind of worldwide collaboration needed—separate, but interlinked, centers of inquiry—to begin building a climate of opinion favorable to adaptive change on a world level.

Ideas of community are closely associated with notions of dignity and equality. To be a participant is to enjoy a sense of status that cannot be acquired by the weaker members of the present *stratified* and *competitive* system of world order. In a stratified system to be weak and poor is to be less significant. Such a pattern is evident in relation to the control of nuclear weapons, the possession of electrical appliances in homes, or the average level of educational attainment. The nuclear powers are eager to prevent the proliferation of nuclear weapons to additional states, but they are not willing to eliminate their own nuclear capabilities or even to reduce their discretion to use nuclear weapons. There is no mutuality of obligation. It is not a world community at all, but a stratified society. The nuclear powers seek to bargain with nonnuclear governments, offering technical assistance to develop peaceful uses of nuclear energy in exchange for their commitment to remain non-nuclear. Bargaining about reciprocal interests goes on in every political community, as interests of actors do differ, but in relation to such a fundamental matter as the retention of the capability and discretion to destroy virtually the entire world there can be no genuine community unless and until *all parts* of the world plan and execute a globally acceptable policy bearing on nuclear weaponry. On a matter of such encompassing concern as the risk of nuclear war, the evidence of history suggests how foolhardy it is to rely upon or tolerate the wisdom and prudence of particular governments. Severe limits are imposed on the potentialities for cooperation in a world that is highly stratified in matters of diplomatic prerogative, military power, technological prowess, and standard of living.

The community approach to world order emphasizes the need for widespread and genuine participation by all those actors who are involved in the work of world-order transformation. The outcome of the community approach is the creation of new bonds of affiliation that embrace man and the natural habitat of the global biosphere; we will need new symbols and myths to proclaim the emergence of vital forms of collective identity. Because of the relevance of ecological considerations the idea of a human community should be expanded to embrace the concerns of both man-in-society and man-in-nature.

———————

The principal efforts to promote a new world-order system appropriate for the nuclear age have embodied values that combine the search for *world peace* with the creation of a welfare program able to overcome *world poverty*. The incorporation of welfare ideals into world-order thought is partly a consequence of the growth of "welfare states" within the advanced countries of both socialist and non-socialist orientation, and as such represents an acceptance of governmental responsibility for the provision of minimum conditions for a tolerable life to everyone. The idea of collective welfare also reflects a certain kind of political expediency as it attempts to defuse potential revolutionary situations which seem present whenever a social group of any size experiences a strong sense of *severe deprivation*. On a global level, then, the introduction of welfare thinking reflects both a growth of ethical concern for the alleviation of human misery and the sophistication of peace thinking about the objective social and economic preconditions of war and peace. A comparison of the League Covenant and the United Nations Charter (and subsidiary instruments) symbolizes the growth of this kind of concern, as does Pope Paul VI's assertion in the encyclical *Popularum Progressis* that "development is the new name for peace."

Also present in this nuclear phase of world-order thinking is the growth of a concern with minimum human dignity, especially in relation to the elimination of racial discrimination. Recalling that slavery was a widespread human institution in almost *all ancient* civilizations, but not always based on race, it is of considerable ethical significance to note this verbal endorsement of the ethics of racial equality. It is equally significant, however, to note the *practice* of discrimination and inequality based on race that persists in all parts of the world. Sentiments of racial equality follow from welfare concepts and involve an appreciation that a minimum basis of human existence involves an affirmation and approximation of racial equality. Again, as with welfare, a gradual and sporadic growth of ethical consciousness is joined to an awareness that the peace cannot long be kept except by terror and oppression in an atmosphere of racial discrimination. Hitler's recourse to genocide, an extraordinary relapse into barbarism, underlies some of the modern insistence that the violation of a domestic population is not one of the rights of a sovereign state. Therefore, a peace system for the nuclear age has come to include *dignity* (or *social justice*) demands, as well as *welfare* (or *economic justice*) demands.

In this sense, contemporary approaches to world-order reform have been enriched by social progress in the leading domestic societies. In fact, the transformation of principal domestic societies may be our most critical world-order learning experience. We may now hope, in similar fashion, that the new rise of public concern with clean air and clean water will lead domestic governments to turn increasingly toward efforts to safeguard their environment against further deterioration. There is reason to expect that domestic environmental concerns, as with economic welfare and racial equality, will be shortly transferred into the mainstream of world-order thinking. Chapter VII attempts to encourage this transfer process by offering a set of recommendations based on world-order needs in the ecological age.

The reorientation of thought is a crucial initial phase of world-order system change, but it is only that. The creation of adequate schemes for world order in the nuclear age did not produce the transformation of attitudes and institutions that were needed to avoid large-scale warfare. The tradition of world-order thought has been detached from the politics of *adaptive change,* from the second phase of mobilization, and from the third phase of transformation. In Chapter VIII we will offer some *guidelines for action* in the ecological age designed to *mobilize* the particular demands that might make *transformation* a political possibility.

World-order thinking in the nuclear age carried forward the basic direction of reform proposed in the pre-nuclear period. A time of transition existed between 1914 and 1945, between the start of World War I and the explosion at Hiroshima, that foreshadowed the inability of traditional ideas of security (national defense, self-help, and alliances) to establish a tolerable arrangement of world society. The nuclear age made this transitional perception more vivid and compelling. The nuclear age has not now ended, but its organizing character is being superseded by the ecological dimension. The *Torrey Canyon* oil spill of 1967 can be understood as the Hiroshima of the ecological age. To describe the present era as the ecological age is only to call attention to a central feature of the situation. There is no pretension that such a label suffices to describe the totality of relevant issues.

VII. Designing a New World-Order System

1. THE IDEA OF DESIGN. René Dubos has made explicit the basic, as yet not generally apprehended, situation of human society: "The ecological constraints on population and technological growth will inevitably lead to social and economic systems different from the ones in which we live today. In order to survive, mankind will have to develop what might be called a steady state. The steady state formula is so different from the philosophy of endless quantitative growth, which has so far governed Western civilization, that it may cause widespread public alarm." There are several key elements in Dubos' position that apply to an analysis of world order: (1) the old system is unable to cope with the mounting pressures on the ecosystem; (2) human survival requires fundamental adjustments, including especially a shift from the infinity-consciousness of a growth mentality to a sense of finiteness associated with operating within a steady state system; and (3) the shift from one system to another, even if it can be brought about by human effort (rather than ecological collapse), is likely to disrupt the beliefs, habits, and expectations of most people to such an extent as to induce anxiety and agitation.

René Dubos believes that "the constraints inherent in the world of the immediate future make ideas concerned with design, rather than accumulation of facts related to growth, the dominant needs in the advancement of science and of technology." Although Dubos is concerned with the adaptation of science and technology to the ecological crisis, his emphasis is precisely appropriate as a guideline for thought about an adequate world-order system. What we need to do is to concentrate upon design, not as a static image of a closed system but as an active process of learning and building; the idea of design includes the process of building over a long period of time, cathedral-building in the sense of sustaining a large vision and embellishing on a basic plan of action as the occasion allows, a continuous process that goes on as long as historical time persists and can hence never be finished.

This approach to the creation of a world-order system seems most responsive to the urgencies of the ecological age. It is an approach at odds with the main traditions of world-order reform, an approach that has not yet even been articulated in a coherent form. As matters now stand, one would search in vain through the world-order literature for a statement comparable to that of René Dubos.

There is a long record of proposals for new systems of world order. Some of these proposals have been put forward by eminent men: Dante, Sir Thomas More, the Duc de Sully, Immanuel Kant, Jean-Jacques Rousseau, and William Penn. Some proposals have been promoted by visionary statesmen: Henry IV of France, Czar Alexander, and Woodrow Wilson. But at no time has there been any very concerted political effort to bring about a drastic change in the structure of international society.

Changes have been made in the rhetoric of world politics. Great conferences of world leaders have been periodically held to the accompaniment of fanfare. New international institutions have emerged with great rapidity over the last fifty years. Nevertheless, the basic attitudes, the controlling patterns of action, and the dominant forms of organization persist. Sovereign states remain the basic organizing unit of international life, and the integrity of world order depends on the fragile and ephemeral stability that reflects the tension between deterrence and rivalry.

Schemes for world-order improvement continue to be put forward by statesmen and scholars. The urgency of world-order imperatives is widely stressed on state occasions, so widely, in fact, as to grow routine and vapid. Even hardened realists have acknowledged the peril of our times, usually by doubting the capacity of the system to prevent a nuclear war. Hans Morgenthau—although once the most influential advocate of a national interest approach to foreign policy—writes that the persistence of state sovereignty "is at odds with the rational requirements of the age. Modern technology has rendered the nation-state obsolete as a principle of political organization." Most recent discussions of world order are post-nuclear and pre-ecological in their emphasis, proposing the creation of a peace system capable of restraining large-scale political violence and generally providing some mechanisms for helping poor countries to develop their economies and some procedures to enable governments to resolve disputes by reliable, fair, and peaceful means.

But new proposals for new systems of world order remain remote from the realm of practical politics. The dynamics and momentum of international rivalry and the immediacy of domestic concerns absorb the energies of ruling groups and cloud over the imagination of the more articulate and informed portions of the citizenry. Proposals for dramatic changes in world-order systems tend to be regarded, to the extent that they are considered at all, as either fanciful exercises in utopia-building or as projections of very unfair structures of power and privilege disguised by very high-minded rhetoric.

It might be chastening to recall that the so-called Great Design of Henry IV (as worked out by the Duc de Sully) created an international police force to keep the peace among the federated states of Europe, but also entrusted the police force with the mission of conquering "such parts of Asia as are most commodiously situated and particularly the whole coast of Africa which is too near to our territories for our complete security." And that Dante proposed, in *De Monarchia,* a world-order system based on world empire and envisioned as essentially an expanded version of the ancient Roman Empire of Augustan times.

The United States proposals after World War II to establish a system of international control over all atomic materials and facilities "meant in effect," as Richard Barnet has analyzed the Baruch plan, "capitalist control of the Soviet atomic industry since the international agency would be made up primarily of non-Communist powers with economic and military ties to the United States for as long as even the most optimistic Marxist could see." In addition, the United States in its proposals retained the authority to decide when to transfer control over its stockpile of atomic weapons to the international agency over which it would exert continuing political control. Finally, by virtue of its leadership in atomic developments, the United States would have a commanding lead in a rearmament race in the event of a breakdown of the arrangements for international control. It is not surprising that the Soviet Union would not entrust its welfare for the indefinite future to the unilateral goodwill and wisdom of its arch-rival in the struggle for preeminence in world affairs.

We find, then, an understandable fatigue greeting grandiose proposals for new systems of world order. The assumption has been that such proposals are either pipe dreams or covert schemes to advance special interests or both. In either form, these proposals are not likely to be taken seriously. It is often supposed that basic changes in world order will become realistic only after the end of World War III. Until there has been a large painful breakdown of global proportions, many hard-headed experts contend, it seems pointlessly idealistic to devote any energy at all to the design of world-order systems. This skeptical attitude remains prevalent among sophisticated observers of the international scene. It produces a sense of helplessness, especially if combined with an awareness of the dangers and inefficiencies of the existing system of world order. There are two fundamental elements that together shape this position of despair: (1) the present system of world order is perilous in the extreme and manifestly unjust; (2) no alternative system will be seriously considered by the political leadership situated in any principal society of the world, nor could it be implemented even if miraculously agreed upon at leadership levels.

In such an intellectual setting a withdrawal of concern often results regardless of a situation of heightened danger. People tend their own gardens, devote their energies to private and immediate pursuits, and limit their image of "the possible" to modest and marginal adjustments. The transnational movement unifying the struggles of the poor peoples represents the most vital cause in contemporary extranational political identifications. People are dying and enduring sacrifice and risk for the liberation cause, its goals make worldwide sense, and its projects have resulted, although not without damaging reversals, in a series of revolutionary successes on the national level. This radical movement is not at all animated by world-order concerns. By comparison with the liberation movement, grand visions of a new world order appear sterile and hypocritical, the word-weavings of affluent and uninvolved patrons of the moralizing professions, who talk of good as a pretext for not doing anything about evil.

This aversion to the grand solution is particularly strong in the Anglo-American setting, where prevailing traditions of action are associated with problem-solving, practicality, and gradualism; more general approaches are discarded as abstract and dogmatic, irrelevant to the real affairs of the real world. There is in this tradition, despite the intrusion of pieties and ideologies (e.g., John Foster Dulles, Dean Rusk) onto the political scene, a step-by-step effort to identify and promote common interests of the big governments as the best and only practical way to improve the quality of world order. Our common-law tradition, a penchant for philosophical pragmatism and operationalism, and the dominance of Protestant forms of worship all reinforce these tendencies to regard with suspicious disdain advocates of a new system of world order that can be brought into being by the voluntary activity of reasonable men. On the other side, Soviet analysis is based on a sense of the contradictions between classes in non-socialist societies and by the international contradictions between socialist and non-socialist societies. Such an analysis tends to regard all apparent accommodations on a world level as unreliable and expedient, susceptible to collapse and repudiation as soon as the balance of political forces shifts.

Since World War II there has been some tendency of high statesmen to affirm the need to bring "the rule of law" as assuredly to international affairs as it has been brought to domestic society. President Eisenhower was fond of saying that "law" was the sole alternative to "force" in world affairs. The Clark-Sohn plan was promoted under the label "world peace through world law." Such pleas for improving the prospects for world peace, building on the imagery of "law and order," were essentially exercises in celebratory rhetoric. On the Soviet side, calls for instant and total disarmament, without offering any serious plan to implement the calls, performed a similar role of appeasing peace sentiments without threatening to dismantle the war system. Both superpowers have been firmly committed to the self-help system and these commitments have never been subjected, so far as we know, to any serious internal scrutiny in either the Kremlin or the White House.

The incantations of leaders on ritual occasions were able to cloud over the subject matter of world order with images associated with "law and order" in domestic society. As these images of law and order have continued to diverge from ideas of justice and dignity, an increasing number of people, especially among the young, have identified law and order with anti-riot squads, police clubs, and the unprincipled behavior of those who benefit from the status quo. To call for a new world order strikes many of the young as a naive, and possibly sentimental, desire to repeat on a world scale all that is most wrong on a national scale. To build up police, government, and central authority invites, in such a view, an even more pernicious concentration of power than now exists, power that would be exercised for the benefit of the powerful, justified by abstract principles, and used to the detriment of the weak and abused. Such radicalism leads toward a kind of liberation-style Leninism, namely a commitment to liberation as a personal ethos and as a national and transnational cause, a commitment that often produces a position of radical opposition to all "establishments," including especially the Soviet form of mind-crushing bureaucratic socialism. Again, these attitudes lead to a rejection of the world-order literature which has indeed been built around the introduction of ideas of national law and order onto the global scene.

These, then, are some of the shadows that fall across

the path of a world-order activist. If, however, we are to have any prospect of initiating a new politics for an endangered planet, then it is essential to proceed with the design of new world-order systems, not as blueprints of the future, but as signposts that give directions and heighten political awareness. In this chapter, we explore some lines of positive development, emphasizing that the future will be a rapidly changing milieu in which goals as well as means will need to be constantly reinterpreted in light of evolving circumstances and consciousness. What follows, then, is an outline of a world-order design for an endangered planet.

In working out this design I will take it for granted that the transition from here to there is achieved by political means. I assume that no apocalyptic bridge to the future will or can be built in the decades ahead. That is, my reasoning supposes that there will be no nuclear war or equivalent ecological catastrophe (e.g., the death of the oceans) and no sudden burst of enlightened cosmopolitanism such as would enable a world constitutional convention to produce a new world-order system agreeable to leaders of all principal societies. Existing risks, structures, and attitudes will persist, and the situation will intensify as the pressures on the planetary life-support systems continue to mount. Positive and negative potentialities will be present in this expected future setting, the destiny of mankind depending on which set of tendencies gains ascendancy. A book of this sort seeks to strengthen the positive potentialities by interpreting storm warnings, building signposts, and sketching some outlines of what the future could become.

In setting forth a world-order design, the ancient Greek idea of *kairos* is taken seriously—that is, the possibility of a historical situation's arising in which an old order is shattered and rapid, dramatic reconstruction can occur. Already we have referred to a wartime sense of emergency as enabling new initiatives not otherwise possible.

By *kairos* we generalize this possibility, and relate it to human consciousness as well as to events and action. It is important to be *prepared* for the *kairos,* ready with a response. To be ready involves a *true orientation* of thought and feeling, strategic points of access, and the capacity for decision. In a sense, to be ready means no more and no less than having an accurate appreciation of relevant reality, unflinching in its integrity and unyielding in its will to promote the survival of life on earth. World-order designs are part of the content of a reality-based hope directed toward past, present, and future that will enable a creative response to a situation of *kairos.* More modestly, sustaining hope induces action and participation, slows down the process of decay, and gains time for the forces of transformation to build up their strength. A positive design that does not overpromise allows action to be grounded on a solid foundation of *hope,* and thereby can act as a breakwater against *despair* and *immobility.* These various counterweights of thought, feeling, and action are together part of the struggle to combat *alienation* in a period of rapid and profound change, change so profound that the status of man himself is drawn into question and the boundary between man and machine grows indistinct. It is truly a time when the need for *refocusing* is intense. In this sense world-order designs can be likened to lenses especially ground to enable perception of the specific terrain of the endangered planet.

2. INFORMING VALUES AND GOALS.   There are several elements of consciousness that help shape the structural (or visible) design of a new world-order system. There is first of all the sense of urgency and concern that arises from the evidence that the planet is endangered in certain specific respects and that existing patterns of political organization and behavior offer no realistic prospect of removing or lessening the danger. The diagnosis and prognosis of the endangered planet suggest certain directions of response. If unregulated violence tends toward ever-widening circles of destruction, then the development of methods to confine and mitigate violence becomes an obnoxious and necessary antidote. If continuous exploitation of nature and its resources pushes human existence up against the carrying capacity of the earth, then an economy based on a steady state rather than growth commends itself. Since population growth leads to crowding and to a variety of pollutions that are likely to become more serious with the passage of time, then the identification of optimum population levels and the search for ways to approximate such an optimum are called for. If vast amounts of resources are spent to construct and replace weapons systems, involving threats of mutual destruction, then the idea of reliable disarmament possesses great appeal. If dignity is besmirched by racial oppression and if oppressors and oppressed are bound in a relationship of violence, then the cause of racial equality and liberation becomes central to the creation of a decent and durable system of world order.

Such directions of aspiration make explicit the values that inform the world-order design. Without these values (reinforced by certain social and political suppositions about cause and effect) it would be tempting to argue in favor of a new system of world order based on universal conquest or empire in which unity of control could be the sole demand of the new system and compliance could be obtained by reliance upon the most modern advances in propaganda, surveillance, and punishment. Such centralization of authority could impose rigid constraints on reproductive, distributive, and consumptive standards, perhaps enforcing childbearing quotas, allocating goods on the basis of need, and stopping the production of all wasteful or dangerous items. A certain macabre realism may make some advocate an Orwellian unification of the world as the only way to avoid an ecological collapse. Such anti-humanistic realism is not a world-order design but a consequence of its failure. Unless we can mobilize a world-order movement based on an ethical vision, then it is likely that we will respond with brutality and desperation to the harsh challenges of ecological pressures or to the devastated conditions that may exist in the aftermath of nuclear war.

The idea of a world-order design is conservative, that is, it aspires to maintain the values of individual and collective human dignity, including the basic demand for a political order that enjoys the support of most of the population most of the time. To put such a design into operation depends on inducing changes in attitudes, behavior, and institutions to assure that *coordination of effort* and *equity of benefit and burdens* can be achieved in the course of adjusting to a condition of planetary peril. If "adjustment" turns out to mean the strong over the weak, the rich over the poor, the white over the black, brown, red, and yellow, or even the socialists over the others, then pursuit of stability and harmony between man and nature also entails acquiescence in exploitation and oppression. The thirst for justice has been spread around the political deserts of the world: it must be either quenched or stifled; hence, we must either make a commitment to social and political justice in the world or be prepared to participate in its demonic denial. The latter course depends on applying ever-larger doses of terror and violence as oppressed peoples everywhere are being more and more awakened to the actualities of injustice. In such a world of aroused awareness there is no middle path. The issues are clear and we must be prepared to choose between oppressor and oppressed. We side with the oppressor whenever we tolerate or cooperate with a political order based on injustice. The relations of the liberal democratic West with South Africa represent a symbolic test of world political conscience. This test is failed whenever the issue is resolved by considerations of profit or convenience.

The design of future world-order systems needs to take account of two principal criteria of appraisal:

—desirability (is it worth attaining?)
—feasibility (can it be attained?)

The design needs to inspire hope and to orient action. What is feasible may shift rapidly with minor shifts in the climate of opinion. In this regard the wild upsurge of American concern at all levels—from Students for a Democratic Society to the Nixon Cabinet—on environmental issues within the last year is indicative of what it is and is not possible to do once a certain threshold of awareness is crossed. Also, efforts to condition the climate of opinion, as by warning about ecological hazards, may be suddenly very influential to the extent that confirming evidence becomes available. Nevertheless, the prospects for adaptive change depend on transforming existing values, attitudes, and institutions by bending them to the new realities rather than starting fresh with the proposal of an ideal arrangement of world affairs.

I support certain world-order goals that give shape and direction to the design of a new system:

1. *Unity of Mankind and Unity of Life on Earth.* The creation of a new system of world order must draw its animating vision from the long and widespread affirmation that all men are part of a single human family, that a oneness lies buried beneath the manifold diversities and dissensions of the present fractionated world, and that this latent oneness alone can give life and fire to a new political program of transformation. W. Warren Wagar, a historian who has studied the history of world-order thought in an excellent book, *The City of Man,* has written of the importance of this vision: "We are in desperate need of a new generation of authentic cosmopolitan voices who can bear witness to the unity of man. The

notion of world order must lose its vagueness and abstractness. We must learn how to move men to make a paramount commitment to mankind: we need an ideology of world integration." Wagar dismisses the principal organized efforts of recent years to promote world peace as feeble, half-hearted, and futile indulgences in naive sentimentalism. Peace groups have not appreciated the radical challenge implicit in their own demands for a new order of things and have not faced up to the opposition that will emerge as soon as world-order reforms are taken seriously in a politically relevant way. Wagar argues that the unity of mankind is a goal worth fighting and dying for, and advances a revolutionary strategy in support of it; anything less than this, he argues, is a waste of energy, not serious, perhaps even harmful as it maintains a weak illusion that wishful thinking and pious resolutions can change the world in fundamental respects. Once more to quote Wagar ". . . the only movement with the moral dynamic to attract great numbers of men of intellect throughout the world will be a movement that seeks to create more than peace (i.e., the absence of war), and more than world federation (i.e., the law of power to prevent war). It will be a movement for the unification of mankind in a free and organic world civilization. It will be a movement that sees in the coming of Cosmopolis an end to the cyclical pattern of world history and the synthesis of a panethnic culture quite different from any present national culture, at once more complex and more inspiring, more highly integrated and more free." Some such ennobling vision needs to underlie and guide world-order thought and action, including the mobilization of a desperate effort to oppose the entrenched powers of dissension and oppression that now control the political life of most people. This vision, or variants of it, needs to be discovered or rediscovered by all principal cultures and ideologies active in the world today. For centuries there have been prophetic voices in every major world tradition urging upon man and society the relationship between the unity of mankind and secular salvation. These voices must be raised anew in the struggle that lies ahead.

The failure to approximate more closely in reality the ideals of human unity that have held sway over the moral imagination of man is one of history's most baffling and profound motifs. I find Manfred Halpern's analysis of this issue suggestive: "The fatal flaw in man's past devotion to holiness has been his inability to sustain the illusion or the reality of sacred connection with strangers. Yet most of us are strangers to each other. As a result, we become mere objects or mysteriously evil or wholly invisible to each other." Halpern, a political scientist who is developing an elaborate theory of social, political, and personal transformation, is referring in the passage I have quoted to the malign dynamics of the Arab-Israeli conflict, but his observations suggest a focus for concern and creativity with respect to the attainment of global unity. We need to study this frightening process of estrangement from what is humanly different and understand that the strangeness of other human beings has been one of the strongest obstacles through the ages in the pursuit of a decent world order, and that now this tendency to perceive strangers as enemies imperils the very ground of human survival.

But to affirm and even to realize the unity of mankind is no longer a sufficiently embracing goal. The entire force of the ecological imperative is to expand prior horizons, especially in Western thought, so as to enable and necessitate a vision of the unity of man and nature. We need to go back to the great traditions of Asia, Africa, and the red people of "the new world" to recover a sense of this wider unity of all life. And these earlier traditions need to be reinterpreted and revivified in light of the specific urgencies of the ecological age. The goal of goals in any adequate world-order design needs to arouse, play upon, and activate both sentiments of solidarity among men and bonds of affiliation between man and nature. Without such a central vision the entire effort to evolve a new system of world order seems incoherent and only abstractly related to the provoking necessities and the positive potentialities of a new political basis for the organization of life on earth.

2. *The Minimization of Violence.* We seek a design of world order that reduces to a minimum the role of violence in human affairs. And by violence we mean the reliance on force and terror by rulers and police, as well as the reliance by governments on military power in their external relations with one another. The design of world order here affirmed does not anticipate the disappearance of conflict from the planet—in fact, biological survival and balance at all levels of life depend on conflict and power—but only the reliance on violence by social, political, and ethnic groups in their struggle for scarce resources, space, influence, or prestige.

On a domestic level it is essential to work toward the elimination of all forms of social, economic, and political oppression that lead governments to require terror and violence as the means to maintain and control their own populations. We are also concerned about the development of effective procedures of social and political change so that opposition and discontented groups have nonviolent options at their disposal.

On an international level, disarmament is an essential aspect of minimizing violence, to reduce the costs of preparing for war, to cut down the risks of war's breaking out, and to moderate the effects of war should it occur. It would be essential to eliminate weapons of mass destruction from the war system and to build up a reasonably effective international inspectorate and constabulary to verify compliance with disarmament obligations, and apprehend and rectify violations. The *size* and *mission* of such an international force would depend on the extent to which a political consensus could be mobilized, on confidence in its fairness and effectiveness, and on the stakes of international conflict's being cut down to a point where issues in dispute were susceptible to settlement by nonviolent means. It is hard to suppose that any conflict of interest between New York and New Jersey, despite their inequality in size, would generate threats or recourse to violence.

3. *The Maintenance of Environmental Quality.* A design for world order can no longer devote its concern only to war prevention and the alleviation of human suffering. The political agenda of the ecological age requires monitoring of the environment for a variety of hazards, early identification of causes, and effective capacities to enjoin or prescribe appropriate action. Equitable considerations would have to be taken into account so that the burdens of maintaining or rehabilitating the environment were borne by *all* relevant actors or shared with the community as a whole. There is thus a need for planetary fact-finding, monitoring, and enforcement capabilities, and for an accounting procedure that reports upon injustices and on the failure of efforts to raise standards of living. We need to assess costs in relation to both the cause of harm and the ability to pay.

The more complicated tasks involve fixing ceilings upon population increase, resource use, and waste disposal. These tasks will have to be performed to safeguard the biosphere against the threats mounting against it.

The exact standards of environmental quality are difficult to express in precise terms. The minimum expectation would involve taking steps to prevent the destruction of the life-support system. More ambitious undertakings would involve protecting the health and longevity of man and other species of animal, as well as taking steps to safeguard natural beauty and wilderness regions. Still more ambitious assignments would involve efforts to improve the quality of life by reducing the contamination of the environment to the extent possible. Of course, almost all human activity is carried on at some cost to the environment, and it would be both unrealistic and unwise to posit some absolute norm of ecological nonintervention. It is appropriate to place the *burden of justification* upon those who would act in a manner that deteriorates the environment, to ensure that costs of environment maintenance should not be passed on to the totality of society but treated as operating expenses of business and government, and that the appropriation of natural resources should be carried on in such a way as to support an indefinite continuation of life on earth.

Given the pressures of poverty in many parts of the world it will be surely necessary to couple environmental maintenance with large-scale economic assistance. It may also be necessary to reshape the economies of advanced countries to assure a life of dignity and plenty for the entire population, but not necessarily of material abundance; such reshaping might have to penalize or prohibit certain types of economic growth and reward producers able to stabilize production and profits and satisfy the society's demands for goods that fulfill basic human needs.

The principal point is that the defense and restoration of environmental quality should be conceived of as an active political task that must be understood as a basic element of world-order design. Such an inclusion is, of course, a break from the tradition of world-order thinking, which has concentrated on war prevention and has expanded in recent years to take into account issues of economic and social welfare and, more recently, human rights. Ecological consciousness must inform every aspect of world-order design.

At present, governments conceive of environmental issues as largely national in their scope. Such a conception is particularly prevalent in the United States, encouraged by the continental size of the country and its "can-do" outlook on problems that impinge on national well-being. In Europe the proximity of countries has given a more regional definition to environmental policy questions. Throughout the Third World there has been an understandable tendency to associate environmental problems exclusively with the side effects of advanced industrialization. As a consequence, environmental issues have, as yet, no serious global constituency. The 1972 UN Conference on the Environment may be expected to stimulate some wider appreciation that assaults on the environment have a potential global impact (and, therefore, no government can afford to be indifferent to environmental abuse by others, even if it occurs at a distance) and that the protection of the environment cannot be undertaken by governments acting singly, but requires global regimes that are in a position to take account of the variety of interests at stake. (Therefore, no government or group of governments can hope to solve the environmental quality problem by its own autonomous plan of action.) These points are difficult to demonstrate by available evidence at the present time, but trends in oceanic and atmospheric pollution assure that the near future will occasion the rapid globalization of environmental concern.

4. *The Satisfaction of Minimum World Welfare Standards.* Implicit in the moderation of collective violence is the achievement of steady progress toward the elimination of mass misery. Any adequate world-order design needs to include as a central goal the provision of sufficient food, housing, health, and education to allow all people the conditions for a decent life. We need specific target figures and dates, as well as a sense of which tactics are effective and just. Part of the task is to give concreteness to the call for minimum standards by outlining their contents. Certainly there exists a need to provide enough food to avoid starvation and malnutrition, enough health facilities to assure human beings

medical help and advice, enough housing to provide tolerable shelter and comfort, and enough education to help an individual use his capabilities. Such a welfare program is closely linked to population control and, eventually, to population reduction. In fact, the requirements of a program considering limitations on the resources available make it essential to establish a ceiling on the maximum world population. Otherwise the dictates of social and economic justice work against an endorsement of the ecological imperative. To uphold life on earth in circumstances of dignity requires us to confront at once all four threats to planetary welfare and survival.

The idea of minimum provision of the necessities of human life should be viewed as a *right* of individuals and societies. The assistance needed by the poor should be accomplished by a taxing scheme that is based upon a redistribution of income and wealth in accord with an ability-to-pay graduated scale. Obviously this kind of redistribution could not be achieved all at once, but serious five-year welfare master plans on a global scale should be carefully worked out in conjunction with population control plans. Obviously the population level and growth rate of richer societies would also probably be subject to control for reasons of environmental defense, perhaps to more rigorous control. What kind of control would depend on the assessment of urgency and the character of the political order, ranging from gentle persuasion to criminal prohibition and compulsory abortion or even forced sterilization. These more extreme responses to overpopulation are not as remote from our world as many people seem to think. Already, for instance, we have reports from certain crowded sections of India that "untouchables" were being pursued by villagers who were determined to have them sterilized. Such happenings, admittedly isolated incidents in the world of today, do give us a foretaste of the ugly politics of response to the problems of ecological pressure. The weak are made to bear the early burdens of response by application of the force available to the state. It is not even alarmist to suppose that subordinate population groups might be "eliminated" to serve "the common good," or that the drive to restrict the rights of minorities may intensify as the dominant group feels the squeeze from population pressure.

The fixing of levels, the allocation of burdens, and the determination of realistic time intervals would be among the most difficult tasks of the welfare sector of a new world-order system. Questions of necessity, of judgment, and of preference are intermingled in the identification of what is a "minimum" standard and in the decision as to what is a "satisfactory" rate of progress. We would suggest a ten-year plan for the elimination of world poverty with a zero growth rate for world population and for country-by-country population. Adjustments on either side could be bargained for by an appeal to special circumstances. In a second ten-year period a search for man/milieu harmony should adopt as a goal a negative birthrate for the world to be maintained until an agreed optimum is reached, probably somewhere between 500,000,000 and 1,000,000,000. It would be important to equalize rates of decrease on the basis of national, ethnic, cultural, regional, and religious considerations.

Some idea of the magnitude of such a welfare project can be obtained from a plan by the Food and Agriculture Organization, a specialized agency of the United Nations, which estimates that $86 billion is needed over the next fifteen years to provide enough food for expected population increases. The FAO analysis, called the World Indicative Plan, carefully examines the food needs and agricultural prospects throughout the world. It notes that demand for food products in the 64 developing countries (65 percent of the total) is increasing at 3.9 percent per year, whereas agricultural output is increasing at only 2.7 percent per year, which means that in the period from 1965 to 1983 it would require $26 billion in food imports to close the gap. Such an amount far exceeds the capabilities of developing countries to supplement their domestic productive capacities through the purchase of imports. High-yield grains and cereals may alleviate some of the pressure on food imports in the period ahead, but will not overcome the shortage of protein in many societies. The relevant point at issue is that an adequate world welfare system depends on a massive reallocation of effort and resources throughout world society, an effort that seems well beyond the capacities of the sovereignty system.

The reduction of material inequality is another aspect of a world-order design that is closely related to minimum welfare standards. To assist the poor it is necessary to take from the rich, especially when further increases in the world GNP are causing the biosphere to deteriorate and are depleting the resource base available for the future. Whether the equalization of living standards is possible, or even desirable, is a question that can be deferred to the future. Some equalization is necessarily present in the duty of the rich to alleviate the misery of the poor, but whether this duty should be understood as also involving the elimination of material disparities is a quite separate concern. Suffice it to say, the minimization of group violence seems to require the removal of both incentives to revolt against the prevailing order and counter-incentives to discourage or defeat revolt by enforcing through official violence the basic norms of that system of order.

5. *The Primacy of Human Dignity.* A world-order design needs also to be built around the values of human dignity. Some of these values are already to some degree contained in the preceding discussion, especially protection against violence, conservation and rehabilitation of the environment, and provision of minimum welfare

for everyone. In addition, however, there exists a painfully evident need to design a world-order system willing and able to uphold personal rights of conscience and autonomy and group rights of assembly and cultural assertion. We live now in a world where brutal abuses of individuals because of their beliefs or their views, and of groups for their ethnic, national, political, or religious identity, are commonplace. Codes of personal and collective freedom need to be embodied in the world-order design, and procedures established to interpret and implement the rules of conduct agreed upon.

But should the system practice "pure" tolerance, even in relation to those members who might advocate and agitate for regressive policies based on warfare, exploitation, and repression? Can the fabric of the new system be woven strongly enough to absorb such attitudes and acts? Or is it probable that the design can never come into being unless a prior transformation of human consciousness occurs such as to make anti-ecological thinking a harmless aberration, the work of soapbox fanatics, somewhat comparable to the professions of faith made by the most fundamentalist of religious sects?

Obviously, the design needs to make provision for resistance against those who would destroy it. Any kind of enforced disarmament process has to have some methods for dealing with violators, most probably a police and constabulary capability. In addition, as in countries valuing political liberty, a large risk can be incurred with respect to speech and advocacy. In a period of stress it seems likely that some abridgment of this liberty would occur, resulting in the impairment of the quality of life. The central point is that a balance needs to be struck between the affirmation of dignity and the security of the world-order system against its potential enemies, and that those who are entrusted with balancing functions— perhaps the judges in regional and world tribunals—need to be provided with constitutional traditions that incorporate the principal, and occasionally conflicting, goals of the world community.

The regulation of technological innovation and scientific progress will have to be undertaken in relation to the values of human dignity. Already we are aware that the scientists who harnessed the energy of the atom and made possible its military use at Hiroshima and Nagasaki are not necessarily benefactors of mankind. Now we seem poised on the verge of discoveries and developments that will enable the construction of machines capable of doing any task, however creative, that man now does. Genetic engineering will soon permit the creation and modeling of men to take place in scientific laboratories. Such breakthroughs have a fundamental bearing on the place of man in the world and should be evaluated by men as beneficial or harmful. The tradition of scientific freedom needs to be reconsidered from the viewpoint of human capacity to put discoveries about the forces of nature to constructive social and ecological use. We may be finding out to our surprise that the dialogue of the seventeenth century between Galileo and the Catholic Church needs to be reopened to confront the urgencies of the Nuclear Age and of the Ecological Age. Putting the issue differently, we may discover that Olympian gods exhibited a reluctant wisdom by chaining Prometheus—the bearer of progress and technology—to a rock so that he might suffer under public scrutiny.

6. *The Retention of Diversity and Pluralism.* There is a tendency toward unification in world-order designs, especially designs shaped by the agenda of the endangered planet. The ecosystem can be sustained only by some method of central guidance. The war system, international stratification, and the competitive ethos all rest upon the divisiveness of political fragmentation. In this sense there is a virtual consensus that nationalism and the associated traditions of state sovereignty remain decisive obstacles to the development of a rational (survival-prone)

system of world order, obstacles that entail apocalyptic consequences in a world of nuclear technology and ecological hazard.

In addition, many of the most persuasive interpretations of international conflict emphasize the existence of incompatible political systems on a national level. If all sovereign states adhered to a common system of government, it would help shape effective responses to shared hazards. In the event that all world society were to become Communist or "Liberal Democratic," then, so this reasoning proceeds, the functional imperatives of survival and welfare could be implemented much more readily. The splits within the main ideological groupings of the present period (e.g., the Sino-Soviet dispute) cast some doubt upon the significance of homogeneous political allegiances, just as earlier periods of international history revealed that more than common religion or civilization was needed to sustain political harmony. World Wars I and II can both be correctly understood as ideological civil wars fought primarily among Western contenders for world preeminence.

Toleration and moderation in political outlook rather than homogeneity of political systems seem to underlie the prospects for a cooperative relationship. Mutual toleration mutually perceived and heightened by common dangers is the most reliable basis for alliances among sovereign states. Such a basis of converging interests would also serve the designing requirements of world order.

Beyond this practical point, diversity and pluralism are life-enriching. Variety is itself part of the splendor of life, and underlies the celebration of the rich variations of flora and fauna that have arisen in the course of natural evolution. The strong movements to conserve endangered animal species and threatened wilderness areas, and to preserve the beauties of nature, is partly an expression of the intrinsic value that many people attach to diversity and pluralism. The positive effects of diversity seem to arise in all realms of human affairs, including politics, culture, and religion. There is a variety of appropriate forms of political arrangements, given only a common framework of affirmation, that is essential for the persistence of individual variation. An ideal design of world order would certainly not call for the elimination of separate languages, myths, or political ideologies.

It will be necessary to reconcile coordination with diversity; not diversity of any kind, but the widest possible diversity consistent with the mandate of the ecological imperative.

7. *The Need for Universal Participation.* The designer of a world-order system must be sensitive to all parts of the world. No exclusions or withdrawals from the system can be tolerated if the fundamental mission of the organizing center of world society is to safeguard the planet as a whole. The exclusion of China from participation in the United Nations since 1949 would be unthinkable in a world-order system designed for the ecological age. Membership and participation need to be given to all principal actors so as, among other reasons, to vindicate and legitimize claims by the central institutions to assert authority and control. Those who do not participate in institutions or in the process of decision tend to feel little or no obligation to give their respect to adverse policies and decisions. It is important to give all actors who have international significance a serious role in the system. Actors other than national governments will have to be included, perhaps on the basis of interest and ethnic groups, perhaps on the basis of transnational actors and movements, perhaps on the basis of regional organizations and ecumenical political and religious formations. Perhaps all these *kinds* of actors need to be represented in a formal and effective manner.

To be excluded from the process of decision relevant to one's own affairs is to be denied any relationship to destiny. Hence, nonparticipation and oppression go together even if "the oppressor" adheres to a benign creed. The Black Power movement in the United States, various drives for subnational autonomy or national liberation, and the hostile attitude of most non-nuclear states toward the retention of discretion by the nuclear powers to use nuclear weapons are some of the various indications that nonparticipation breeds resentment and opposition.

But suppose a government that is rich or repressive or alienated voluntarily adopts a policy of nonparticipation and seeks to withdraw from the system? Suppose that the choice for a portion of the world is understood to be either withdrawal or acquiescence in a position of inferior status? It is difficult to appraise such questions. Their relevance depends on the climate of specific opinion, on the perception or not of planetary peril, and on a determination by the representatives to the central institutions to organize a response that protects and benefits all, even as it imposes burdens and requires sacrifices. If world conditions are correctly assessed, then it is difficult to imagine the adoption by any major government of a voluntary preference for nonparticipation. Of course, the bargaining for *degrees* and *modes* of participation will underlie the designing process. How can a design assure meaningful participation for principal actors and yet restrain their participation in relation to actors who are weak, poor, or small? If population level is the principal gradient factor, then the association of population size and potency is sustained at a time when it should be dissolved. In this regard it is deeply troubling to take notice of the efforts of Prime Minister Eisaku Sato, the cautious leader of Japan, to encourage an increase in the birthrate of the country. The main explanation for this curious reversal of Japanese population policy in mid-1969 was that labor shortages in certain sectors of the economy were threatening to erode the fantastic rate of economic growth that has been attained recently in Japan.

At present, Japan has a population of 102,000,000 squeezed into a land area smaller than the state of Montana, with 1,333 inhabitants per square kilometer of cultivable land; Holland, the second-most-crowded country in the world, has, by comparison, 565 inhabitants per square kilometer of cultivable land. If area is made the principal determinant of relative influence, then it collides with common sense (Canada outranking India, Japan, Germany), and it associates participation with territory (rewarding expansion and neglecting nonterritorial actors such as regional institutions and multinational trading entities). One recent proposal has been to grade states by a combination of population and per capita energy consumption. It will probably be exceedingly difficult to find *weighted standards of participation* that combine considerations of realism (and allow for expression of unequal influence) with those of acceptability (and stimulate even the weaker actors to join in the new enterprise). Agreement on what are suitable *standards of participation* is likely to be an extraordinarily difficult task. If the design of a world-order system rests on distinct levels and forms of structure, then it may prove desirable to decide upon a variety of participatory forms to satisfy the variety of participatory needs.

In outlining these value orientations it should be understood that they are intended to give a sense of direction to the work of world-order design and to set limits on what should be done to bring about a new world-order structure. There is nothing fixed or disembodied about these value orientations. Indeed, they might be thought of as elements of the sort of political *consciousness* required to design capably and imaginatively a preferred world-order *structure.*

For instance, the idea of ecological unity—the planetary scope of crisis—works against the adoption of partial structures of domination as a basis of world order.

Structures of control such as *Pax Americana* or con-dominium (a possible world system managed through the coordination of Soviet and United States interests) are ex-pressions of pre-ecological thinking. Anachronistic proposals for world order often result from a failure to appreciate the emerging centrality of ecological factors. A typical pre-ecological outlook is contained in the following depiction of the future shape of international stability. "On the sound assumption that in a decade or two China is likely to become increasingly capable of undermining the delicate balance of power between the U.S. and the Soviet Union, policy makers in both Wash-ington and Moscow increasingly make common cause in creating a more stable world in which their own power is not necessarily dissipated and Peking is not made the benefactor of these differences." There is a common fallacy that persists in world-order thinking that *prudent management* of military power is *sufficient* for the needs of mankind and that the transfiguration of power in human history is neither possible nor necessary. The design of a world-order system for an endangered planet affirms both the necessity and possibility of change, and concentrates on identifying the most favorable circum-stances for promoting its vision of the future.

3. STRUCTURAL DESIGN. The external features of new systems of world order, including institutions and ac-tors, constitute its "structure." There are several plausible structures of world order that are worthy of consideration. The uncertainty of the future makes it impossible to make any very confident predictions about the limits of political feasibility. It seems highly probable that the competitive character of international society will persist indefinitely, at least until a catastrophic breakdown. This competitive-ness of states may be susceptible to various kinds of small-scale subversion and thereby made more sensitive to the meaning of planetary storm warnings and early ecological

fissuring. In this section I shall discuss the structural building blocks of a new system of world order, appreciating
that it may turn out to be possible to evolve a whole series
of arrangements, depending both on the nature of world
conditions in the future and on the level of political consciousness. The purpose here is not to develop a blueprint for the future for world order, but to provide the
contours of a vision and some ideas to put it into practice.
Any more directive proposal has a comic quality in the
Bergsonian sense of mechanism encrusted on life.

More serious failure from our point of view is that an
overspecific solution for world-order difficulties allows
the reader to evade any personal responsibility by inviting him, in effect, to join in the fantasies of the author. A current political slogan captures this concern:
"If you are not part of the solution you are part of the
problem." The urgency of the present situation arises
from the need for each person to begin *now*, emphasizing
the seriousness of the world-order situation through *work,
feeling,* and *action,* and even *sacrifice.*

The discussion that follows rests upon the analysis of
Chapters III and IV and the informing values outlined
in the preceding section. It builds toward a coherent
image of the future, by selecting some trends in the present world-order system and exploring their positive potentialities. It also seeks to emphasize changes in the
structure of world society that can be made compatible
with one another. But the chapter does not purport to
offer a new or rival plan for a world-order system that
could be compared to the one put forward by Grenville
Clark and Louis Sohn or that of the Chicago group put
together originally by Robert Hutchins, and continuing
its efforts under the guidance of Elizabeth Borgese Mann
at the Center for the Study of Democratic Institutions at
Santa Barbara, to produce a world constitution that meets
the needs of the day. Instead, the design of structure
put forward here develops a different way to think about

creating a new system of world order, especially through an emphasis on the *dynamics of structural growth* and a consequent neglect of the *statics of constitutional arrangement.*

There are seven structural elements of the existing world setup that seem relevant to the promotion of a new world-order design. The practical prospects for progress depend on accentuating their importance within world affairs by rapidly accelerating their present rate of growth and by expanding the scope of their concern. All seven elements are present, to some degree, in the existing system of world order. If these positive elements can be expanded beyond a certain point, major transformations of power, values, and institutions may take place. Also the convergence of positive tendencies, if interpreted persuasively, may help to form a new understanding of world-order prospects and draw further developments within the horizon of possibility. The structural characteristics that contain positive potentiality with respect to a world-order design are the following:

1. Central political institutions of general authority.

2. Specialized agencies devoted to specific tasks of international coordination.

3. Informal and tacit patterns of coordination among principal world political actors.

4. Transnational and multinational actors and movements devoted to special tasks or to social and political change.

5. Regional and subregional organizations performing certain tasks of a cooperative nature in relation, especially, to economic and security subject matter.

6. Changes in outlook and organization within national societies involving the emergence of more cosmopolitan ways of perceiving the world and a

greater insistence by domestic public opinion upon a restrained foreign policy.

7. Subnational ethnic and religious movements seeking to break off from an established sovereign state or to establish zones of genuine autonomy within the setting of an existing state.

1. *Central Political Institutions.* The political organs of the United Nations—the General Assembly and the Security Council—taken in conjunction with the UN Secretariat and the International Court of Justice amount to a skeletal outline of world government. We have already discussed in Chapter V the very limited roles played by the United Nations in maintaining international security in the present world-order system. The UN has neither the resources, the consensus, nor the prestige to impose its will on its stronger members, and its missions must proceed cautiously on the basis of an unusual and, very possibly, evanescent agreement on means and ends between the United States and the Soviet Union. But it is most often the case that these two states need to be contained within the limits of the UN Charter. The Soviet Union and the United States have acted militarily beyond their own boundaries for purposes other than self-defense, and have claimed the right to station military forces on the territory of allied countries.

Besides the weakness of the UN, it suffers from a tendency to reach decisions on the basis of political, ethnic, and religious affiliations rather than through an interpretation of the facts of a dispute in relation to the norms and procedures of the Charter. The Middle East conflict illustrates the tendency of UN organs to ignore the merits of a particular complaint in order to promote a political preference and express the ethnic solidarity of the Arab countries, the religious solidarity of Islam, the ideological solidarity of the Third World and of the socialist group of countries.

In addition, the organization suffers from its inability

to achieve universal participation. The most glaring absences are mainland China and the divided states of Germany, Vietnam, and Korea. The constitutional absurdity of allowing the Chiang regime in Taiwan to occupy China's seat in the Security Council, although it lost complete control of the mainland in 1949, greatly impairs the seriousness of the organization as a central political institution with authority to regulate sovereign states. In essence, the UN functions as an *instrument* of international diplomacy rather than as an *alternative* to it, especially in matters of war and peace.

The prospects for encouraging peaceful settlement of disputes have been greatly diminished by the recent experience of the International Court of Justice. The World Court in The Hague greatly shocked world opinion in 1966 by its refusal, after five years of expensive litigation, even to determine whether South Africa was violating the terms of its mandate to administer South-West Africa by extending *apartheid* to this former German colony that it had been administering since 1920, supposedly as "a sacred trust" for the international community. The agreement establishing South Africa's rights of administration specifically gave a right of recourse to the World Court to states that were part of the organized society of sovereign states —initially the League of Nations and later the United Nations. At a time when the international community could agree on little else than that official racial discrimination of the type practiced in South Africa is evil, the ostrich-like refusal of the Court to decide the controversy and promote the cause of social justice seemed to confirm the worst fears and biases of the Asian and African states, an interpretation made more plausible by the fact that all the European judges sided with South Africa in the close final vote. The World Court's prestige has also been diminished by the failure of important UN members, notably France and the Soviet Union, to comply with the advisory opinion given in the dispute over whether peace-keeping expenses incurred in the Middle East and the Congo by authorization of the General Assembly could be charged to the membership; the General Assembly refused to apply the sanction in Article 19 of the Charter and deprive members of their vote who failed to pay their assessed share of the costs in accordance with Charter procedure. These two controversies, both of political significance, reinforced the worst skepticism about the worth and effectiveness of the Court as a central judicial institution. Additionally, the inability of the Court to secure much judicial business of any kind suggests that national governments are unwilling to settle international disputes by recourse to such cumbersome judicial machinery. As a consequence, it is not surprising that at the present time only one international dispute awaits determination by the Court.

Despite this assessment, there are certain reasons to expect a greater relevance for the United Nations in the near future. First of all, the United Nations may become a principal arena for spreading word and concern about the endangered planet. As we have already mentioned, that organization serves an indispensable function by providing a communications link among its membership that remains intact during periods of crisis and even warfare. Such a network of communications links may encourage a more even awareness of the impinging and urgent relevance of various ecological issues to the traditional concerns of world diplomacy. To the extent that the endangered-planet position becomes convincing to a variety of governments, it is also likely to increase the receptivity of countries to proposals for central coordination of international life. Therefore, unlike national governments, the UN stands to gain from publicity and evidence supporting the endangered-planet thesis.

Second, the UN is likely to move toward universal membership in the years ahead. The symbolic value of having all major governments participate in the work of the UN may enhance its authority, especially in relation to Afro-Asian conflicts. The inclusion of China would be a major step toward correlating the *formal* operation of the political organs of the UN with the *actualities* of political power. As universality is approached, the UN may be in an improved position to evolve world standards of behavior on the basis of the will of its membership, depending on whether areas of agreement can be identified and developed.

Third, the emergence of common interests among the superpowers may induce a greater reliance upon the conflict-moderating capacities of the UN. If the Soviet-American objectives are to maintain control over the spread of nuclear weapons and to retain political preeminence, then their incentive to cooperate without seeming to bargain at the expense of other states will probably encourage recourse to the United Nations as a mediating body. Such recourse (as an alternative to either confrontation or partnership) may also in time allow the organization to build up some military and financial capability of its own.

These considerations may also prompt principal states to enter into substantial disarmament agreements in order to maintain both political stability and hierarchy. One likely effect of substantial disarmament would be to expand the functions and salience of the United Nations in the field of peace and security, either by way of supervising compliance or by way of increasing the peacekeeping role of the organization. Such reductions in national military capabilities and discretion would tend to give central political institutions more scope for action, more autonomy, and more claim on public support and confidence.

Fourth, the exploration and exploitation of the oceans

for mineral resources seem to require some kind of centralized regime to allocate rights and duties and to assure that the wealth of the oceans is used to help poor countries as well as to augment the GNP of the rich. The UN may very well be a partial beneficiary of some kind of taxing or licensing authority that diverts some of the profits to the world community, thereby achieving somewhat greater independence for the organs of the UN in their relations to large member states. A related role in financing economic development may emerge for the UN or its subsidiary agencies as a response by rich countries to the allegation that their bilateral foreign aid has often been interventionary.

These prospects, and others, make it possible for the UN to move well beyond its current image of impotence and virtual irrelevance. A significant success in preventing or resolving a single world conflict could change popular, and even governmental, expectations about the UN very rapidly. Institutions in the early years of their existence are often looked upon with scorn, but when circumstances make possible a signal success, then suddenly a new image of potency emerges. A single leader of a powerful country who places great trust in the UN may, if he influences public opinion or other governments, alter the whole climate of confidence, and convert the UN into a major world arena. Furthermore, there are no inherent restraints on the development and growth of the organization. The Charter is a flexible constitutional instrument that can be adapted, as it has already been in several respects, to serve the wishes of its political majority; and although its revision may seem difficult to negotiate, it may also be quite unnecessary ever to do so.

The UN structure is probably most durably deficient in its identification of Permanent Members of the Security Council (especially England and France), and in its reliance on a one-state/one-vote principle in the General

Assembly. Voting rules are difficult to revise by common consent, especially if corrective action takes voting power away from actors whose consent is needed. Despite these difficulties, the UN is likely to play a very much more important role in world affairs once the evidence of the perils to the planet begins to be understood. As a structure, the organization provides a flexible framework for transforming values and adapting behavior to altered and altering attitudes and expectations. Short of a major sudden breakdown of the state system, no set of circumstances can be foreseen that will convert the UN into a structure of world government by voluntary, explicit acts. Such a drift may, nevertheless, occur if a strong, central response to the agenda of the endangered planet is sought in common by principal states, especially the U.S., U.S.S.R., China, India, Brazil, Nigeria, and Japan. This drift toward world government may be darkly disguised, even denied, depending on the political requirements of national life. Nevertheless, such a drift, once initiated, can gather a momentum that builds up higher, more ambitious expectations about the roles and functions that can and should be entrusted to the UN. Once bureaucracies begin to expand their activities, the logic of further expansion becomes ever more convincing at each stage along the way until some countertendencies initiate a period of contraction or stasis. The destiny of the UN involves the consideration of social and institutional growth processes.

2. *Specialized Agencies.* There is a type of world-order thought that contrasts the impotence of the *central institutions* with the realistic promise of *specialized agencies* that are entrusted with specific and technical tasks such as labor, health, weather, maritime affairs, food, postal service, airline safety. According to David Mitrany, all efforts to achieve world government and grand political solutions for the problems of mankind have led into "ide-

alistic blind alleys with no effect on public action at all." The development of specialized agencies, based on *function* not *politics* and *diplomacy,* has been "earthbound, concrete, and wholly practical"; these agencies of international cooperation "have grown quietly and undramatically, yet in ways that hold the promise of a true international community." In 1969, the International Labour Office (ILO) was awarded the Nobel Prize for Peace, recognizing the contributions of this most respected of specialized international agencies to the cause of world order.

There are many international agencies performing a variety of tasks. A central distinction is between intergovernmental organizations in which the membership is constituted by governments (so-called INGO's), and international organizations that are made up of other kinds of nongovernmental actors (so-called NGO's). It is evident that with the increasing interrelatedness of international life there is an ever-growing need for functional and technical cooperation. The dynamics of reliable contact presuppose the need to sustain links of communication and transportation, to establish safety standards and rescue procedures, to conserve endangered species of marine life, to manage the resources of the oceans, and to regulate activity harmful to the entire community such as piracy or "pirate" broadcasting.

In addition, there is a wide range of subject matters pertaining to health, labor standards, food and agriculture, culture, and science, where the exchange of information, the avoidance of expensive duplication of effort and facilities, and the formulation of common standards would seem sensible and valuable. Where the subject matter is highly technical, such as controlling the spread of contagious disease, the common interest is so obviously paramount that effective cooperation has been generally present, whereas in other areas, such as securing minimum working conditions or achieving freedom to form trade associations, the task has impinged so directly on political values as to be incapable of any serious implementation beyond that slight level of coercion that may flow from adverse publicity.

The most extravagant claim made on behalf of a functional approach to world order is that it builds up a new system of relationships without appearing to detract from the role of the sovereign state. Once these relationships are institutionalized and staffed by professionals—international civil servants—then the logic of cooperation and the spirit of internationalism are supposed to assure a constant expansion of activities and a performance that is genuinely *international* in character as its officials build up loyalty to the agency rather than take orders from their governments. A heated debate has arisen among the leading students of these specialized agencies. Ernst Haas has ably demonstrated that the growth of these agencies depends on their *politicized milieu,* not, as Mitrany would have it, on the exclusion of politics from their work. Haas has shown, especially in relation to the ILO, that the extension of the agency's work depended on political majorities pressing for action against a particular target, creating a new standard by this pressure that became generalized as a precedent for action of very different character from that which motivated the original claim. For instance, the West wanted to embarrass the Soviet Union for its failure to allow workers freedom to form trade associations and so it conducted inquiries, reached decisions, proposed action. Later the Afro-Asian group used these precedents to mobilize action against Portugal for its treatment of labor in her African colonies. It is, then, the political confrontations, and not their avoidance, that move these specialized agencies out beyond their habitual zones of endeavor.

There is also some reason to question whether specialized agencies—at least in the pre-ecological age—could ever seriously alter the basic contours of a state-dominated world-order system. Technical cooperation might enable certain kinds of action to be taken in the common interest, but it would never be able to erode, let alone override, the basic competitive impulses to maximize national wealth, power, and prestige. The contention that if technical cooperation proceeds beyond a certain point it will *spill over* onto the political terrain does not seem well supported by available experience, especially the history of European economic cooperation in the period since World War II. The spillover effect has not materialized; instead we notice a buildup of political resistance once the process of technical cooperation progresses very far. If political pressures underlie the growth of these institutions, then these pressures have always stopped short of reshaping the basic prerogatives of national governments. In essence, functionalism and state sovereignty are compatible, hence the development of the former does not pose a threat to the latter; in contrast, central political institutions become incompatible with state sovereignty as soon as they achieve any tradition or capacity for independent action. For these reasons it would seem naive to suppose that issues of war and peace could eventually be assimilated into the network of specialized agencies. The international experience that we have to date suggests that the existence of these agencies has not in any way reduced the fundamental zone of discretion claimed by states. The very achievements of these institutions may be in part a consequence of their *irrelevance* to the high matters of state that cause wars, repression, and misery—the issues that have given rise to the frantic search for effective and just central institutions of authority and control. These specialized agencies are helpful for regulating the daily routine of international life, but they are not very significantly related to the main task of building a new world-order system.

But such an analysis of the pros and cons of functional

approaches to world order is cast in a pre-ecological frame of reference. The agenda of an endangered planet is likely to give a heightened potency to functional dimensions of world order. The multiple needs for ecomanagement create a set of demands for central and regional institutional development to obtain access to information, set agreed standards, and establish some procedures for implementation. There are already plans underway to establish a global warning system on environmental issues. Twenty stations are contemplated in critical settings around the world, equipped with computers able to process information quickly and sound alarm bells whenever necessary. The project is being planned mainly by Soviet and American scientists and it is unclear whether the system will be administered by a group of cooperating states as a joint venture or will be placed under some kind of global auspices that would manage the enterprise for the benefit of the whole earth. In matters of resources, population, pollution, and technical hazard there is already much gathering and disseminating of information on an international level; the prospects for policy coordination will improve as the dangers of decentralization grow more evident. Functional agencies may well provide governments with a rational strategy of adjustment that does not involve any dramatic explicit transfer of sovereign prerogatives. At the same time, the critical bearing of endangered-planet issues on wealth, health, and survival is likely to make the responsible agencies very significant international actors who would need a growing budget of major proportions to cope with likely assignments. The functional alternative would also allow a measure of divided authority on the international level, thereby avoiding the Frankenstein problem—that is, being associated with building up a single monster organization that might become capable of imposing some form of tyrannical rule on the entire world. Whether this divided authority would be compatible with an ecologically minded administration of the planet is doubtful if we bear in mind the central tenet of interrelatedness.

Large-scale functionalism could also emerge in relation to the transfer of capital from rich to poor countries and through the creation of an international agency for the management of ocean resources beyond the twelve-mile limit. Significant arms control and disarmament measures would also be likely to rely upon specialized regimes of control that were somewhat detached from the central mechanisms of the United Nations. As these and many other managerial activities expand in *scope* and in *significance*, a growing civil service of heightened stature will emerge as a separate labor force with political influence in world affairs that might be expected to stand somewhat above the parochial perspectives of national governments. This separate identity is likely to be encouraged by the sharpening awareness of the endangered-planet situation and of the obstacles to response posed by traditions of nationalism and sovereignty. Therefore, specialized agencies are likely to play a crucial role in the design of a new world-order system, not primarily because of the past experience and record of these agencies, but because of their greater relevance to the expected needs and conflicting value concerns of the near future. In fact, the specific design of specialized agencies and their interrelation is itself a form of advance planning for the problems of an endangered planet. Serious consequences can result from failures of distinct agencies to coordinate their missions: it is plausible to imagine medical and agricultural specialists introducing cheap insecticides to combat disease and starvation, without being sensitive to the effects on the pollution of the oceans. Clearly a central ecological study center will be needed to gather and disseminate information; in addition, political questions will arise from the crisscross of different priorities (e.g., food versus ocean quality) and from the need to allocate burdens and costs among a variety of potential actors on the basis of a variety of different principles (e.g., should the country itself bear the increased expense of nonpersistent insecticides? How should the costs of protecting the biosphere be shared?).

The need for coordination will become obvious in the years ahead as the boundaries of the earth close in on man. The success we have in evolving voluntary patterns of cooperation will help defer catastrophe and gain time to allow the designing process to go forward without great trauma and suffering for mankind. Fritz Baade, in a relatively optimistic book about the future, perhaps overstating the technological opportunities for meeting the basic needs of mankind, writes as follows about the Malthusian dimension: "At first glance the task of providing food for 6.5 billion people within the next four decades seems staggering and overwhelming. Within the context of these figures, however, we can envisage it as a task easily soluble *if* tackled not by enemies split into opposing camps, but by mankind as a whole working in single-minded and perhaps even fraternal cooperation." Even without reference to ecological dangers and constraints, Baade, writing in 1962, believed that mankind could sustain itself only if utterly unprecedented levels of international cooperation were achieved in coming decades. What may strike a reader as Baade's technological optimism is, then, completely undercut by an overwhelming political precondition. In general, the design of world order is caught in a terrible crossfire between the technical need for *international cooperation* and its political *unattainability*. There seems to be no prospect at all for direct transfers of the central authority and resources now under the control of national governments. These governments are geared up to compete with one another in the sparsely regulated world jungle. The

emerging scarcity of renewable and nonrenewable resources—including space, wilderness, and light—accents
the need for coordination to set and implement allocational rights (so long as abundance could be presumed, it
was possible to regulate only to achieve some level of
compatibility for multiple usage). But these objective
conditions favoring cooperative relations do not seem
strong enough to overcome the competitive habits and
vested interests.

3. *Concert of Principal Actors.* In addition to federalist and functionalist ideas about improving the quality of
world order, the idea of a concert among governments
(of more or less informality) seems promising as a building
block for the future. The notion of "concert" implies a
regular membership, periodic meetings, common policies,
and the absence of any permanent secretariat or organizational center. In European diplomacy through the eighteenth and nineteenth centuries a concert among sovereign powers operated to maintain prevailing ideas about
world order and, especially after the Congress of Vienna in
1815, to arrange for joint action to maintain dynastic legitimacy in countries threatened by liberal democratic revolution. Historical experience with the diplomacy of the
concert suggests potentialities for coordinating policy and
practice to serve perceived and shared common interests.
The use of the concert approach to world order might be
limited to obtaining a minimum consensus among principal governments on basic rules of conduct and the main
line of response to central issues of international public
policy. A concert setting might be less vulnerable to diplomatic maneuver and less unwieldy than the present setup
provided by the United Nations; such a setting might also
simplify the process of broadening the agenda of international diplomacy—perhaps initially relying upon secret
sessions to explore frankly the various interests at stake.
An effective concert would have to achieve universal

participation, but perhaps not only by representatives of national governments. Perhaps the five most populous countries would be represented on a standing committee of the concert, together with representatives of regions and major cultural and ethnic groupings. Such universality would be needed to overcome the suspicion that this body was being used as a covert instrument for bipolar or great-power domination as well as to solicit the participation of more militant actors (perhaps China) in the work of ecological defense. The mission of the Concert of Principal Actors should primarily be to act as an *organ of oversight* to grasp and help organize responses to the endangered-planet emergency. It would be desirable to coordinate the efforts of the concert with the United Nations by means of some resolution of endorsement by the General Assembly. A separate identity for the Concert of Principal Actors would still be needed, however, to allow for *representation* by actors other than states—probably principal specialized agencies should participate as members and perhaps major transnational cultural and political movements as well—and operate in an atmosphere of relative privacy and informality. Such meetings should be regularly scheduled for one week each year and at such other urgent times as are deemed necessary by three or more constituent members. Presiding officials, perhaps even heads of states and other participating organizations, might be committed to such an annual obligation, with lower officials meeting in continuous sessions to carry out the recommendations of the principals, to issue authoritative reports with policy recommendations and action proposals, and to prepare agenda and background material for subsequent meetings. It would be desirable if the Concert of Principal Actors would establish as a matter of principle and tradition that its concerns were limited to ecological issues of planetary dimension and that its activities were not to be regarded as a substitute for, or even a complement to, either traditional diplomacy or the efforts of the United Nations and other constituted international institutions to deal with world conflicts or threats to world peace.

Naturally, the composition and function of such a Concert of Principal Actors would be subject to intense preliminary negotiation and bargaining. Perhaps, too, some prior steps would need to be taken before such an arrangement could become even a serious possibility—the admission of China to the United Nations, the settlement of the Sino-Soviet boundary dispute, the termination of U.S. military intervention in Asia, and the widespread dissemination of a generally convincing account of the endangered-planet argument. Although the idea of concert may seem premature, given the preoccupations of traditional diplomacy, it still deserves careful study and articulation by any student of the potentialities for peaceful change in world-order systems.

Having satisfied these preconditions, however, a Concert of Principal Actors might emerge, initially perhaps on an *ad hoc* basis relating to a particular issue or series of issues, and subsequently taking on a more regular range of topics, at fixed intervals, with specified members, a permanent staff and, possibly, a central headquarters. Such a concert could consider problems of planetary dimension in open and closed session, and seek to achieve among its membership positions of minimum consensus and to identify ways to compromise differences in perceptions and priorities. As with all complex proposals for international coordination, the originating ideas and prescriptions would be basically altered by experience.

4. *Transnationalism.* The interrelatedness of economic, social, cultural, scientific, professional, and political activity has caused an increasing sphere of transnational activities. The characteristic identifications, events, and activities of the age are carried on across national bound-

aries, joining together the efforts and experiences of men and women situated in different parts of the planet. Transnational reference groups—groups with which one identifies—are becoming dominant centers of loyalty and affinity for many critical sectors of national society. It has been observed, for instance, that scientists from the United States and the Soviet Union who attend the Pugwash meetings to discuss disarmament prospects increasingly have more in common with one another than they do with the leaders of their own countries. The same observation is more dramatically true of students and militant minority groups; the youth of the world has a transnational quality, reacting to many of the same abuses in their societies, reading many of the same radicalizing books, and sharing many of the same heroes and villains. Rock festivals around the world—whether held in Woodstock, New York or on the Isle of Wight—exhibit a similarity of dress, life-style, and ambience that bears witness to the transnationality of many cultural trends. The appeal of liberation movements to those who regard themselves as dispossessed by the structures of power also makes political activity take on a transnational character. The Black Panthers are not just opposing racism and exploitation in America, they conceive of themselves as part of a worldwide movement that is going forward in Latin America, Africa, and Asia. The Black Panther Party newspaper reports on all of these international struggles against oppression, and urges upon its readership a sense of an ongoing world revolution of which the cause of black revolution is just one instance. Even business groups, through the rise of the multinational corporation, are conceiving of their interests in transnational rather than in national terms, and in some cases are becoming centers of power and authority that either ignore or compete with the governments of states. Finally science and technology, by scale and scope, are pushing the organization of many phases of life into transnational modes of operation. It is superfluous and costly to duplicate computers, reactors, or transport equipment in each state. Satellite TV and radio broadcasting also has an obvious transnational character, allowing people situated throughout the world to participate simultaneously in new experiences, such as the latest space exploit or the most recent ecological disaster. As with other forms of transnational activity, both good and bad ends can be served by the erosion of national boundaries. The distinction between good and bad that is relied upon here is whether the activity carried on illuminates or obscures the endangered-planet crisis. The economics of modern industry tend to undermine the organizing significance of national boundaries. The movement of highly skilled labor—the whole phenomenon of the "brain drain"—further illustrates the transnational underpinnings of modern life. In fact, national bureaucracies are exceptional entities in the modern world by their exclusionary rules that limit eligibility to members of the national population group.

The growth of transnational operations and identifications tends to produce a wider sharing of perceptions about events and problems throughout world society. This transnational character of experience promotes a convergent awareness of what is taking place in the world. Especially among those segments of national society that are seeking drastic social change there is emerging an awareness of the unity of mankind and of the world as potentially constituted by a single family of man. This movement toward wholeness and unity, which is at the very center of various transnational expressions of the ecumenical movement that has been so successful in various Christian religions, builds a climate of attitude that is consistent with the needs of an endangered planet. If the development of awareness is the critical need during the immediate years ahead, then these efforts to light the same candle in many parts of the world complete the difficult and critical task of reaching agreement on the seriousness and nature of the threats to human welfare and survival, and of their underlying causes. Such an affirmation of the unity of mankind is not to be identified with either enthusiasm for or aversion to world government or federation. The Soviet Union and in fact most governments remain skeptical about, and even ideologically opposed to, building up central institutions in a divided world. Yet an awareness of the unique historical setting of danger emergent in various forms throughout the world, and linked by transnational affiliation, would seem to encourage the growth of a sense of the urgent need to transcend present rivalries so as to be able to coordinate policies and an acknowledgment that shared problems and aspirations should take precedence in political consciousness. At the same time, the positive side of diversity could itself become, as we have suggested, a guiding policy of world-order design. It is not *diversity* of language, culture, ideology, or race, but a structure of *stratification* based on these diversities that produces misery and assures the persistence of the war system. Transnational movements and actors may work toward an idea of diversity without stratification, and hence diminish the dependence of national governments upon threats and uses of force. Such a reorientation of attitudes would seem to condition the design of a new world-order system in a highly beneficial way.

Transnationalism as a set of trends, and as a potential way to assemble an album of pictures of the endangered planet, provides the world-order designer with promising opportunities. Enlightenment, even if it comes to certain national groups, is not likely to produce adaptive change unless there is across the earth a widely shared sequence of occurrences and of interpretations. The discovery of the endangered planet needs to take place—possibly in a variety of guises—in many different national settings at about the same time, and these separate discoveries of roughly the same phenomenon will have to be linked up with a planetary movement calling for drastic change. The transnational domain provides linkages on many levels of social, political, vocational, and economic existence, and may even eventually provide the psychosocial groundwork for reorganization of international life. As such, transnationalism provides a principal building block for a world-order designer; the precise future of transnationalism is, of course, impossible to foretell, being contingent on many related happenings. At each stage of the future, however, the positive use of transnationalism can improve greatly the prospects of adaptation to the urgencies of the ecological age.

There are, however, three qualifications on this affirmation of transnational trends. First, the conversion of rather narrowly conceived transnational groupings into pressure groups lobbying for a new system of world order may be an exceedingly slow and uneven process. Given the urgency of our needs, we may not be able to mobilize these transnational groupings rapidly enough. Second, the cosmopolitan ethos that often accompanies transnational growth is likely to be atypical for the wider social and political setting that exists at local, national, and regional levels; as such, this kind of wider endorsement of what Anne Morrow Lindbergh has called "earth values" may be confined to arenas dominated to various degrees by the counter-culture. Third, the conflicts over agenda and priorities—what shall be done first? how important are various claims for action relative to one another?—would be likely to surface even among transnational groups as soon as *action,* rather than mere *sentiment,* was at stake. There are differences in perception, feeling, and interest that make equally sensitive individuals understand the endangered-planet argument in a variety of ways.

5. *Regionalism.* In recent decades there have been many experiments with economic and political organization on a regional or subregional scale. Countries have sought to pool their capabilities or to lighten the burden of competitive relationships. Regionalist plans seek to develop political actors that are larger than the state, but not universal. On the basis of homogeneous political and economic systems, shared history and traditions, physical contiguity, common levels of industrial development, and common interests, there could develop in theory at least, step by step, a supranational approach to the organization of collective life. Some of these regional units could compete with the more powerful states and could save smaller nations from a permanent situation of impotence within the world system. Also within the regional framework, violent conflict might be less likely to occur in the event of a dispute, and limited peace systems might begin to emerge. Another kind of regionalism has been stimulated by the effort of a superpower to organize an alliance system (e.g., NATO or the Warsaw Pact), or to disguise its position of dominance beneath an elaborate set of legal forms (e.g., the Organization of American States). In the 1950's, John Foster Dulles carried this kind of spurious regionalism (spurious in the sense of the absence of any wider community base) into a series of security arrangements that had treaty status but no implementing will or organization (CENTO, SEATO, ANZUS).

There are many kinds of regional organizations, although most of them either emphasize *inward-looking* arrangements to promote security and prosperity or *outward-looking* arrangements to guard an area against a perceived military threat from outside. The Organization of African Unity is primarily inward-looking; NATO and the Warsaw Pact are primarily outward-looking.

Some experts on regionalism have speculated that if economic integration proceeds by way of erecting common external tariff barriers and eliminating internal barriers to trade, then it is likely to be followed by political integration, including the gradual displacement of sovereign centers of power by supranational institutions. The European experience has been the principal testing ground for theories of integration, illustrating both the strength of economic integration as carried on within the framework of the Common Market and the resistance of national governments to developments that impinge on traditional prerogatives of sovereignty in matters of high politics. General de Gaulle slowed down the process of economic integration in Europe by his opposition to any displacement of sovereign authority, and there does not appear to be any likelihood in the near future of achieving a federation of Europe like that envisioned by such original champions of regionalism as Jean Monnet.

In certain situations a regional organization may provide a mantle by which to disguise exertions of power by strong states against weak ones. Certainly the regionalism involved in the action by the OAS against the Dominican Republic in 1965, or against Cuba at various times since the Castro Government took over on January 1, 1959, served mainly as a transparent cloak thrown across the unilateral policies of intervention being pursued by the United States Government. A somewhat equivalent form of spurious regionalism has been relied upon by the Soviet Union to justify its periodic impositions of military control upon the states of Eastern Europe, including its military occupation of Czechoslovakia in 1968.

In other circumstances, as with the Arab League in the Middle East or the OAU in Africa, the regional grouping functions in part as a belligerent party in relation to some common enemy present within the area—Israel in the former instance, and the white regimes of southern Africa in the latter instance. In these settings, the regional actor is a competitive focus that helps organize a common front in a situation of intense conflict, but may also disclose the limits of effective cooperation even in the presence of a strongly endorsed common policy objective. The domestic and national preoccupations of most governments become very apparent in these experiments in regional cooperation. Such preoccupations are particularly prominent with respect to nations threatened with civil strife—there are no surplus energies or capabilities available for external projects, however worthwhile they may appear to be.

Regional organizations may have distinct modes of conduct, depending on the setting. The OAU has acted on a number of occasions to moderate or settle conflicts among its membership, as in relation to the Algerian-Moroccan and the Somalia-Sudan boundary disputes. Similarly, the OAS in the summer of 1969 acted as an effective intermediary in bringing the soccer war between El Salvador and Honduras to a rapid and constructive end.

The world-order appeal of regional organizations is twofold: first, to move foreign policy and international relations beyond the nation-state, and second, to create a stepping stone to an overall or central organization of world affairs. In the first context, the regional buildup dilutes the affiliation of the individual to the national government, and thereby weakens traditional forms of patriotism, softens interstate rivalry, and may diminish the prospects of warfare. In the second context, the experience with supranationalism supports the further transition to a more unitary system of world order.

Some of the environmental problems of the endangered planet can be dealt with through regional planning, thereby enlarging the scope of political authority and responsibility. Also, if the traditions of nationalism tend toward competitive patterns of behavior inconsistent with the cooperative and overlapping needs of the nuclear age and the ecological age, then a regional interlude might provide a breathing spell to enable still more adaptive patterns of behavior to take hold. Therefore, a world-order designer might stress the positive potential of regionalist developments to gain time in which more fundamental shifts in political attitude could take place.

Regional units could also be entrusted with a variety of tasks in the new world-order system, including the protection of diversity while achieving coordination on a planetary level. To overcome the tensions caused by distrust and interest on a global level, regional actors might be assigned certain tasks involving peacekeeping and disarmament relevant to their area of concern. Similarly, regional tribunals and councils of conciliation could give expression to distinctive patterns of beliefs and values, although still fulfilling the need for community-centered procedures of dispute settlement to supplant the self-help procedures of the present competitive and coercive system. Shared values and physical proximity provide individuals with a supranational identification that poses less of a threat that special features of national culture and history may be overlooked than does entry into a world system. For this reason, regional movements, if oriented toward the actualities of an endangered planet, could make a significant contribution to the creation and maintenance of a new world-order system.

In this regard, it may be important to notice the creation by NATO of a Committee on the Challenges of Modern Society that is being entrusted with turning the attention of this security arrangement to welfare and survival threats arising from environmental decay and adverse social effects of modern technology. Daniel Patrick Moynihan, speaking before the North Atlantic Assembly in October 1969 on behalf of the United States Government, went a long way toward acknowledging an endangered-planet situation. Mr. Moynihan said:

> Just as advancing technology has given rise to the central social vision of our age, so also has it be-

come the central problem of the age. In massive and dominant proportion, the things that threaten modern society are the first, second, third or whichever order effects of new technology. . . . For a quarter century now, mankind has lived with the possibility of the ultimate technological disaster, that of the nuclear holocaust. But more recently, it has come to be perceived that this would be only the most spectacular of the fates that might await us. The perils of the modern age are wondrous and protean, and if anything, accumulating. An ecological crisis is surely upon us; and developing at quite extraordinary rates.

If awareness needs to emerge in many places and at various levels of organization, then these early signs that regional organizations are being expanded beyond their original undertakings to encompass ecological issues, represent a welcome recognition that these matters belong on the agenda of all actors in world affairs. Of course, recognition of a problem is not to be confused with undertaking a solution or even with a willingness to pay the price of a solution. In fact, being noticed may be an *alternative to,* rather than an *adjunct of,* serious social and political action. As governments become centers of public relations activity, there arises a serious danger that the language of concern and urgency will become totally divorced from realms of feeling and action. The right words and images massage the public conscience without disturbing the demoniac processes that are busy completing their destructive work.

The conditions for regional growth are better in certain parts of the world than in others, and within each region it may be more feasible to progress with certain issues than others. The boundaries of some regions are drawn by ideological affinities rather than by reference to geographical or historical-linguistic units. Such variability merely stresses the need for flexibility in thinking about the relevance of regionalism to world-order design models. Within particular international settings, certain kinds of regional growth may contribute to or retard an overall strategy of response to the endangered-planet analysis. Regionalism by itself cannot achieve the kind of planetary coordination needed to overcome patterns of competitiveness and combativeness, nor can it compromise and bargain toward an agreed global set of priorities and policy goals. Regional organizations are quite likely to evolve, in fact, as more effective units of competition to represent the interests of weaker, poorer countries within still wider arenas of interaction. Even European economic integration is based, in part, on the need to aggregate separate national capabilities so as to enable the region to compete in world markets and to operate at the frontiers of technology, given the immense capacities of the United States and the Soviet Union. The multinational corporation is a nonregional actor that may try to undermine this regional objective by maintaining extraregional control of economic activity, although such control will be centered in a corporate board of directors rather than in a government. Both Latin American and European observers have emphasized the prospects of regionalism for enlarging capabilities, and the vulnerability of these regional developments to various forms of penetration from outside the region. Regionalism is a two-headed beast or, more positively, regionalism resembles the Roman god Janus by having two heads, one looking backward, the other looking forward.

In relation to the war system, the development of serious regional competence also possesses a double possibility. Regionalism may provide a useful way to shield from UN scrutiny claims to use violence in international relations, and it also may deter recourse to unilateral claims to use force, or discourage large states from relying upon superior force in settling disputes.

Again, as with other world-order building blocks, but perhaps to an ever greater extent, the potentialities of regionalism depend upon an analysis of the mix of negative and positive factors present in any given situation. The regional movement is definitely an upward spiral in world affairs—exhibiting an upward curve despite numerous setbacks and disappointments—but its role in future world-order systems is made problematic as a consequence of its ability to recondition traditional modes of competitive behavior and its opposite capacity to guide and hasten transition toward greater cooperation and coordination of national efforts.

The regionalist sentiment does not accentuate the wholeness of the earth, nor is it likely to be sensitive to the urgent need to associate this wholeness with the limits of life-support on the planet. In this respect, the development of regional institutions as a partial world system can never supplant the case for unitary images and action that arise out of the basic understanding of the mounting threats to human survival and the habitability of the planet by all forms of life.

6. *National Reform.* The development of world-order system designs remains largely under the control of national governments. The attitudes that prevail within governmental circles have a critical bearing upon the possibilities for modification of the behavior and structures of international society by peaceful means. Reorientations of national political consciousness in response to the endangered-planet situation offer the greatest of all targets of opportunity, and if there is a failure to achieve any reorientation on critical national levels, then almost any other effort at transformation will be nullified. If world-order redesign is going to be accomplished by nonviolent and nontraumatic approaches, then the receptivity and capability of governments to proceed by way of self-modification are essential.

Such an observation is obviously correct if related to the transfer or curtailment of military power at the national level. Disarmament, arms control, arms policy, and the creation of an international police force are measures that each presuppose the assent of the participating governments. Unilateral action is possible, and in some instances desirable, but usually some kind of agreement or interactive process of behavior is needed. An arms race is a sequential spiral of interactions among governments; it might be possible to initiate a reverse or negative arms race by successive reductions of defense appropriations accompanied by expectations that if other relevant governments took comparable steps, then still further reductions would take place. The underlying issue of orientation involves the pattern of dominant values and attitudes held by political and parapolitical elites, an orientation that may be difficult to alter very much by evidence and rational argument because it is maintained by bureaucratic structures to protect the interests and honor the beliefs of the rich and powerful groups in a particular society. Most elites are socialized into an acceptance of a world view or cosmology that reinforces the existing setup of power and privilege. The idea of undermining or transforming the status quo does not normally even occur within such a socialized personality.

Only an emergency or a revolution creates enough bureaucratic latitude to modify basic modes of behavior; the outcome of elective politics in the major liberal democratic societies of the West, including the United States, may occasionally cause 5 percent shifts in policies, priorities, and behavioral patterns, but can never involve reorientations of 25 percent or more. The 25 percent figure is the lower level of response called for by an endangered-planet context. Where there are no elective politics, the difficulties of achieving an orientation seem to be at least as great, the longer-term bureaucracy having an immense capacity to slow down any disposition toward change; and even if a drastic break with tradition occurs at a given point there is likely to follow a reassertion of the slowly evolved bureaucratic values after a lapse of some years. Bureaucratic nullification can block new policy paths merely by the battling of different agencies of government to preserve or enlarge their sphere of influence and authority. Neo-Stalinism has succeeded in restoring many of the constraints upon Soviet life that were supposed to have been repudiated along with Stalinism. The effort of Pope John XXIII to revitalize the Catholic Church as a source of social and political inspiration has been almost nullified during the papacy of Paul VI. The alluvial course of value-change within large governments makes it very difficult to penetrate a major bureaucracy with a new set of interpretations leading to the adoption and implementation of new priorities.

As I have had occasion to point out above, there are many indications that top governmental leaders in the United States appreciate the dangers and costs of continuing population growth and pollution increase, and are giving dramatic expression to this appreciation in their speeches, especially on ritual occasions, as before national and international bodies. But, it is also true that no commensurate redirection of resources and energies has taken place to enable a serious effort to mitigate the worst effects of these ecological strains. Furthermore, on such flashpoint environmental issues as the SST and the regulation of offshore oil drilling, the policy momentum has remained with advocates of GNP, competitive industrial development, and the maximization of profits for industry and revenue for government. The very strong case for withholding further appropriations from the SST was brushed aside by President Nixon's assertion that the United States needed to develop a supersonic commercial aircraft so as to sustain the superiority of its aircraft industry vis-à-vis other countries. Even more symbolic of the continuing dominance of old priorities was the outcome of the ABM debate, resulting in the policy of proceeding to develop and deploy the Safeguard System. In this debate, supporters of the ABM put forward one unconvincing argument after another, and yet managed to carry the day against the mobilized opposition of moderate liberal opinion, bolstered by very solid support from the scientific community. The inability to prevent, or at least to defer, the introduction of MIRV systems into weapons planning also exhibits the control over national security policy that is retained by the military-industrial-congressional establishment, and also discloses the failure of this establishment to shift their own perceptions on issues of national welfare.

A final illustration drawn from American experience arises from the country's role in the Vietnam War. Despite the failure of American policy on almost every level over a period of many years, despite the startling disclosures of the brutality and corruption of the Saigon regime, despite the lack of battlefield progress, and despite the ugly disclosures of the Songmy massacre and other war crimes, there is a continuing insistence on peace with honor and an insistence that honor for the United States consists in maintaining the Saigon regime as the legitimate government of South Vietnam.

Soviet behavior in Eastern Europe and the Middle East exhibits the same disregard of the new urgencies of international affairs, and a steadfast dedication to the old modes of violent and competitive politics.

The old values and interest groups remain ascendant. Challenges are being mounted by critics situated within all principal sovereign states. As a result these issues have assumed greater prominence, and there is a widespread public reaction against polluted skies and waterways, evidenced in the outcry against further encroachments upon the environment of power plants, dams, bridges, highways, jetports, or mines. It may be that the popular mood can be catalyzed into a strong popular movement if several more disasters on the scale of the Santa Barbara blowout occur in the next few years and are perceived and interpreted as part of an underlying and deteriorating situation of ecological imbalance. The widespread, if provisional, enthusiasm accompanying the Environmental Teach-ins of April 22, 1970, is suggestive of a new national mood on these issues, although it may encourage a false complacency about what is being done.

In modern society broadcast media play a great role in preserving and altering the field of public awareness. Those who have privileged access to the media are also those who benefit from the existing arrangement of power and policy. Both the rich and the poor suffer from pollution (just as they both suffer—although less evidently and less equally—from war); there is reason to hope that as the evidence of impending disaster continues to gather, more and more effort will be made to educate and arouse the public. A climate of serious concern might emerge rapidly under such circumstances, especially if coupled with a strong expression of governmental concern and action. If these spurts of awareness occurred in different parts of the world, then further momentum would be built, and positive feedback might lead to even greater levels of aroused concern. In such a context, proposals for significant change tend to get a better hearing, and might even be well accepted.

A reorientation of consciousness on the national level may lead to an entirely new conception of social and political progress. Progress may become associated with *stability* of economic output rather than with its *increase;* with a steady-state economy rather than a growth economy. The national interest may become associated with a reduced domestic population and with diminishing rates of world population increase. Structural changes would have to be introduced to make these pursuits a central activity of government.

Another area of reorientation could involve subordinating foreign policy to a common set of legal rules and procedures. Such patterns of national restraint—if imposed on the behavior of principal states—might moderate international politics to a point where arguments for large defense spending would grow weaker and weaker in relation to other claims on resources.

Most specific steps toward a redesign of the current world-order system depend on modifying the character of national governments, their characteristic forms of aspiration, perception, and action. These modifications will vary greatly from state to state depending on the kind of political system, the stage of development, the extent of national wealth, the degree of cohesion, the extent of perceived external threat and conflict, and the moral and cultural traditions of the society. In the Soviet Union, for instance, changes in the character of the Communist Party may represent the only means by which to reorient political consciousness, whereas in the United States such a reorientation of political consciousness may depend on convincing heartland and blue-collar America that its bounty is being depleted by the GNP-mindedness of its governing groups. Some specific steps toward reorientation will be discussed to illustrate more concretely the direction of effort that is needed at the present time to awaken ourselves to the realities of the ecological age.

The national setting as a world-order building block is both a decisive arena and one in which prevailing habits of mind and action continue to work against adaptive change. But it should be understood that shifts in national consciousness, especially a realistic appreciation of endangered-planet threats and timetables, and an understanding of the inability to deal with these threats by relying upon the existing weak traditions of coordination, create receptivity to world-order redesign, or at least lessen the resistance to the search for new forms of order that might be able to sustain the habitability of the planet.

In a sense, the initial task of the world-order redesigner is to demythologize the state, especially in principal societies where continued reliance on its benevolent character is unwarranted. When people have their needs more or less satisfied, there tends to be a diluted attachment to any particular form. The fact that New Jersey has lost most of its "sovereignty" to the federal government is a development of no consequence to most of us at the present time, but it was of great moment toward the end of the eighteenth century when the federal constitution was being drafted and ratified and almost foundered in the conflict between the demands of bigger states for *proportionate* interest and of smaller states for respecting their absolute sovereignty. The Senate and the House represent the resulting and largely successful compromise. Most of us have shifted our identification from the original sovereignty of the era of confederation to the enlarged sovereignty of the United States, and identify our political destinies with the main governmental center of power, wealth, and prestige in Washington, D.C. Such a shift may be less successful for most white residents in Alabama or Mississippi. The federal government may become associated with alien values and hostile traditions that undermine regional forms of social and economic order that probably would have been safeguarded had effective sovereignty been left to reside in state capitals.

To some extent the same possibility of reattachment can take place in relation to political communities wider than the nation-state. The transnational links of the Arab and the Jewish communities in relation to the Middle East conflict illustrate one variety of political attachment that cuts across state lines. Regional communities drawing on common bonds and benefits might also be able to attract and retain the political faith of many individuals who now identify primarily as nationals.

To claim an identity based on being a citizen of the world or a denizen of the city of man has struck most people as vague and unreal—as an exotic piety that makes no political sense, given the current level of political consciousness. There is no world structure that is able to exalt or protect its adherents, there is no enemy against whom to organize, and there is no widespread sense of bondage to men who belong to a different tribe or speak a different language or worship different gods or believe in a different kind of political system. Suddenly, at some propitious moment in the future, a world consciousness may emerge with great force. It lies latent in all the great cultural and moral traditions that guide the destinies of separated national societies. To bring out these convergent tendencies and to associate them persuasively with human welfare and survival may be the great educational challenge of all time. Because of the present strength of national governments, this effort at persuasion must first of all be directed at the populations living within nations.

But will governments preside over their own dissolution? They have no choice, if the basic line of analysis is correct. The available choice is whether governments will assist the process of dissolution and guide the outcome in a constructive direction or will resist dissolution and help induce a catastrophic breakdown in world affairs. If universalizing tendencies are resisted, then ecological pressures will cause destruction. The state mechanism must either facilitate a political mutation or accept an ever-growing prospect of inviting destruction. Political elites tend to be rigid, favoring destruction over accommodation, and therefore every hopeful prophecy strikes most of us as unconvincing. Nevertheless, those who believe in overcoming the obsolescent ethos of nationalism, at least for the nuclear superpowers, must keep testing the political potency of our position with all the ingenuity and strength at our disposal.

There are two final observations that apply particularly to the situation that prevails in the United States at this time. First of all, we should adopt a skeptical attitude toward the remarkable upsurge of recent interest about environmental and ecological issues in the United States. In an inconceivably short period of time—a matter of months—the quality-of-life issues became prime political subject matter. President Nixon's 1970 State of the Union Address is probably the high point in this apparent awakening of politicians to the centrality of a set of issues that had been left for decades in the hands of the sanitation engineer and the amateur bird watcher. My suspicions are that the special appeal of these environmental issues at this time is to take the mind and eyes of the country off the barbarism of American war policies in Vietnam and the disgrace of its urban ghettos. In this sense, and the redirection of energy may not even be a calculated effort by the men who are advising the President, the discussion of pollution almost further pollutes the environment of politics. We cannot purify our land and water and air at the same time as we destroy crops and forests in Vietnam—and in the process commit a new war crime that has been called "ecocide"—nor can we make purity and quality the goal of domestic politics when we abide rat-infested dwellings for our urban poor. Politics is not often based on consistency of feeling and action, but there are levels of contradiction that raise issues of pathology and that cast doubt upon the sincerity of professions of concern. These doubts are reinforced by the failure to back up the alleged concern of government with resources (i.e., budgetary appropriations) and with specific decisions (i.e., SST, ABM, offshore oil drilling). The politics of ecology is very complicated because there are mixed motives present and because there is a genuine, as well as a spurious, upsurge of public concern. Whether ecological fanfare acts as goal or Miltown will be decided by the swing in the political balance within the United States.

My second observation concerns the lost awareness in America of our revolutionary tradition. Early in 1969, a survey was taken of a cross section of Americans at an air-force base in Germany. The Preamble of the Declaration of Independence was circulated without identification on an ordinary piece of paper to discover how many citizens would affirm the sentiments contained in this fundamental statement of the American creed and mission. Of the 252 persons interviewed, only 68 (or 27 percent) would sign the document, and only 41 identified it correctly. Of the sample surveyed 59 percent positively rejected the ideas contained in the Declaration of Independence. Among the responses to the document reported are the following: some called it "a lot of trash"; one man asked if the document "had anything to do with the Communist Party of America"; another responded, "Who wasted an afternoon writing this?"; another said, "It sounds like long-haired kid stuff"; and four said (possibly because the surveyor was black) that the document was a plan to set up a separate black state. Such a range of responses suggests how far the mood of the country has moved away from its inspirational beginnings. Such a shift is not necessarily wrong, but when the shift involves the denial of attitudes toward drastic change that are desperately needed in the ecological age, then it becomes a proper cause for concern and regret. If we could revive a sense of our revolutionary origins, and discover their relevance to the agenda of an endangered planet, then we could use our past as a foundation of strength upon which to build a future. Our historical past provides an indispensable source of support for a new tradition of radical politics.

7. *Subnational Militancy.* A final world-order building block has been created by the rise of pressures throughout the world to grant subnational autonomy to groups within a state that do not identify fully with their central government. Whether it be the Catholics in Northern Ireland, the Scottish or Welsh in Great Britain, the Kurds in Iraq

and Iran, the Naga in India, the Ibos in Nigeria, or the Turks in Cyprus, there is a demand, periodically backed by violence and a willingness to risk death, for subnational autonomy. Such enclave movements work against the centralizing tendencies of the modern state and incline toward a refeudalization of international society. To the extent that large-scale states are broken down or preoccupied with domestic issues, there is some moderation of a competitive approach toward international relations. The rise of subnational militancy, especially in modern societies, reflects in part a sense of the declining utility of the state mechanism as a method for organizing security and economic well-being. Perceiving this decline unleashes group grievances that have been dormant during the decades when, by and large, the state seemed like a necessary and inevitable political form. With nuclear weaponry, with a growth in transnational identities, with regional economic communities, there also comes about redefinition of the principal locus of political loyalty and personal identification; such a redefinition is further encouraged by real or imagined inferiority of status experienced by subordinated groups within the nation. Thus subnational militancy is a latent tendency in all large multi-ethnic states that compose the contemporary system of world order in a fashion that complements the latent supranationalism and even universalism contained within most national belief systems. The strengthening of subnational enclave consciousness can be used as an instrument to combat the larger, more dangerous enclave consciousness that undergirds and gives vitality to the one universal political energy, namely, "nationalism."

The relevance of subnational militancy to the redesign of world order is difficult to assess, but important to notice. It has become a very pervasive social and political force in recent years and is likely to remain so for some time. The character of subnational movements varies greatly from place to place. Like other world-order building blocks, subnational militancy is capable of inhibiting, as well as facilitating, world-order system redesign. An adequate sense of design will seek to discover the specific conditions under which subnational militancy is a positive development in international life. The one general comment that seems valid is that widespread subnational militancy would tend to lessen the predominance of the state as a center of political life; such lessening would, in turn, be likely to open the way to somewhat more functional solutions of problems impinging on the quality of human life by its tendency to demythologize the sovereign character of the state.

The larger political communities that bring together in one nation several—often many—disparate militant groups depend for their cohesion on more mystical or coercive strategies of government. Repression may foreclose subnational militancy, as has been the case, by and large, with respect to black separatist groups. Smaller, ethnically homogeneous states tend to have more spontaneous bases for constituting themselves as a political community. Therefore, there is less need to "impose" peace within national boundaries and less temptation on the part of outsiders to "intervene" to uphold the rights of the suppressed group. A natural definition of the political unit is more likely to result if the groups espousing subnational militancy achieve autonomy; of course, the more extreme subnational claims formulate their goals in terms of secession and separate statehood. As we have earlier noted, Leopold Kohr argues persuasively in *The Breakdown of Nations* that such fragmentation would produce a more moderate world-order system, less disposed toward violence and less endowed with capacities to wage protracted warfare backed up by huge population and industrial bases.

Smaller state units would tend also to lean toward a more cooperative set of economic and functional relationships. The interrelatedness of modern life requires larger, not smaller, units of policy-planning and execution; larger *function units* would, of course, be consistent with smaller *political units*. In conceiving of the reconciliation of the values of unity and diversity in an endangered planet, it is entirely reasonable to support both subnational militancy and supranational institution-building.

In this chapter we have tried to assemble the principal elements of world-order design that will be useful in the decade ahead. These elements concern both matters of *consciousness* and matters of *structure*. The design model has been given an overriding special task of adapting political organization on the world level to the urgencies of the ecological age.

These elements of design are thought of as building blocks more or less useful depending on how the future unfolds. The redesign of a world-order system is conceived of as a *process*, not an *event*. As such, the design will have to be constantly reinterpreted in light of the policies which themselves are subject to modification through further experience. The fixed point of analysis is a search for modes of human existence that assure the survival and welfare of mankind and other forms of natural life. By welfare is meant the conditions of healthy autonomy. For man this means food, housing, hygiene, education, beauty, as well as a measure of respect and rectitude. Our world ideals call for a vision based on *harmony within limits*, harmony among human groups and harmony between man and nature. Without some political approximation of both kinds of harmony we have no hope of avoiding much longer some awful day of reckoning. In the next chapter I shall examine some specific proposals that build toward world-order redesign.

VIII. World-Order Activism: First Steps

The tenor of earlier chapters has emphasized that a new system of world order is needed and is exceedingly difficult to bring about. Quiet persuasion, the appeal to reason and reasonableness, have little ability to çhallenge and transform structures of power. And we must make no mistake, adaptation to the ecological age presupposes new structures of power and the elimination or bypassing of old structures. The *first law of ecological politics* is that there exists an *inverse relationship* between *the interval of time* available for adaptive change and the *likelihood* and *intensity* of violence, trauma, and coercion accompanying the process of adaptation. Put in a less technical but no less precise way: the sooner the better, especially if the survival of democratic society is a positive goal to be secured. A study made available in February 1970 reports that atmospheric lead concentrations in San Diego have reached as high as 80 percent of the safety limit proposed by government health agencies, and that these concentrations have been increasing at a rate of 5 percent a year. Time is running out for even positive responses that are based on gradualism and largely voluntary compliance by the principal agents of harm, in this case the use of lead as a gasoline additive for automobile fuel.

We are in a position, then, where the precariousness of the human situation is becoming increasingly evident and where the period of time available for peaceable adjustment is being steadily shortened. Certain modest adjustments—arms control agreements, increased expenditures on population control, or prohibitions on particularly dangerous pollutants—may lengthen the period within which to evolve a new politics grounded in an appreciation of the endangered-planet situation, but there are many other events and processes at work that seem to hasten the end of human history by shortening the probable interval between the present and the ecological future. Underlying such a discussion is our uncertainty as to the exact character of risk or the point of virtual irreversibility. For instance, movements throughout the world to suppress dissent and to polarize principal national societies, to revive traditional values, and to reassert the drive toward competitive advantage over national rivals in the international system are some of the more discouraging features in the present situation. I have no idea as to exactly when to expect an ecological collapse, but I know it could happen even now, and that it is even more likely to happen with each passing hour of insufficient response.

In such circumstances, the need for action is great. How can we act to transform political power in a direction that accords more closely with our understanding of the needs of mankind and of the earth itself?

We who are convinced of this emergency setting are at present virtually powerless. We need, first, to spread the word of truth and then organize for its realization in the politics of principal societies. We need to begin as if learning to walk; walking commences with steps. A movement to reconstitute world order also needs to commence with steps, perhaps tentative ones, that can be retraced. Learning to walk involves persistence, many trials and errors, patience with failure, even some wisdom about disappointment. The virtues of persistence and fortitude are essential for the man who would promote by *action* the redesign of world order.

In this chapter we shall set forth a wide variety of specific proposals, projects for possible first steps. Within any large national or regional setting we need many steps taken from different points of departure, but moving in the same general direction. Within the world these first steps need to be taken in light of the concrete circumstances of time, place, and value as they exist in various places. To build a world-order movement capable of shaking the foundations of the present system, it is essential to depart from many shores, in many ships, with a variety of destinations, united despite their separateness by a concern for planetary survival. Despite these many distinct voyages it will be possible to identify a common purpose that will produce an increasing momentum and bring about unexpected confluences of these superficially separate voyages.

In setting forth some first steps an effort is made to parallel the earlier organization of the book. Hence, attention is given first to proposals that would heighten awareness of the overall planetary crisis, its unity and totality, and, then, to the fourfold threat that cumulatively produces the endangered-planet situation—namely, the war system, population pressure, resource depletion, and environmental overload. This chapter is a sequel to Chapter VII, in which the positive tendencies present in world political behavior were discussed as the basis for bringing a new system of world order into being by peaceful means. Political arenas and contexts—such as regionalist or supranationalist tendencies—were looked upon as containing positive potentialities. Here in Chapter VIII some specific areas of political action are proposed to heighten the prospects for political transformation of a desirable and necessary variety.

1. THE OVERALL FOCUS ON THE ENDANGERED PLANET: SOME FIRST STEPS.   My purpose in this section is to excite and arouse people to commit their careers, their resources, their energies and, if necessary, their well-being to solving the problems of an endangered planet. These proposals, then, are hortatory in intention, but seek to put concern with survival in its correct overall context.

Declarations of Ecological Emergency. National governments and other principal bureaucracies become receptive to adaptive change when their existence is threatened in fundamental respects. Periods of war, civil strife, or economic depression are the most notable examples of governments' acting in the spirit of emergency consciousness. Responses to disasters—floods, earthquakes, and hurricanes—also indicate that governments can act quickly and responsively under the pressure of a specific occurrence that jeopardizes the total or partial well-being of the community. The ecological crisis, aside from sporadic outbreaks, has had a cumulative, quiet, barely noticeable character to date. We do not even know whether thresholds of virtual irreversibility have already been crossed. This quietness needs to be challenged by strident displays of the objective circumstances. One such display might involve persuading national governments to declare with fanfare and prominence the existence of a National State of Ecological Emergency, and to undertake annual reports on efforts to meet the threats. The United States, as the most affluent and world-pervasive state, could exert great leadership by being the first country to make such a declaration. It would then become natural to discuss ecological issues as affairs of state and to shift the concerns of government toward the agenda of the endangered planet. President Nixon emphasized environmental issues throughout his 1970 State of the Union Address, but barely touched upon these issues in his far lengthier State of the World Message. At present, the flurry of interest in questions of pollution is not altogether reassuring. The prevalent call for response seems to believe that environmental challenges can be handled within the existing framework of government, that the overriding locus of these challenges is domestic, and that the market system powered by GNP-mindedness and depending for "stability" on perpetual growth can incorporate the costs of environmental protection. This represents a very misleading interpretation of the circumstances of planetary danger.

Parallel efforts to alert world consciousness to the full extent of the danger should be made in international institutions, especially in the political organs of the United Nations, the General Assembly, and the Security Council. It is most important to introduce the ecological perspective into the center of world-order activity and thought. Resolutions by the organs of the UN could work toward building a world consensus on the nature of the endangered planet and on the proper character of response to it.

The effort to solicit such declarations of acknowledgment would serve as a valuable educational instrument, since the challenges of persuasion and response would encourage both advocates and opponents to mobilize the evidence, and the debate would lend prominence to the entire issue, as well as sharpen perceptions as to the real character and complexity of the situation.

The International Law Commission (ILC) might also be asked to prepare a Draft Declaration of Ecological Rights that would serve to complement the Draft Declaration of Human Rights agreed upon in 1948 within the

United Nations. Such a document might represent an initial effort to formulate a set of goals that would command widespread assent among the peoples of the world and could express clearly the need for the coordination of human behavior to maintain the quality of life on earth. Such a Draft Declaration could itself become an educational instrument for the promotion of awareness, as well as provide a preliminary test of whether it will prove possible even to *formulate,* much less *implement,* a common set of ecological policies. Of course, the more specific the provisions become, the more difficult the negotiation of acceptable standards will become, especially on matters for which different priorities exist among states. It would also be desirable to expand upon the suggestion of Professor Arthur Galston, a biologist at Yale, and urge the ILC to prepare a Draft Convention on Ecocide to complement the Genocide Convention. This Draft Convention could then serve as the basis of a world conference of governmental representatives convened to agree upon the text of a treaty making it an international crime to commit ecocide.

The Establishment of Survival Universities or Colleges of World Ecology. At all levels of our educational process we need to introduce the ecological way of thinking, feeling, and acting. One way to begin might be to establish one or more universities devoted to the training of specialists in problems of world ecology. The curriculum might follow closely the principal concerns of the early chapters in this book, the faculty could be of rotating multinational composition, and the students drawn from everywhere. Humanities and the arts could be related to exploring the character of human survival. An ecological theater could flourish in which new kinds of social conflicts were portrayed. Henrik Ibsen has already demonstrated how effectively issues of social and political conscience can be dramatized, and recent experiments in guerrilla- and street-theater show the possibilities for enacting in theatrical settings ecologically oriented parables of the endangered-planet situation. I could happily imagine minstrel theater groups touring the land in association with national church organizations to bring people a closer awareness of their dependence on nature and the precariousness of the human situation at the present time.

It has also been proposed by Professor Theodore Brameld of Boston University that there be established "Experimental Centers for the Creation of World Civilization." These centers would be concerned with action, as well as contemplation, and would study and take part in efforts to foster the growth of a world civilization as part of the reorientation of consciousness that must precede a new system of world order. Professor Brameld's ideas are largely conceived of as a response to the urgencies of the Nuclear Age; it is, then, a conception that in terms of this book is pre-ecological in its definition of the problem of world order. Nevertheless, his sense that man must discover ways to realize and give meaning to the unity of mankind, and thereby avoid World War III, applies even more directly to the dangers of ecological collapse. Such centers need also to concern themselves with the precarious condition of man's relations with nature, which in essence involves a true appreciation of man's dependence upon the intricate network of ecosystems that together support life on earth and an understanding of what it means to exist within a limited domain, a tiny island enclave in the vastness of the universe.

An associated educational idea already underway is a Peacekeepers Academy that would train students in the arts of peace, preventing and moderating group violence, the causes of conflict, nonviolent alternatives to war, peacekeeping, mediation and arbitration, and so on. The Academy has an international faculty and has held its initial session in Vienna, selected because it is considered a politically neutral site; it has obtained students from many countries, and hopes to develop attitudes associated with world citizenship. Although the focus has been on war and peace—a Nuclear Age project—its scope and intention could be easily expanded to fit within the consciousness of the ecological age.

Radical students are increasingly bringing environmental issues into the inner orbit of their concerns. A survival fair was held at San Jose State College in February 1970, the principal event being the burial alive (i.e., with engine running) of a brand-new car. Local residents protested against the act and accused the organizers of the fair of trying to destroy the capitalist system; in other words, they got the point.

On a more prosaic level, it would seem helpful to organize debates and discussions in town meetings and schools that look beneath the surface of environmental problems. We need to debate, in particular, such questions as the following:

—Can the capitalist system and the market economy be adapted to the needs of the ecological age?

—Can a national government succeed with policies on environmental issues without securing the cooperation of other principal governments?

—Can the demands of social justice be fulfilled if we substitute "steady state" for "growth" as the primary test of economic well-being?

—What is the optimal population for this town, this city, this section, this country, this region, and this world, and how can it be achieved within a decade?

—Is it possible to respond to the demands of the ecological age without destroying the foundations of political liberty and the spirit of democratic society?

Conventional research efforts can also be redirected in a variety of ways to be responsive to the present situation of danger. A publication of "radical" demographers (reacting to the traditional distancing and aloofness of population specialists from social issues) came up with

new research suggestions that appeal to me because they
cut through the niceties of professional jargon:

1. Participant observation studies of the board of
directors and top officers of major industrial corporations such as General Motors. Individual re-
searchers would be assigned to individual company
officials to follow them through their activities for
several months. How do these companies combat
attempts to regulate their pollution? How do companies decide which rivers to pollute and where to
build their smokestacks?

2. Intensive interviews with state legislatures and
the national Congress on why they are raising their
salaries but cutting back on welfare payments. Find
out how public officials feel that lower incomes will
help the poor raise themselves out of their current
status.

3. Surveys of French military leaders in the 1950's
and American military leaders in the 1960's about
their successful population control program in
North Vietnam. Find out what techniques were used
so that North Vietnam has the lowest birthrate in
Asia.

Probably these research suggestions will strike many
readers as strident and embodying prefixed conclusions.
Perhaps, but the need to get at the roots of the crisis
makes it vital for us to be receptive to novel ideas that
would not prevail in more normal times.

Shifting educational moods and arenas, it would be
desirable to develop on a crash basis a new program of
studies for children at all ages—a program oriented
toward the affirmation of earth values. Images should
emphasize man's kinship with and dependence upon
nature. The whale might become a basic symbol of survival studies. Man the hunter and the destroyer imperils
his own species as well as every other form of life on

earth. Earlier lost ideas about the music of the spheres, the Edenic vision of men and animals in harmonious play, and Eastern visions of man as a participant in life processes, not as an exploiter or master of nature, could be introduced at all stages of human development. Notions of play could be informed by earth values—peace games could become part of a child's earliest learning experience. Increasingly, we know that the first years of life largely determine subsequent attitudes toward life. Although it is late in the day to begin with young children, the effort to create a whole new climate should be initiated at all possible levels.

World Political Party. Another initiative would involve an effort to organize a world political party dedicated to the redesign of world order and guided in its endeavors by some version of an endangered-planet analysis. The creation of branch parties in different parts of the world would help also to shape a consensus as to the priorities and goals of world-order change, as well as mobilize nongovernmental sectors of society. To some extent, at least, government officials can be reached through the United Nations, but the world political party will engage the attention of the ordinary world citizen. During early stages of its existence, at least, membership in the world political party would be fully compatible with membership in national political parties. Also in the early stages a large measure of ideological decentralization would be accorded party branches and there would be no headquarters and no very specific schedule of goals or universal platform. To assure some degree of coherence, a common statement of goals would have to be agreed upon by a world board of party founders. But these goals can involve only the commitment to organize within national societies a political party branch dedicated to the elimination of the war system, tne achievement of a minimum quality of life for all people, the conservation of resources, and the protection of the environment against abuse and collapse.

In some societies, a world political affiliation is looked upon as a betrayal, and can result in pressure and even punishment for involved individuals. The Soviet Union and states closely associated with it have often described even professional associations dedicated to world government or a new system of world order as guilty of an ideological heresy known as "mondialism," that is, as taking a position that supports the possibility of a stable and just system of world order on some basis other than the triumph of socialism in all national societies. A world political party could not presently, at any rate, hope to operate on a formal basis anywhere within the Sino-Soviet sphere of existence. At the same time, the participation of representatives from these countries, even if only in an informal consultative capacity, is essential, to inhibit the impression that the world political party is a new means by which to subordinate one part of the world to another.

There is also a need for a variety of local forms of party operation, depending on the political setting, the numerical and financial strength of the party, and its capacity to bargain for influence within the councils of conventional political parties. In liberal democratic societies the closeness of elections may induce major concessions to the viewpoint of the world political party in exchange for its endorsement or its agreement not to run a candidate of its own. To be effective, a world political party initially would have to attract enough loyal followers to establish itself as a serious political presence to be reckoned with; in essence, this presence can be secured if membership and financing proceed well. It would be important to have a very effective leader as the head of the party branch in each society who would command respect and attention throughout the society as a whole and be in a position to arouse enthusiasm, confidence, and loyalty among followers.

Again, the effort to initiate a world political party, if undertaken seriously, would contribute to the education of publics throughout the world and make many more people responsive to the ecological imperative. It is not now too soon to bring people together who would be interested in such a project, although it is important to prevent the world political party from acquiring the pale cast of the typical peace group activity. Perhaps, a close tie with individuals concerned on a transnational basis with issues of racial and economic justice, with environment defense, and with political change might assure a concreteness of commitment that is an essential ingredient of political vitality. World federalist projects, by the very abstractness and remoteness of the objectives, have tended to attract apolitical personalities.

Warren Wagar, sensitive to these issues, has warned that "To suppose that educators or publicists or businessmen can, through an apolitical process of propaganda and enlightenment, achieve so thoroughly political a goal as world federation is quite simpleminded." Wagar asks us to "remember that we are speaking of a revolutionary transformation of the structure of world politics. . . . Establishing a world political party is only very remotely analogous to establishing, for example, a party for the prohibition of alcoholic beverages in the United States, or a political organization such as the Christian Democratic International." A world political party will have to work for the *transformation of beliefs as well as the redistribution of political power* within the present world political structure. It is not enough to work for a change in the *policy* of the state (or states) or even in the *authority pattern* (replacing the existing elite by a new one). It will be necessary to develop, whether by conscious design or uncoordinated evolution, far greater centralization of control over the world ecosystem than now exists. As I have argued in Chapter VII, such centralization does not imply a buildup of a tight system of world government. The central guidance mechanism may be a composite of many elements, but it must have the overall ability to evolve and maintain a dominant structure of planetary cooperation to take the place of existing international structures whose origins and functions concentrate upon the competitive drives of sovereign states. As the ecological crisis widens in scope and deepens in impact, there is likely to be great political turmoil and confusion in many societies, and very possibly a bloody series of confrontations between the forces of change and the forces of reaction over world-order issues.

Wagar's Ark of Renewal. Taking full account of the obstacles to world-order redesign and of the prospects for ecological collapse and for Armageddon, Warren Wagar has proposed the formation of a semi-conspiratorial worldwide group with its "headquarters and most of its membership in the least ambitious and most geographically isolated countries of the world." Wagar's idea is to plan for survival of a nucleus of individuals who are dedicated to the establishment of world government when the opportune moment of history presents itself: "There is no more opportune moment for radical change than in the aftermath of a world catastrophe. It could well be mankind's literally last chance." According to Wagar, "at war's end" this survival colony "would emerge as a conspiratorial task force dedicated to persuading the other survivors throughout the world to form an indissoluble world union as man's last hope of preventing complete extinction or reversion to savagery."

In his later formulation, Professor Wagar gives a more specific character to his proposal: "The colony would consist of perhaps one thousand persons, including technicians in all fields and specialists in public relations and management. It would have to possess a considerable fleet of light aircraft and a microfilm library of the world's books and works of art and technology: it would be an ark of civilization. Building and sustaining such a colony and keeping its personnel and facilities up to date might involve an initial investment of $100 million and yearly expenditures of $20 million thereafter—certainly no less." Perhaps we need to go about the work of establishing such a colony, or several of them, and thereby build up an underground cadre of revolutionaries committed to a radical change in world-order design. Wagar's emphasis is placed explicitly, and perhaps somewhat prematurely, upon the desirability of and necessity for a world government presiding over the unified "city of man." As indicated in the previous chapter, I am less convinced than Wagar that the only type of adaptive world-order change involves the political and cultural unification of mankind in a single cosmopolis. Such unification is consistent with the image of a limited and enclosed earth, of wholeness, of man as part of a family, but it also may, in certain circumstances, stimulate the rise of a new center of demoniac power with the potentiality for tyrannizing and exploiting mankind. Those who would enter Wagar's ark must themselves form and reform the goals of their enterprise with every twist and turn of world history.

Wagar is among the first writers on world-order problems to demonstrate an awareness of the mounting ecological crisis, but the main tenet of his proposal may be too closely tied to earlier thinking about world government as a war-prevention system. I wonder whether there is any hope of surviving ecological collapses by locating in any place remote from industrial societies. Do ecological sanctuaries exist on earth? Is there a place to hide? It is no longer clear that it will be possible, even if one guesses correctly about where to wait out the catastrophe, that there will be any escape from a major nuclear holocaust. The earth is enclosed, as well as finite, and as a whole dependent upon certain conditions of harmony among its parts. One image that has been used recently to summarize our plight is to think of the earth as an automobile parked in a closed garage with its engine running.

The conspiracy needs to embark upon its mission on the basis of the departure point: the endangered-planet situation—and chart its course toward some kind of destination that involves the harmony of man and nature as well as man with himself and with other men. The colony needs to be set up at once and to operate initially as a map room that keeps recharting the voyage of the world-order ark in light of new discoveries and the quickly changing, totally unfamiliar setting. The ark can be better thought of, perhaps, as a center of awareness than as a place of refuge or a sanctuary during the period of catastrophe.

But Wagar's type of proposal is thought-provoking in the best sense, and here again the effort to carry it out and understand its seriousness may well induce a deeper awareness of the real situation of danger that all of us are part of, including especially our children born at this strange and possibly decisive moment in world history.

2. COUNTERING THE WAR SYSTEM: SOME FIRST STEPS. In this section we shall set forth a variety of current proposals that are designed to moderate the dangers associated with the war system, especially to reduce the magnitude and pervasiveness of war in the course of human affairs. No claim is made to present all, or even most, of the proposals worthy of consideration. Instead, this section gives a sampling of constructive thinking on "first steps." First steps are designed to lead to next steps, to a series of continuations that build momentum, cut resistance, and finally attain the acceleration needed for "takeoff."

No First Use of Nuclear Weapons. It would seem desirable to induce all governments to commit themselves not to initiate the use of nuclear weapons against an adversary. Nuclear weapons could still be stockpiled to the limits of national policy for potential use against any adversary who violates such a nuclear agreement.

The immense and indiscriminate destructive power of nuclear weapons, their inevitable obliteration of any distinction between military and nonmilitary targets, their extraordinary capacity to inflict enormous and possibly irreversible damage upon civilian populations, their uncertain genetic impact upon future generations, and their highly disruptive effects on the ecosystem, are among the many reasons favoring the prohibition of these weapons. It can be safely said that no political objective can vindicate the initiating use of nuclear weapons and that the insistence of a government to reserve the discretion to use nuclear weapons wounds further the tender fabric of mankind.

The General Assembly of the United Nations, in Resolution 1653 (XVI), passed in 1961 by more than two-thirds of its membership (but over the opposition of the United States and the United Kingdom), notes that "the use of weapons such as nuclear and thermo-nuclear weapons is a direct negation of the high ideals and objectives which the United Nations has been established to achieve . . . ," and declares that:

"(a) The use of nuclear and thermo-nuclear weapons is contrary to the spirit, letter and aims of the United Nations and, as such, a direct violation of the Charter of the United Nations;

"(b) The use of nuclear and thermo-nuclear weapons would exceed even the scope of war and cause indiscriminate suffering and destruction to mankind and civilization, and, as such, is contrary to the rules of international law and the laws of humanity;

"(c) The use of nuclear and thermo-nuclear weapons is a war directed not against an enemy or enemies alone but also against mankind in general, since the peoples of the world not involved in such a war will be subjected to all the evils generated by the use of such weapons;

"(d) Any State using nuclear and thermo-nuclear weapons is to be considered as violating the Charter of the

United Nations, as acting contrary to the laws of humanity and as committing a crime against mankind and civilization . . . ."

The nuclear weapons now in the arsenals of major powers possess 50 to 1,000 times the explosive power of the atomic bombs used in World War II against Hiroshima and Nagasaki. Five survivors of Hiroshima brought suit against the Japanese Government for its waiver in the peace treaty of the damage claims of its nationals against the United States. The decision in the case of *Shimoda and others v. Japan* is an important document of the nuclear age since it examines in detail the legal status of nuclear weapons in the context of the attacks on Hiroshima and Nagasaki. The Tokyo District Court, with a notable touch of irony, handed down its decision on December 7, 1963, deciding that these attacks did indeed violate international law, although the complaining parties, for reasons of procedural capacity, could not recover damages. The entire proceeding, which examines facts and law with scrupulous fairness, is heightened by the drama of having the Japanese Government in the peculiar position of defending the legality of American attacks against its own cities. We have already had occasion to refer to the lingering effects of these atomic attacks in 1945 on the political consciousness of Japanese society. As recently as November 1969, the Japanese Premier, Eisaku Sato, while on a state visit to negotiate the return of Okinawa to Japanese sovereignty, was reported to be very cognizant of the strong persistence of the shadow cast by the 1945 attacks: "Mr. Sato was said to have referred to the continuing but diminishing 'nuclear allergy' that Japanese have to all things nuclear arising from memories of the atomic bombing of Hiroshima and Nagasaki." Of course, to characterize this consciousness as a "nuclear allergy" is to express a very conservative perception of this abiding concern, but it does serve to underscore the unique horror associated with these weapons, even compared to such other horrors of modern warfare as the obliteration bombings of such cities as Dresden, Rotterdam, and Tokyo during World War II by so-called conventional weaponry. It is their special capacity to destroy, their potential danger to later generations, and their scarring effect on the imagination that establishes the case against tolerating the legitimacy of nuclear weapons as instruments of national policy. In such a circumstance, to "ban the bomb" is to struggle against the gathering forces of doom and, in a positive sense, to work for the beneficial survival of the planet.

The Soviet Union and China have frequently proclaimed their own intention never to use nuclear weapons first. In contrast, the United States and its principal NATO allies have indicated that the defense of Western Europe, at least, rests on the credibility of a nuclear response to a potential non-nuclear provocation. Such a reliance on nuclear weapons for the defense of Europe is not essential, given the balance of capabilities in that part of the world. And furthermore, there does not appear to be any significant intention on the part of the Soviet Union to promote a westward expansion of her sphere of influence by either direct or indirect military pressure. In these circumstances, and because of the overwhelming importance of discouraging use or acquisition of nuclear weapons, it would be a momentous event in human history if the United States, acting alone or in conjunction with other nuclear states, would issue a Declaration of No First Use of Nuclear Weapons (or possibly join with the Soviet Union in promoting a treaty prohibiting the threat or use of nuclear weapons).

Such a declaration, if appropriately explained and coordinated with other security policies, would also establish a certain kind of *reciprocity* between nuclear and non-nuclear states with respect to nonproliferation of nuclear weapons. Such reciprocity, in fact and in spirit, is an essential precondition for further disarmament progress. It discourages the impression that the super-powers are managing international arms control negotiations to perpetuate their domination of world society. A no-first-use principle would in many ways help the overall effort to raise the nuclear threshold above the reach of international conflict.

The process of persuasion needed to gain political support for a Declaration of No First Use of Nuclear Weapons would itself serve valuable educational purposes. The justification of the proposal would include plans to meet security requirements by non-nuclear means. Such a declaration, even if universally adhered to, would not be a guaranty that nuclear weapons would never be used first; no simple prohibition can ever expect to secure absolute compliance under all circumstances. However, such a declaration, combined with some adjustments in defense planning that would probably accompany an initiative of this type, would inhibit reliance on nuclear blackmail and threats, and would probably lead to the removal of nuclear weapons from battlefield deployments. The prohibition on initial use might also swing the balance against a governmental leader who contemplated starting nuclear war during a period of crisis. We have no reason to think that the further withdrawal of legitimacy from nuclear weapons would encourage governments to rely upon subnuclear means of military expansion.

Declaratory steps require no inspection, no action, no complicated process of international bargaining and negotiation. A declaratory step is simple, a matter of words expressing a fairly clear intention. At the same time, such a Declaration of No First Use of Nuclear Weapons may have a profound effect upon world political consciousness by altering attitudes toward weapons of mass destruction. In any event, a declaration would give a certain legal form to a preexisting moral consensus with respect to nuclear weapons. Building on such a consensus would also help create some sense of human solidarity, which is itself a vital aspect of the struggle against alienation in the ecological age.

On November 25, 1969, President Nixon made an official declaration of United States policy on biological and chemical weapons that proceeded in a desirable direction. With respect to biological weapons, the United States issued a Declaration of Renunciation (which includes even a repudiation of any right of retaliatory use). Mr. Nixon's declaration relied upon the following formulation:

> The U.S. shall renounce the use of lethal biological agents and weapons, and all other methods of biological warfare.

> The U.S. will confine its biological research to defensive measures such as immunization and safety measures.

> The D.O.D. [Department of Defense] has been asked to make recommendations as to the disposal of existing stocks of bacteriological weapons.

In the same statement President Nixon reaffirmed on behalf of the United States "its oft-respected renunciation of the first use of chemical weapons" and extended on the occasion "this renunciation to the first use of incapacitating chemicals." Such recourse to declaratory arms control measures by unilateral initiative serves as a direct precedent for the kind of action that seems so desirable in relation to nuclear weapons. From every human perspective, ranging from prudence to decency, the case for repudiating the first use of nuclear weapons is at least as strong as it is for biological and chemical weapons. Certainly the characteristic effects on man, society, and nature are at least as malignant if nuclear weapons were to be used against men or earth. Since

these weapons are being stockpiled and constantly re-
fined by all principal states, it is of particular importance
to achieve the moral strength and political wisdom needed
to renounce the right to use or to threaten to use them.

Strategic Arms Control. The most characteristic first-
step gestures involve limitations on the quality and quantity of strategic armaments, negotiated by governments
and formalized in an international treaty. The treaty
method has been used to exhibit agreement on ways to
impede the arms race (for example, the limited test ban,
separate treaties on the demilitarization of outer space
and Antarctica) and to make the existing arms setting
less prone to *unintentional* collapse (for example, an
agreement to establish a "hot-line" or telecommunications link between the White House and the Kremlin).
The draft treaty on nonproliferation of nuclear weapons
combines both features—attempting to prevent a nuclear
arms race among the principal non-nuclear states in the
world and seeking to avoid the added danger arising from
an increasing number of independent centers of governmental authority having access to nuclear weapons. Arms
control aims to stabilize the existing international order,
to moderate tendencies toward destabilization, and to
enhance tendencies toward stabilization. In this sense,
arms control is *conservative,* as its principal objective
is to *conserve* the existing system rather than to *change*
it into something else. Disarmament is more radical,
since it contemplates a shift in the distribution of power
and authority within the world system, and in most forms
entails the creation of new actors in world affairs that
would have to be beyond the reach of even the most
powerful national governments. Again, it may make the
point concrete to think of the relationship between the
state of New York or California and the federal government in the United States. Except for police and riot
control functions these units have yielded control over
the military instrument to the central government. Never-

theless, the character of the federal compact depends on voluntary patterns of allegiance under most circumstances. There is little disposition to impose central policy upon subordinate units except with respect to matters of vital principle, and even then the overriding tendency is to secure voluntary compliance, with enforcement an option of last resort.

Arms control measures are not normally taken as steps toward a new world order, although their proponents may contend that they are. Support for gradualism rests on the notion that the war system can be dismantled (or at least tamed) piece by piece. The evidence for this possibility is not at all convincing. Arms control measures are usually justified to save money, to avoid provocative nuclear postures, to increase stability during periods of world crisis, to maintain existing patterns of military preeminence, and to avoid deterioration of health and environment (as by testing in the atmosphere).

But the shift from a state-oriented world system to a community-oriented world system will not result from the adoption of a series of arms control measures. Besides, national officials act basically to preserve and enhance, not to replace and diminish, the role of the national bureaucracies that they serve. In addition, bureaucratic politics, including interagency and interservice (army, navy, air force) rivalries, set sharp limits on what a major government can agree to do at any given moment.

Therefore, if adaptive change on a global level is the objective, then an arms control approach can come to serve as a *substitute for,* rather than a partial *realization of,* that goal.

If arms control measures, however, have the effect of lengthening the probable interval before the occurrence of a nuclear catastrophe, then they tend to support the early tasks of transformation. Opportunities are created to initiate more basic processes of change. Each measure needs to be assessed on its own and in relation to these more fundamental objectives. In most situations, the prospect of reducing the likelihood of nuclear war or deferring its probable occurrence is a sufficient intrinsic good to warrant support for the measure even aside from its contribution to the struggle to redesign world order. The rationale for such support seems especially strong in light of how little we know about how to redesign the present system of world order except that we need as much time as possible. The point here, then, is a double one: first, to be clear that limiting the further expansion of the war system or preventing its catastrophic collapse is not to be confused with bringing about its demise; second, to appreciate that gaining time is very helpful for those who are working to build a new world order. It is true that if arms control measures provide reassurance about the adequacy of the present system, then it may work against a full *acceptance* of the endangered-planet argument; acceptance implies appropriate action, a sense of urgency, and an understanding of the scope of required change. People interviewed throughout the United States in 1963 after the signature in Moscow of the Limited Test Ban Treaty seemed often to believe that the treaty had virtually eliminated the danger of nuclear war; in fact, of course, the test ban had no bearing at all on the will or capacity to use nuclear weapons and did not even involve any commitment to bring the arms race to a halt. Underground weapons testing continued to be permitted, and has facilitated the rapid evolution of new, more lethal and accurate nuclear-tipped missiles, as well as such entirely new weapons systems as the ABM and MIRV. Such naive public attitudes, although perhaps not typical or long-lasting, help explain why so few people in a community react to *danger* until it is impressed upon them by a *catastrophe*. Even educated people in positions of responsibility tend to resist evidence of a danger that calls for major revisions of structure and beliefs. Arms control is such an ambivalent approach to world-order redesign because it offers the kind of palliative that allows the most alarming evidence of mounting danger to be brushed aside by those who are eager to retain the present military-security setup.

An International Peace Force. Most proposals for reforming world order have been built around the twin pillars of drastic disarmament at the national level and a world police force at the world level. For instance, in the proposals initially submitted by the United States Government to the Eighteen-Nation Disarmament Conference at Geneva in 1963, this combination of national disarmament and world police was at the core of the plan. The U.S. proposals set forth a series of objectives including the following one: "To ensure that during and after implementation of general and complete disarmament, states also would support and provide agreed manpower for a United Nations Peace Force to be equipped with agreed types of armaments necessary to ensure that the United Nations can effectively deter or suppress any threat or use of arms." In the final stage of the disarmament process this UN Peace Force would have been given "sufficient armed forces and armaments so that no state could challenge it." The United States and the Soviet Union jointly affirmed these objectives in the seventh principle of the so-called "McCloy-Zorin Agreement," officially known as the Joint Statement of Agreed Principles on Disarmament Negotiations, dated September 20, 1961. The Soviet draft treaty on general and complete disarmament, put forward at Geneva in 1962, is ambiguous and vague about the nature and capability of a world police force, and seems to envision nothing more reliable than earmarked national military contingents available to carry out UN peacekeeping operations as authorized by the Security Council.

The Clark-Sohn plan for limited world government is the most carefully elaborated conception of the coordination between disarmament and world police concepts.

In the Clark-Sohn plan, which relies heavily on a very intricate system of checks and balances, a world police force would have a *standing military* complement of between 200,000 and 400,000 and an additional reserve component of between 300,000 and 600,000. The effort to assure that the UN would enjoy a preponderance of force in a disarmed world also led Clark and Sohn to provide this UN force with access to nuclear weapons under certain exceptional circumstances of actual or imminent use of nuclear weapons by a breaker of the peace. Such access by the world police would require a prior declaration by the main political organ of their re-vamped United Nations that "nothing less than the use of a nuclear weapon or weapons by the Peace Force will suffice to prevent or suppress a serious breach of the peace or a violent and serious defiance . . . of the United Nations." In the Clark-Sohn plan, nuclear weapons are placed under the custody of a UN Nuclear Energy Authority and can be transferred to the Peace Force only in accordance with procedures designed to minimize the prospects of unauthorized threat or use. In this entire discussion of world police concepts, there is a subtle and delicate balance between assuring that the world community evolves an effective capacity to maintain peace in a disarming and a disarmed world and safeguarding against the emergence of some central tyranny that could dominate and enslave the whole of mankind.

In the situation at the start of the 1970's, there is no longer serious discussion of drastic or general disarmament. Such a goal is viewed by men of influence as totally unrealistic. Instead, most efforts are being directed toward reducing the rate of arms competition among principal states and in moderating the spread of high-cost arms competition to the poorer countries of Asia, Africa, and Latin America. The idea of a peace force to accompany disarmament has no real relevance to the present scene, certainly not as a first step toward a new system of world order, although such an idea will regain importance if a few other disarmament steps are taken. One such step that is a modest precursor to a world police system would involve the establishment of a UN Peace Force of between 15,000 and 50,000 men trained and equipped to deal with small-scale threats to and breaches of international peace under current world conditions. At the present time, the UN has to improvise a peace force on each occasion in which it is called upon to act. The delay in response, even assuming the presence of the political will on the part of principal members, often makes it difficult or hazardous to carry out the UN task effectively. Often, too, disagreement among big governments is exhibited by contradictory expectations as to what is the appropriate UN role in a particular situation. The Congo Operation between 1960 and 1964, although undertaken by the UN in response to an appeal by the Congolese Government, and in an atmosphere of agreement on the part of Security Council Members, illustrated the vulnerability of such an undertaking to the cross-pressures of world political conflict. The experience shook the confidence of the organization in its capacity to maintain a consensus of its members or to impose its own autonomous views upon the organization. As already discussed in Chapter V, the post-Congo UN has appeared timid and intimidated, unable to mount any response to such major and bloody disruptions of world peace as the Indonesian upheaval of 1965, the Nigerian Civil War that ended early in 1970 with the collapse of Biafra, and the continuing destruction of Vietnam by the use of advanced American airpower year after year.

Some of these difficulties might be mitigated by the creation of a small UN peace force, sustained by a regular budget and staffed by states who were *not* Permanent Members of the Security Council. A trained, available peacekeeping force could act on short notice, with personnel trained for the special requirements of small-scale, quick-reaction missions. Such a police force might reduce the incidence and duration of both internal and international warfare. If reasonably successful, a somewhat reduced prominence of the war system in the overall scheme of things might result, as well as the elimination or containment of some international flash points that impede agreements among rival governments. No direct attempt would be made to curtail the size and activity of national military establishments, but the functioning of such a UN peace force might *illustrate* the economic and political advantages of relying for national security more on the activities of international institutions. This development could help build up a climate of opinion inclined to dismantle the national security state and to transform the war system, especially given the rising pressure of domestic priorities and of demands not to waste resources on obsolescent military equipment.

The idea of a peace force, then, is to take a first step toward building up capacities, experience, and attitudes that would favor strengthening what I called in Chapter V the Charter conception of international relations (collective security) at the expense of the Westphalia conception (national security). The creation of a small UN peace force seems to be a definite improvement on the prospects for peacekeeping in the world today. Because of the political conflicts going on within the UN, it is often impossible to mobilize any response to the outbreak of warfare. Sometimes this failure to act creates a pretext for alleged "peacekeeping" by the superpowers: the U.S. in Lebanon (1958), the Dominican Republic (1965); the U.S., United Kingdom, and Belgium in the Stanleyville Operation (1964); the U.S.S.R. in Hungary (1956), Czechoslovakia (1968). It is a hopeful sign that Charles W. Yost, the U.S. Ambassador at the UN, observed in early 1970: "Even those great powers who have been most skeptical may at last be coming to recognize that multilateral peacekeeping by an international organization is infinitely less dangerous and less costly, and in the long run more effective, than unilateral intervention."

It is, of course, true that such an initiative may encounter severe difficulties under present world conditions. As has been stressed previously, as long as China remains outside the organization it is unconvincing for the UN to claim universal authority. Furthermore, so long as governments such as South Africa, Portugal, and Israel pursue policies that are at odds with the preferences of the overwhelming majority of the membership, the UN is itself in an adversary relationship to certain sovereign states, and these states would certainly resist UN claims to exert authority over their affairs. The UN is, in effect, allied with one part of the community in a conflict with another lesser part. Such a role of partnership is common enough for national governments, but it is likely to impose a considerable strain on the UN, since so many governments among its membership, including especially the Soviet group, remain extremely sovereignty-oriented and, hence, are wary of community claims that might be directed at a future time against their own patterns of political control. It is the fear that peacekeeping precedents established against pariah states might be turned against other states on future occasions that probably explains, as much as anything, the failure of the UN to build up through all of these turmoils any standing peace or police force, or to make use of the machinery for peacekeeping set up by the Charter in Articles 43–47. In fact, the Charter framework as it now exists could be activated to support the creation of a standby international peace force of almost any size. Article 43 is particularly central to this unused potential of the organization:

Article 43

1. All Members of the United Nations, in order to contribute to the maintenance of international peace and security, undertake to make available to the Security Council, on its call and in accordance

with a special arrangement or agreements, armed forces, assistance, and facilities, including rights of passage, necessary for the purpose of maintaining international peace and security.

2. Such agreement or agreements shall govern the numbers and types of forces, their degree of readiness and general location, and the nature of the facilities and assistance to be provided.

3. The agreement or agreements shall be negotiated as soon as possible on the initiative of the Security Council. They shall be concluded between the Security Council and Members or between the Security Council and groups of Members and shall be subject to ratification by the signatory states in accordance with their respective constitutional processes.

These arrangements were supposed to be coordinated by a Military Staff Committee (Articles 46 and 47) consisting of the Chiefs of Staff of the Permanent Members of the Security Council (or their representatives) and the Security Council. This quite elaborate potential apparatus has lain dormant since the creation of the UN in 1945. Why? The split among the Permanent Members of the Security Council is the most evident answer, but it is far from the only answer. Indeed, the fundamental explanation would seem to be that the Charter mood did not take sufficient account of the bonds of affiliation that reinforce the ideals and practices of national sovereignty. The Charter itself, then, was in 1945, and still remains, a somewhat utopian instrument, given prevailing patterns of world-order beliefs. The failure to carry out the Charter intention to create a peacekeeping force by voluntary arrangements among UN members illustrates how difficult it will be to bring about the kind of changes in world order needed if welfare and survival are to be upheld in the ecological age.

Most discussions of peace and police forces have presupposed a war/peace setting, but it is quite possible that in the future the tasks of a world police mission would involve the regulation of behavior deemed detrimental to the environment, such as large-scale use of hard pesticides (DDT family) to increase agricultural yield, in continued defiance of community prohibitions. Or one could imagine a call for UN action to prevent nuclear weapons tests or to assure compliance with safety conditions in an offshore oil-drilling operation. The possible assignments that might be given to a world peace force should be reconceived in light of the entire agenda of an endangered planet; world-order thinking can no longer be confined to the problem of war within and between states.

Demilitarizing the Large Sovereign State. The most difficult and significant first step to end the war system would involve various possible approaches to the demilitarization of large states. These states, especially the nuclear superpowers, have built up huge stockpiles of weapons and militarily oriented bureaucracies, and have spawned large supporting interest groups spread throughout the society. Major weapons systems are funded so that virtually every Congressional district in the United States has a financial stake in going forward with the development. Richard Barnet has succinctly depicted the political tie-in by quoting from Carl Vinson who, as Chairman of the House Armed Services Committee, was presenting a billion-dollar military construction bill to the House of Representatives: "My friends, there is something in this bill for every member." Barnet goes on to say: "Defense contracts are currently distributed among 363 Congressional districts. Congressmen have the incentive to vote right on military appropriations. Many are cultivated with junkets, testimonials, and other diversions." To justify the continuing expansion of military capability far beyond any purpose of defensive security, there is a tendency to exaggerate the capabilities, intentions, and hostility of a potential adversary. Since it takes several years to make a weapons system operational, there is also the lead-time problem that makes military planners devise a security system that five years hence will be able to cope with whatever the other side does. This interactive process is what produces arms races at fantastic cost to all parties, no increase in security, and great wastage in resources and talents. The psychology of the arms race breeds tensions that make it more difficult to proceed with all forms of international cooperation, and sustains the competitive approach to international relations in a setting wherein competition is becoming more and more self-destructive with each passing year. Sectors of the government try to make a case for new military expenditures, as with counterinsurgency warfare—by showing that real missions exist or will exist in which the military hardware can be put to use. Therefore, the foreign policy of a government may be led in the direction of acting up to its capabilities.

The evidence is becoming very strong, also, that excessive profits arise out of the tie-in between government and the defense industries. Cost overruns, subcontracting pyramids with profit margins at each level, and the development and construction of elaborate weapons systems that will be virtually obsolete by the time they are available for use, are all part of this national security system. Powerful Congressional allies of this relationship between government and industry have generated what has come to be known as the military-industrial-legislative complex, giving momentum to the arms race, creating a powerful lobby against arms control, and spreading its influence out over the society in such a way as to give the impression that local prosperity and employment depend on giving political support to a larger and larger defense budget.

The intricacies of the military-industrial-legislative complex need to be understood in undertaking any curative action. Many large corporations depend on defense contracts for a large percentage of their business; some companies, such as General Dynamics and Lockheed Aircraft, are almost totally dependent on the continuation of high rates of defense spending. A.T.&T., Ford, General Electric, General Motors, and Ling-Temco-Vought are among the ten top United States defense contractors. Many top executives in the defense industries are retired military officers apparently hired either because they delivered contracts in the past while working for the government, or because their contacts and experience in the Defense Department made them likely to secure contracts in the future. Richard Barnet, writing in 1969, says that "a recent report of the Joint Economic Committee of the Congress reveals the 2072 retired military officers of the rank of colonel or Navy captain are now employed by the 100 top defense contractors." Barnet gives many telling examples of the dynamics of these relationships: "Major General Nelson M. Lynde, Jr., helped arrange a 'sole source' contract with Colt Industries to procure the M-16 rifle. Five months later he was working for the company. Lieutenant General Austin Davis retired from an important role in the Minuteman missile procurement program to a vice-presidency of North American Rockwell, a prime contractor in the same program."

The profit system operates in such a way that everyone but the taxpaying public benefits from increasing costs. What Barnet calls "military socialism" means that the competitive system does not operate to keep costs low. Inefficiency is rewarded, and many considerations enter the negotiation of a defense contract other than the quality and the price of the product. Barnet puts the situation in sharp focus:

> The essence of the free enterprise system is competition, but 57.9 percent of all defense procurement is negotiated with a single contractor and only 11.5

percent through formal advertised competition. Under the capitalist creed only the efficient survive, and those who can neither provide quality nor control their costs fall by the wayside. Defense industry, as we shall see, is shielded by the government from the harsher realities of the competitive system. It is relieved of the obligation to be efficient and is protected by the government from most of the normal risks of doing business for profit. Government and a vast dependent industry have struck a bargain under which industry has surrendered a few management prerogatives to the Pentagon in return for substantial subsidies.

The economics of the defense contract are a story of shocking abuse. A large contract is subcontracted to a group of firms that in turn subcontract for parts and subsystems. At each level profits for the whole effort are taken, including those portions that are performed for profit by subcontractors. Joseph Goulden and Marshall Singer explain the special profit structure governing military procurement:

The trick is that in military contracting a fair profit is figured as a reasonable percentage of a contractor's "costs," but the contractor includes in his costs not only the expenses of the work he did, but also all the payments he made to the next contractor down the line. Thus the government ends up paying profits on the work done on trailers by Fruehauf but also to Douglas for the "cost" of paying Fruehauf to do the work, and to Western for the cost of paying Douglas the money with which to pay Fruehauf. The government pays profits three times, once for the work and twice for the mere service of passing the government's money on down the line. Even stranger is the fact that Douglas's "cost" is larger than Fruehauf's because it also includes

Fruehauf's profit; and since Douglas's cost is greater, its profit is larger, even though Fruehauf did the work. Western, the farthest removed, has the greatest basis for profits, since its costs include the profits of both subcontractors: Profits on profits on profits.

These ideas of contract result in fantastic profits being earned in relation to very little work being done. Goulden and Singer give many examples. A particularly extreme one illustrates the situation. The Consolidated Western Division of United States Steel delivered 1,032 missile launchers to the United States Army at a price of $73.5 million. Douglas Aircraft made plastic raincovers for these launchers at a cost of little more than $3, totaling $3,361. According to Goulden and Singer, "Douglas then took a profit on that amount *plus* the Consolidated price, a profit of $1.2 million on $3,361 work, the percentage: 36,531 percent. Western Electric, which gave each missile a quick $14 inspection at the base, proceeded to pyramid its return on top of Douglas's. Western did not do quite as well, coming out with a mere 6,684 percent return on its effort." Such a structure of profits is tantamount to authorized theft, and can occur only because the military budget is subjected to so little scrutiny. Recent developments suggest that such an extreme situation is unlikely to persist for very long, as the evidence of abuse begins to become a matter of public, documented information. At the same time, power structures do not often dissolve as a result of close scrutiny, but are more likely to reorganize themselves to respond to particular criticism and yet carry on as before. Large organizations, whether governments, churches, or corporations, display an extraordinary rigidity and tend for this reason to find it difficult to adapt to changed conditions.

It is important to clarify the ways in which this militarization of society operates, and to identify the groups in society that are paying these large defense bills and those that are benefiting. It is also crucial to demonstrate that we are wasting resources needed for other social purposes by investing so heavily in superfluous items of defense. Japan and, to a lesser extent, West Germany, are good examples of states that have, largely because of their defeat in World War II, spent a smaller percentage of their GNP on military purposes and have, in partial consequence, experienced very rapid increases in overall economic growth. These examples are helpful demonstrations of the proposition that a healthy capitalist-style economy does not need to rely on war industry to experience economic growth, especially if it can rely for security on powerful allies. The United States military budget, even aside from Vietnam, has climbed up above seventy billion per year, an extraordinary figure that is larger than the entire GNP in 1966 for such important states as India, Poland, Australia, or Brazil. Most of these resources put into defense are wasted or worse, and are diverted from applications that would bring people a better quality of life.

All these reasons make it very critical to begin reducing the power and influence of the military establishment, especially within the United States. A first step in this direction would be to cut the military budget by a fixed percentage—say 5 to 10 percent—each year for an indefinite number of years. The Defense Department would tend to redesign its capabilities within these new boundaries of available resources. Other efforts in the society would be needed to relocate and retrain labor, and convert defense industry to other ends. Such efforts would have special value if they were guided by endangered-planet considerations. It may also be worth considering John Kenneth Galbraith's suggestion to convert those corporations that depend on defense contracts for more than half of their revenue into public corporations. The creation of public corporations, while adding to the already dangerous concentration of economic and political power in the federal government, would at least eliminate the profit motive from the operations of the defense industries, and would formalize and expose the already close links between government and those large corporations now engaged almost totally in defense work.

Underlying these suggestions is the basic assertion that far too much money is currently being devoted to defense expenditures from the perspective of either the national or the world interest. Far too many weapons are being purchased to serve the needs of national security, even given the postulates of the war system. The United States and the world would be much better off with a minimum deterrent posture, with overseas bases abolished, and with American troops withdrawn from foreign countries. It might then become possible to evolve a foreign policy based on the combination of national defense, in the sense of discouraging hostile governments from attacking the homeland, and assisting in the effective development of genuine collective security arrangements carried on within the framework of international institutions. Such a reorientation of defense policy would have as its explicit objective a strike force of between 100 and 400 missiles deployed in such a fashion so as to have the maximum prospect of surviving a first strike and being able to deliver a retaliatory second strike on target. Such a posture—what has sometimes been called "a minimum deterrent"—is difficult to spell out in any precise way, but such a missile force, combined with a standing army of no more than 500,000 (as compared to the present total of 3.5 million), would need less than 25 percent, and quite possibly only 10 percent, of the present military budget. Even 10 percent of the present United States military budget is a larger amount than any country now spends on defense, except for the Soviet Union. We propose here the rationality of scheduled reductions of defense ceilings at the rate of 5 to 10 percent per year, adjusted upward or downward, depending on whether or not reciprocal behavior by other states results, and by the general climate of international society. The objectives of these steps toward demilitarization would be not only to devote resources to more useful ends and to lessen the influence of the military upon domestic society, but also to induce *a negative arms race,* that is, a beneficent spiral of inter-actions between governments that would encourage one another to match respective reductions in defense spending. In initiating this process, especially in its early stages, warning of betrayal and prophecies of doom would be made by die-hard Pharisees, adherents of the outmoded system. Prudence demands that such unilateral initiatives be carried on with due caution, including constant re-appraisal in light of the behavior of others. The new defense posture should be undertaken in a gradual, tenta-tive, and contingent spirit. It would be especially con-structive to obtain at least *tacit* assent to the pursuit of a negative arms race by most world leaders. It would also provide reassurance if some of the reductions could be undertaken as a consequence of negotiated agreements.

This course of demilitarization of national society is unlikely to occur in the near future unless provoked by major social and political convulsions. Although some shift in resources—or some new schedule of priorities—is likely to take place in the United States during the next five years, it is likely to be on a small scale, and to involve very little explicit or actual downgrading of military in-fluence within the society. And, in fact, the very power of the military makes it hard to move effectively toward demilitarization; the effort itself creates some risk of polarization and may carry the drift toward militarization to an even further extreme in the United States. If this should happen, it would tend to lead foreign societies in the same direction. The issue is, in essence, a political one, a struggle to alter the ratio of domestic resources and energies. The struggle, although political, has strong ecological overtones, since it involves a determination on several levels of consciousness as to whether to presuppose the primacy of the destructive forces, and thereby resign ourselves to their supremacy within the existing order of things, or begin the long job of reconstruction by presenting evidence that demonstrates the need to *defend* the environment against ourselves. In this security crisis the enemy is truly lodged within our midst, and we may indeed need a witch hunt to identify all these wrongdoers—many innocently unaware of their roles—who have subverted the structure of our political and economic organizations and have attained high office and public eminence.

The war system can also be eroded by making it somewhat more difficult for the head of state to commit the society to a war policy. The United States enjoys a long constitutional tradition of restraint on arbitrary action by the President or Congress. To reinvigorate this tradition is particulary important now because the present world role and capabilities of the United States give it a stake in the outcome of political conflicts going on almost anywhere in the world. The long American involvement in the Vietnam War has revealed, among other fallibilities of our foreign-policy process, the extent to which the United States can adopt a war policy without at any point making a solemn and considered political decision to do so. The Tonkin Resolution of August 1964 involved a Presidential effort, in misleading circumstances, to obtain the endorsement of the Senate for a very specific military response to a staged military provocation attributed to North Vietnam, that is, repeated torpedo attacks upon two United States destroyers engaged on a peaceful mission on the high seas. These official contentions have been undermined by the following disclosures: The destroyers were operating within twelve miles of the North Vietnamese shore (the width of North Vietnam's territorial waters); and their mission included giving artillery support for landing and surface military operations by

South Vietnamese armed forces in combat on some islands in the Gulf of Tonkin. The evidence increasingly supports the conclusion that there was either no attack on the American warships or only a slight skirmish. It is also now clear that the so-called "provocation" relied upon to justify an air strike in "reprisal" against the port facilities in North Vietnam of these torpedo boats was part of a general plan to initiate a limited war against North Vietnam, which only became fully manifest some months later, in February 1965, when the bombing of North Vietnam became government policy without any further effort to obtain the approval of the United States Senate. Lyndon Johnson also later alleged that bombing of North Vietnam staved off the collapse of the Saigon forces in late 1964 and early 1965. Such an experience, culminating in a bloody and destructive war that had at its peak over 550,000 Americans fighting in South Vietnam and many additional thousands engaged in the naval and aerial bombardment of North and South Vietnam, demonstrates the need to erect higher political, bureaucratic, and legal barriers between war and peace within the United States Government. In small part this can be done by rebuilding the legislative role, especially by a reaffirmation of the constitutional requirement that an American fighting involvement in a foreign war requires a Declaration of War by the Congress. This declaration can be limited in objective, scope, and duration, but it would allow policy to be planned on a joint basis and might foreclose a slow, incremental increase in military involvement that started out as an assistance program and ended up as a full-fledged combat role. Obviously, the reassertion of a legislative role is no guarantee of restraint—under certain circumstances of congressional militancy it can even hasten or accentuate a war policy—but it is part of a general assumption that publicity tends to have a moderating impact on foreign policy, especially if achieved early in a conflict situation bearing only a remote relationship to the security of the homeland. There is a danger that in reacting to the Vietnam experience a new idea of constitutional structure will emerge that might contribute to some future involvement in war, if the political context was one in which the President was more cautious in approach and the Congress or public more bellicose. Despite this real danger of designing against a recurrence of a past mistake, it seems persuasive, on balance, to encourage full discussion and disclosure as the bases for a valid decision by a government to embark on war as a decision of national policy. In the event that a sharp division in public opinion resulted, then this itself would serve as restraint, especially if the decision preceded the major commitment of force and the loss of American lives.

In the same spirit of early restraint, it would be very useful to insert into the governmental process a public official whose job it would be to clarify the link between a proposed or actual course of foreign policy and the requirements of international law. Such an official, serving either as a kind of Ombudsman for World Affairs on the White House staff, or as an Attorney General for World Affairs, would be chosen for his professional competence, and would be given a tenure that was not coincidental with Presidential terms of office. This official would be expected to respond to inquiries on legal status directed at him by Congress, the courts, and independent agencies, as well as by the President's office and, under certain circumstances, by members or representatives of the public. No pretense would be made that international law *clearly* identifies what can and cannot be done—even the Supreme Court rarely attains agreement among its justices on close questions—but an expert interpreter would at least have access to the highest policymakers and have a duty to point out what clearly cannot be done, what is problematic from a legal perspective, and what constitutes legal arguments for and against a proposed course of foreign policy. Such an exposure of legal doubts would ideally be tied in with a greater reliance on third-party procedures to settle international disputes and to resolve difficult legal questions. The Ombudsman or Attorney General for World Affairs might, for instance, be given the competence to insist that a controversial action of government, if appropriate, be referred to the International Court of Justice or otherwise to a panel of impartial and non-national experts in international law for a definitive opinion. There is no reason to urge unilaterally upon the country a *legalistic* foreign policy, but the more serious consideration of a nonpartisan presentation of the legal status of various actions, especially in the area of war and peace, might help to reorient national policy in the direction of global considerations and, by example, to elevate the status of international law elsewhere in the world. If so, then international law could help clarify the dangers inherent in the present close association between the ideology of sovereignty and the war system.

In the same spirit, domestic courts might be encouraged to pass judgment on legal issues, even those involving matters of war and peace. Individuals asked to enter the armed forces or to pay taxes should be entitled—according to the precepts of a liberal democracy—to find out whether they and the society are victims of arbitrary action by the government. The whole idea of war crimes is premised on the opportunity to test the legality of "superior orders" so as to mitigate the choice between contradictory legal imperatives. The United States took the lead in the United Nations to have the General Assembly affirm the judgment against German war leaders at Nuremberg. The General Assembly affirmed Nuremberg by unanimous vote at its first session in Resolution 95 (I), which instructed the International Law Commission to formulate the principles underlying the Nuremberg Judgment. These principles were formulated by 1950, and provide a useful summary of the relevance of international law to the responsibility of an individual whenever he believes that his government is waging a war of aggression or is relying upon illegal battle-field practices in the course of a war. The Nuremberg Principles make it clear that an individual cannot excuse his illegal and criminal acts merely by proving that he was carrying out the "superior orders" of his government. Of course, the Nuremberg idea could be used in the future as a *sword* to punish leaders of a defeated enemy alleged to be guilty of initiating a wrongful war, or it can be used as a *shield* by individuals seeking to interpose the require-ments of international law—or often only their beliefs about these requirements—between themselves and the demands and policies of their own government. It is in this latter usage that the Nuremberg argument has been relied upon by young Americans who hold the conviction that the United States involvement in the Vietnam War is immoral and illegal. Incidentally, this conviction is shared by many international lawyers of eminence in the United States and elsewhere in the world. Courts have disallowed this argument, not on its merits, but because it threatens an infringement on the powers of the Presi-dent in the area of foreign affairs. Courts have relied on the Doctrine of Political Questions to justify their proce-dural rejection of these legal attacks upon the war, pre-supposing that Presidential decisions on war policy are matters of pure discretion outside the ken of law. Such an attitude of judicial deference is an exhibition of pre-nuclear thinking on issues of war and peace, and fails even to take account of the legal prohibition on nondefensive uses of force embodied in such international treaties as the Kellogg-Briand Pact of 1928, and relied upon at Nuremberg to convict the German leaders of war crimes. It was the United States that championed this legal pro-hibition upon the discretion of governments to wage war in the period after World War I. Given the advent of nuclear weapons and the reaffirmation of the prohibition on aggressive war, it becomes most important to take seriously these efforts to outlaw war, at least to the extent of applying in our own courts the world-order ideals we proclaim so insistently as essential for a durable peace. Domestic courts have always been the principal judicial tribunals for the application of international law, and it has generally been conservative jurists who have argued on behalf of judicial activism in international law cases, especially insisting that domestic courts in the United States should hold void under international law confiscatory decrees of expropriation of foreign governments directed against foreign investment. Despite the vagueness and dubious character of international law rules applicable to the expropriation of foreign investment, American courts have been directed by Congress to apply these rules to controversies brought before them. There is no inherent reason of judicial practice why courts should not be encouraged to assert their relevance to the legal issues raised by a controversial decision of the President to use Americans to fight a war. The most basic rights of an individual are at stake, including those that protect an individual against unwarranted risk, in a situation where the contention is that the government is engaged in "illegal" action. An individual has no means other than by judicial redress to put forward his contentions. During the Korean War, the Supreme Court held by divided vote in the *Youngstown Sheet and Tube* case that the President lacked the constitutional authority, despite an official proclamation of "a state of national emergency," to "seize" the steel industry in order to facilitate the war effort. If property rights are entitled to such legal protection in a period of emergency, then courts can surely discover a judicially acceptable manner to protect the rights of individuals. Domestic courts could also provide very important educational forums, as they have in the areas of civil liberties and the administration of criminal justice, within which to clarify the relevance of legal restraint to foreign policy.

Such an educational effort would assist the overall objective of demilitarization of domestic society. It is of particular importance to detach the idea of citizenship from traditional images of national patriotism, especially blind adherence to the will of government: "my country right or wrong, my country," or its modern expression in the slogan "America—Love Her or Leave Her" invoked by those who would silence critics of America's role in the Vietnam War. The survival and welfare of the country depend on early, effective adherence to minimum standards of international conduct, including a gradual renunciation of violence and its threat as the way to resolve international disputes. The prospects for reducing the role of international violence are also related to the reduction of the extent of international stratification, especially as exhibited by the extremes of jet-set affluence and mass misery. A first step in the present period would involve acquainting young children with the images and attitudes of world community thinking as the basis for national citizenship. Earth values should be placed at the center of a citizen's creed from the earliest exposures of infancy: the objectives of citizen and state alike should be thought about as the discovery and maintenance of harmony between human society and the natural environment. Such a broader framework of identification would help tear out the psychological roots of human support for the war system on the national soil of the most powerful states. As such, it would help create a climate for the progressive demilitarization of the nation-state. The recruitment and training of men for military careers should also be informed by earth values. Such an educational reemphasis would involve adjusting the awareness and vocation of the military profession to the new objectives and conditions of our times. If the "military mind" could begin to evolve a constructive response to the special urgencies of the ecological age, there would be a far greater prospect of building a positive consensus as to the ends and means of social changes. Demilitarization of nation-states represents a critical first-step target in the search for a sequence of adaptive change bearing on the political organization of the world and on man's total relationship with the earth.

3. MODERATING POPULATION PRESSURE: PRELIM-INARY GESTURES, GUIDING IDEALS, AND THE REDISCOVERY OF EQUILIBRIUM. The danger of population increase has been widely appreciated in many parts of the world, but there has been a failure to take effective action in most societies.

A great deal of confusion and controversy also results from interpreting demographic trends and, in particular, predicting the length of time needed to achieve a zero growth rate. Governments are hesitant to interfere with the family and to establish standards regulating family size. Individuals often view their decision about family size as so minute in relation to the aggregate problems as to be of only trivial social consequence (the paradox of aggregation) and restrict family size, if at all, for selfish reasons. It remains uncertain whether "the perfect contraceptive society," that is, a society in which everyone could avoid having all unwanted children, would offer, if combined with educational and propaganda effort, a solution to the problem of population increase in various parts of the world within the next decade. Where educational levels are low and little industrial development has taken place, it may take many years— more than we can reliably project—to bring population growth to zero, partly because of the obstacles to a perfect contraceptive society, and partly because the desired family size averages far larger than replacement.

In thinking about population policy there are several

distinct points to make. First, there are fundamental uncertainties about future behavior that arise from our inability to predict with confidence. Second, there are now too many people on earth to establish a tolerable world social system, and the prospect assures the presence of many more people each year for the indefinite future. Third, the problems associated with world population are aggravated by the distribution of people and by the continued growth of national population in high per capita GNP countries. Fourth, the longer we defer effective action to prevent further population growth, the more difficult it becomes to avoid adjustments based on coercion and catastrophe. Fifth, purely voluntary approaches to family planning, even if expanded rapidly and dramatically, do not promise to prevent further population growth or to reduce the average number of children per family in most national societies to the replacement level of 2.1–2.3 within the twentieth century.

Concrete proposals for action need to be taken in light of this situation of mounting danger. Demographic trends present cumulative behavior that shows what large consequences can result from small upward or downward shifts in growth percentages. Such shifts can be induced, however, only by coordinated efforts at all levels of government, in all parts of the world, although some societies are more critical than others, because of either their size, their present population density, their poverty, or conversely, their wealth.

Replacement Ethos. International institutions and national governments might, as a first step, be induced to commend small-sized families as serving the national and the world interest. In fact, if citizenship as a whole could be persuaded to adopt a *replacement ethos*, then the increased availability of family planning assistance would begin to create some prospect for a self-adjusting aggregate population. Such a replacement ethos would have to be urged on the populace with real conviction by the leaders of society, and presented as a matter of urgent national welfare and personal dignity. Failure by individuals to conform, more or less, to the replacement ideal should be viewed as antisocial behavior, comparable to polluting a river or lake.

The campaign should start out as a call for voluntary action, but a call being made at a time of national and planetary emergency. Individuals who seek a large family should be encouraged, and perhaps given monetary or social rewards, to adopt children beyond the second. Families who have, either accidentally or not, more than two children should be encouraged to offer for adoption their "surplus" children. The ideal of replacement would need to regard additional children as in some senses "surplus." Obviously the parent, not the innocent surplus children, should be made to feel responsible for his antisocial action, and, preferably, the main efforts should be preventive.

Leaders of government, especially in the poor and crowded lands of Asia, Africa, and Latin America, must alert their societies to the prospects for more drastic action if voluntary campaigns to reduce family size are not carried out. These leaders *must limit* the size of their own families so as to set examples and to be in a position to speak on the subject with moral authority. Already voices are heard urging rich countries to apply the principle of *triage* by refraining from giving food or economic assistance to societies that have no realistic prospect of meeting the challenges of the future. Paddock and Paddock, for instance, urge concentrating all efforts at aid by rich governments on those specifically named societies that do have some hope. Such an approach involves a radical repudiation of human solidarity, requires standing aside while millions perish, and naturally induces the society called "hopeless" to adopt the most desperate strategies of self-preservation. As such, *triage* is dangerously naive about political consequences. The willingness of responsible leaders of opinion and students of the subject to propose *triage* is itself a storm signal that should help restore the seriousness and sense of urgency that are indeed needed to bring the situation under control. Proposals to require sterilization of certain social classes or ethnic groups, or to put contraceptive chemicals in water supplies, also reflect attitudes of despair that surround conventional approaches to the population problem. Those who are offended or shocked by these drastic solutions are challenged to make effective the effort to instill rapidly the replacement ethos in the moral and political consciousness of mankind.

There are several steps that might be taken. First, it would be desirable to encourage global awareness of why we would all be better off, independent of food supply, with a smaller population, and why we all would be worse off if the median projections of 6 to 7 billion for world population in the year 2000 are realized or exceeded. Such an awareness, if depicted in many distinct world settings by governments adhering to different, and often rival, political systems, could help produce a really significant expression of world public opinion. Second, it would be desirable to provide as a public service the technical means and medical hygiene needed to achieve effective and safe contraception, including fully legalized and free abortion. There also should be an effort to reach these ends without interfering with normal sexual relations. Third, it would seem desirable to encourage voluntary sterilization after two children, perhaps subsidizing the decision with a payment or an exemption from tax obligations for a year or two. Fourth, it would be desirable to grant approval to social relationships that do not involve building a conventional family, including consensual homosexuality. Fifth, progress reports should be issued by the United Nations every three months to report national, regional, and world changes in family size. In a world that measures and appraises change by reference to minor shifts in statistical growth curves—especially GNP statistics—it would be effective to express the struggle against population growth in comparable terms, relying on standard accounting procedures, and highlighting the application of more or less effective approaches to the regulation of family size. The costs of data-collection could be minimized by relying upon the basic profile of national population—now available in the UN Demographic Yearbooks—and adjusting it in terms of basic birth and death figures for the interim period. Many governments collect basic data of this kind, and material on changes in family size is at least as important, at this point, to economic, social, and political health, as that on changes in labor force, in unemployment levels, or in the cost-of-living index.

Population Plans and Goals. It is essential also to supplement the encouragement of small-sized families with social models of optimum populations, in terms of aggregate numbers, age structure, and geographical distribution. Such a projection of overall goals would help to define the magnitude of the adjustment now called for, and the progress, or lack of it, in relation to these goals. The effort to fix optimum populations for local, national, regional, and world communities would be part of the reeducation in finitude that is needed all over the world, and especially in those societies that accept the prospect of limitless economic growth, an ever-expanding GNP. The level of world population selected as tolerable and as ideal would itself be a measure of national and global understanding of the overall situation of an endangered planet. It is possible to discuss population levels only after an assessment has been made as to the bearing of interrelated issues of environmental quality, life-style, food supply, resource base, standards of living, and political stability. Only after such an assessment does it make sense to identify thresholds—

say 1, 2, 3, or 5 billion—as the maximum tolerable or optimal world population. Fixing such thresholds also makes evident the importance of time scale. Other considerations aside, the longer the time scale, the lower the level of aggregate world population. Thus, for instance, Richard Watson and Phillip Smith, using a time scale of hundreds of thousands of years, argue that 500 million is the maximum sustainable world population, given the need to recycle resources used by economies. Others whose concern is limited to the generations of their children and grandchildren would tend to adopt more "realistic" figures of 3 to 5 billion, which assume the ability to sustain at least the present world population for a reasonably long time if growth pressures can be successfully resisted in the near future. If the problem is conceived only in Malthusian terms, then the problem of an aggregate world population may be denied altogether, as has been done by Colin Clark and Jean Mayer, or put off far into the future. Paddock and Paddock, Borgstrom, Ehrlich, and others, have, in fact, assumed that the Malthusian analysis alone justifies the effort to cut back the existing world population. To take national welfare seriously, then, involves projecting goals that envision scheduled population reduction and devising policies for their realization.

It is obviously premature to select a figure—say 2 billion—as representing an appropriate planetary ideal. At present, it would be desirable to study the various consequences of different world populations upon world-order issues such as survival and welfare, environmental quality, the habitability of the planet, and the prospects for future availability of resources and energy. If one objective is to maintain a rough equilibrium between what man consumes and what the earth possesses, then the optimal population would be lower than if short-run ecological balance were the criterion. The closer to equilibrium the lower the population and the simpler the

economic lives of the inhabitants. A second variable
of great importance is the standard and style of living
enjoyed by the average world inhabitant. A pastoral
commune can maintain an equilibrium with nature for
an indefinite period that is impossible to achieve in a
modern urban milieu.

Therefore, the process of fixing population thresholds
raises profound issues of life-style and time scale. One
can imagine different sorts of judgments being reached
in different societies: the United States, let us say,
adopting a low-population, high-consumption model of
accommodation, with India adopting a high-population,
low-consumption model of accommodation. If we do
indeed affirm a world of diversity, then these kinds of
cultural choices should be protected and understood.
At the same time, we need a concept of equality so that
some parts of the world will not be expected to endure
a disproportionate share of the burden of accommodating
to the endangered planet. Various parts of the world
need to produce assessments of the situation and proposals for adjustment that rest on ideas of maximum
and optimum threshold, time scale, life-style, and satisfaction of minimum human needs. The step proposed,
then, is the projection of a "world plan" and a schedule
for its fulfillment.

**4. PROTECTING THE RESOURCE BASE: A FURTHER
REDISCOVERY OF EQUILIBRIUM AND LIMITS.**   As
in other sections of this chapter, the idea here is to
sharpen awareness by discovering limits and proposing
limitations. In this respect, a first step is to repudiate
a sense that things can go on indefinitely as they have
been going on. Such a repudiation is especially relevant
to the stockpile of world resources. After fears of shortages in earlier centuries, there has developed in modern
man a sense of unlimited prospects for expanding the
material base of human society. The large impoverished

sections of the world have embraced this prospect of limitless expansion with particular enthusiasm as offering them an eventual way to deal with the misery and suffering of their peoples. The technological mentality envisions an ever-expanding horizon of opportunity that will provide man with a supply of resources that grows more rapidly than demand. Expectations of this sort are especially associated these days with the mineral wealth of the oceans and the seabed, and with prospects for the successful exploitation of low-grade ores, even those contained in ordinary rock. As was discussed in Chapter IV, these visions of indefinite abundance amount to fantasy. Shortages of critical minerals, even at present levels of consumption, are likely to emerge in the next few decades, causing modern economies to experience serious dislocations. It is simply not correct that mineral substitutes always exist, or, if the cost of the substitute is greater, that this can then be shifted to the market.

As with population policy, we need to evolve a resource policy sensitive to the issues of survival and welfare. Such a policy would rest upon a model of equilibrium, and would depend on underlying decisions relating to population size, defense budget, life-style, environmental quality, and economic goals. Obviously, if the war system could be abolished, or greatly diminished, resources could be redirected toward more productive use. Present arms-spending projections call for the expenditure of $4 trillion on the world military establishment during the next decade alone. The continuing efforts by governments to stimulate and promise economic growth at all levels of society, including those social classes already affluent, assure growing per capita resource pressure in a context of population growth. Even with developments in recycling and reclamation of waste products and scrap, there is likely to be a steady deterioration in the resource base of mankind.

In the period just ahead it is essential, first of all, to

gather information on expected resource supplies and demands. This information will tend to identify which shortages are likely to emerge in the years ahead, and might have an impact upon those societies that are highly industrialized and have high per capita consumption. More broadly, an awareness of emerging shortage will confirm the need for *limits* and for new structures of *organization and management* in human affairs. The market mechanism cannot hope to protect human interests in a world of dwindling resources. The proposal made here should reinforce the effort to stabilize a world population as soon as possible. Complex issues of life-style, time scale, and economic priorities bear centrally on resource needs within a given society.

In addition, it would be desirable to shift some control over resource policy beyond the reach of national governments. Private corporations, especially given the device of the multinational corporation, are increasingly significant actors in world affairs, and need to be regulated from a *world*, as well as from a *national*, perspective. If left free to operate virtually as they please, then profit-making drives will subvert the effort to reverse the pressures building within the world ecosystem.

Resources are also required to satisfy basic human needs. A resource policy will have to be developed in conjunction with the creation of a world welfare system, allocating resources to enable poor countries of the world to alleviate the misery of their populations. At the same time, the global ecosystem, including its resource base, would not be able to accommodate a solution to world poverty that proceeded by way of society-wide industrialization of the sort that has occurred in North America and Western Europe. We encounter here once again a fundamental dilemma, namely, that the effective pursuit of economic and social progress on behalf of the poorer countries hastens the ecological collapse

of world society. At the same time, to thwart the alleviation of misery is to induce political disarray and extrem-
ism, which is likely to culminate in some kind of general
war. A rational use of the resources of the world will
have to take account both of the basic needs of mankind
and of establishing an equilibrium between human
consumption and the capacities of nature. Such rationality has profound implications both for resource priorities and for distribution patterns. We are already
beginning to discover that the rich economies of excess
consumption are engaged in highly self-destructive
patterns of existence. The urgent immediate needs
are to gather evidence so as to substantiate the dangers
that now exist, and to plan the sorts of adjustments
that are called for to adapt to the urgencies of an endangered planet.

5. PROTECTING AND REHABILITATING THE ENVI-
RONMENT: INITIAL STEPS.

Many signs of planetary
danger are disclosed in the most industrialized societies
by a variety of environmental breakdowns and stresses.
There are numerous specific responses to the almost
countless variety of environmental issues. The newspapers carry reports of efforts to ban DDT and fungicides, of dangerous concentrations of lead and other
chemicals in the air, of rivers and lakes that have become
like cesspools, of oil spillage, of garbage disposal problems; the list of urgent environmental issues is constantly
being extended. There is also an increasing understanding of the need to deal with environmental issues
in a coherent fashion, and a dawning appreciation of
the fact that pressure on the environment is itself tied
up with resource and population issues. Congressman
Henry Reuss organized a set of hearings in September
1969, under the title "Effects of Population Growth
on Natural Resources and the Environment," that amply
document the contention that the environment cannot

be protected without at the same time establishing policies bearing on per capita GNP, life-style, population growth, and the operation of a market economy in a setting of increasing scarcity.

In this regard, the proposal for the creation of a National Conservation Bill of Rights is an important political initiative. The proposal has so far taken the form of a Joint Resolution introduced in the House of Representatives (H. J. Res. 1321) on June 13, 1968. The Resolution, which does not at present have much prospect of success, contains only a single four-part provision, worth quoting in full:

Article—

Section 1. The right of the people to clean air, pure water, freedom from excessive and unnecessary noise, and the natural, scenic, historic, and esthetic qualities of their environment shall not be abridged.

Section 2. The Congress shall, within three years after the enactment of this article, and within every subsequent term of ten years or lesser term as the Congress may determine, and in such manner as they shall by law direct, cause to be made an inventory of the natural, scenic, esthetic, and historic resources of the United States with their state of preservation, and to provide for their protection as a matter of national purpose.

Section 3. No Federal or State agency, body, or authority shall be authorized to exercise the power of condemnation, nor undertake any public work, issue any permit, license, or concession, make any rule, execute any management policy, or other official act which adversely affects the people's heritage of natural resources and natural beauty, on the lands and waters now or hereafter placed in public ownership without first giving reasonable notice to the public and holding a public hearing thereon.

Section 4. This article shall take effect on the first
day of the first month following its ratification.

Such a provision would give constitutional status to environmental claims, although it represents, as its very
name suggests, too narrow an approach to environmental
defense. At the same time Section 1 is capable of unlimited legislative development, and the entire Congressional
initiative represents a belated, if not all-encompassing,
acknowledgment by the federal government that the
maintenance of the environment no longer can be taken
for granted as a source of permanent support for social
and political action, nor can lesser subunits of government be expected to act effectively to uphold the quality
of the environment. The Santa Barbara blowout illustrates
the opposite tendency, namely, the unwillingness of the
federal government to protect a local community against
damage when substantial federal revenues and vested
private interests are at stake. The effort symbolized by
House Joint Resolution 1321 provides citizens with a right
to initiate legal action against environmental abuse and
to call the government to account. Many of the most
serious hazards to environmental quality are government-
caused, arising from weapons tests, giant nuclear power
complexes, military installations, projects of the Army
Corps of Engineers, and many other activities that are
governmental in character or sponsored by government.
Above all, then, we need to make governments more
sensitive to environmental pressures caused by their
own drive for maximum power, wealth, and prestige.
Once again we come up against the inadequacy of the
state system to cope with the problems of our time.
   More specific proposals have been designed to improve
water quality by persuading, subsidizing, and coercing
industrial firms and municipalities to treat their prodigious
waste output with modern sewage systems. As matters
now stand, the cost of waste disposal is covertly trans-

ferred to the larger community rather than dealt with as a matter of responsibility at the source. Economists talk of "negative externalities" to describe those costs that are borne by the community but not reflected in the price of any specific product. Pollution is a principal illustration of a negative externality, whether the perspective is the operating budget of an individual homeowner, an industry, a municipal council, or a more central unit of governmental operations. In early 1970, Senator William Proxmire proposed a system of effluents charges to be levied on industrial firms that discharge wastes into water, varying with the strength and toxicity of the waste, and assessed on a per-pound basis. Such a system of charges could devote the revenue collected to the large job of improving sewage systems at the municipal level. This system would also underscore the point that "the cost of a product includes the expense of getting rid of wastes in such a way as to avoid pollution or of paying the community, in effect, if the wastes are not discharged in a harmless fashion. Such an approach may deteriorate the market position of some products, but such a deterioration would seem consistent with real cost which needs to include some charge for environmental damage." The idea of an effluents-charge system could be broadened to assess all entities and individuals an amount equal to, or somewhat in excess of, their damage to the environment. Another advantage of this approach would be to encourage greater investment in antipollution equipment. A system of environmental user charges would also make society as a whole grow more aware of the economic and social sacrifices that may be called for to protect the quality of life and to uphold earth values. The idea of damage assessment contrasts with the long tradition of economic development in the West—a tradition that persists to this day—that regards the land and rivers and skies as dumping grounds for human waste and subject to exploitation. In fact, the treatment of the Red Indians by early settlers in America illustrates the same exploitative attitude toward a human group as is present in the abuse of the physical environment. The colonial mentality violates the larger spirit of the ecological imperative. Or rather, the earlier exploitation of human groups is an early case study illustrating the same disregard of the conditions of human welfare that is now also present in relation to the environment.

Finally, it would help improve our prospects of avoiding environmental deterioration and catastrophe if nations and regions established large-scale research and policy institutes to provide early warning and public notice of potential environmental danger. These National Institutes of Environmental Policy should, if at all possible, be financially and politically independent of any government, as governments may be among the principal wrongdoers and, therefore, may possess an interest in suppressing or understating environmental dangers that arise from their activities. Also, these institutes would need experts of real stature whose recommendations would carry weight if it became necessary to reach decisions or to take action on a crash basis. A special function of such an institute would be to monitor and detect the deepening of quiet crises, such as slow or rapid buildups of poisons in land, water, or air dangerous to man and other animals, temperature and climate changes, or impending geological hazards arising from subsurface nuclear explosions or offshore mineral prospecting and drilling. Such institutes could provide reliable information, focus public and bureaucratic awareness on environmental issues, identify the world dimension of environmental problems, and offer recommendations for action. They could also furnish legislative committees, governmental agencies, and citizens' groups concerned about environmental questions with educational materials and expert advice on specific questions, including research facilities and guidance.

The National Institute would issue a public annual report on environmental quality each year, containing specific proposals for curative action; the director of an institute might also be given the duty to submit confidential material to the President.

A World Institute of Environmental Policy should also be created to focus upon the distinctly international aspects of environmental quality. The existence of such a World Institute might be associated with the United Nations, but its operations should be as autonomous as possible. The World Institute should, of course, coordinate its various monitoring and information-gathering activities with those of National Institutes. A special concern of the World Institute would be to identify problems of multinational concern, and to propose both supranational (institutional) and international (coordination of governments) arrangements for the consideration of governments. Governments, organs of the United Nations, and other world actors should be able to request advice and counsel from the World Institute. A special task of the World Institute might be to train experts in various branches of environmental studies to staff National Institutes and to organize conferences and specialized programs to bring together experts from the latter institutes. The World Institute would also make a public annual report to the United Nations, and might also have the right to submit confidential memoranda to the Secretary General of the UN.

The various proposals bearing on the environmental dimension of the endangered planet are meant to be *illustrative* of the kind of first steps that could be taken at the present time. Their overall purport is to gain time, increase relevant awareness, lessen the risk of disaster, and avert irreversible catastrophe. More fundamental reorientations of belief and organization are, however, dependent on inducing basic changes in the relationship between man and nature, especially with respect to the avoidance of applying technological know-how without greater concern for environmental effects.

Environmental issues have a wide potential popular appeal. There is some reason to hope that a political movement built around the desperate need to protect the air, water, and land might sweep across the country and the world. Everyone is hurt by the decay of the environment; everyone benefits by its rehabilitation. At the same time, there are many centers of economic power that operate profitably only so long as the environment provides a cheap dumping ground for their activities or their products. A first-step measure, however, is to introduce environmental issues into the platforms of major political parties and publicize the voting record and positions of candidates for elective office in relation to environmental policy. The Swedish interpreter of the world survival scene, Rolf Edberg, states: "We can foresee the day when the problem of man's natural environment will be recognized as by far the most important of all social problems—locally, regionally, nationally, and internationally."

The specific proposals discussed in this chapter have been mainly concerned with encouraging a better awareness of the endangered planet, which means, above all else, the discovery that human civilization is pushing up against the limits of the earth's carrying capacity. Some proposals also have been put forward because they are the kind of first steps that might induce a series of beneficial adjustments. Still other proposals were suggested to try to gain time, by making the current world system somewhat less catastrophe-prone during an interim wherein a new ecological consciousness might form and gain access to power and influence throughout the world. Part of this new consciousness will involve a linkage in the imagina-

tion of men between sensitivity to environmental destruction and compassion for human suffering. The wholeness
of the earth, the cause of restoration, and the impulse
to live and endure will have to produce both a repudiation
of the destructivity of the war system and a concern for
the elimination of mass misery from human affairs. I
believe that we can rescue ourselves from our current
precarious situation only by solving the *whole* problem of
the endangered planet and by evolving the social and
political forms that are compatible with a solution.

IX. Two Images of the Future

Speaking on May 9, 1969, the Secretary-General of the United Nations, U Thant, expressed his belief in the existence of a planetary state of emergency: "I do not wish to seem overdramatic, but I can only conclude from the information that is available to me as Secretary-General that the members of the United Nations have perhaps ten years left in which to subordinate their ancient quarrels and launch a global partnership to curb the arms race, to improve the human environment, to defuse the population explosion, and to supply the required momentum to world development efforts." Suppose that the President of the United States happens to read U Thant's gloomy timetable for the future in a reflective moment of self-doubt. The President ponders the situation and decides that it is essential to find out more about the gravity of the threats to mankind. Accordingly, he summons five of the most thoughtful and creative men in the country and asks them to assess the alleged crisis as soon as possible. In a month they report back in a manner that upholds the analysis and presentation of Chapters III and IV. Our agonized leader grows convinced of the urgency of the situation, but is baffled as to what can be done.

He is able to arrange through private channels an urgent, top-secret meeting of the most trusted single advisers in each of the world's top ten governments and at that meeting will present these advisers with the results of his own inquiry. He asks these advisers to go back to their own governments and study the situation for themselves. Six months later the ten Heads of Government agree to meet to discuss whether human survival and the habitability of the planet are in jeopardy, and what can and should be done about it. They all agree that the situation is desperate and that basic processes of change in the operation of world society must be initiated within the next decade.

Before this fantasy becomes a plausible project it is necessary to draw some sketches of the future. The first sketch is written on the assumption that nothing much is done to change the awareness level, the organizational features, or the behavior of leading national governments. The second sketch is written on the assumption that the political leaders of the world cooperate to save mankind from the present prospect of annihilation. In a sense the negative sketch is a summary of the endangered-planet argument and the positive sketch is a recapitulation of the world-order argument.

To explore this fantasy should not be understood as a cheap way to bypass the existence of separate sovereignties. As matters now stand there is no prospect that any world leader is prepared to propose drastic action to save mankind and the earth, much less that the principal leaders of the world can agree on a common definition of the situation. The discovery of a conspiracy among the leaders of the ten biggest countries would almost surely result in their removal from domestic control, and might stimulate their critics and enemies to question their loyalty or even their sanity. Harold Lasswell, who has written widely on the behavior of political elites, considers national leaders to be "entrapped" by the expectations of their followers and virtually helpless to give effect to any counsel of reasonableness that involved a drastic change in international relations:

> I have no doubt that no matter how meager or extensive their training may have been in science and technology, Mao, Johnson, De Gaulle, Wilson, or Kosygin are convinced of the catastrophic potential of nuclear weapons and of an anarchic world system. However, I suggest that even if they wanted to agree with their opposite numbers to establish a new system of effective public order, they would refrain from taking crucial steps for fear of losing support in the arena of internal politics. A power relation, after all, is a two-way affair: to have power is to be empowered. Any rumor that top political figures are planning to put the body politic in a position where it can be surbordinated to a foreign coalition of powers is bound to stir personal, factional, party and governmental participants into action. What statesman can be blamed if he comes to the conclusion that political suicide contributes neither to his career nor to the fundamental reconstruction to which he may be personally devoted?

Of course, we find ourselves caught within a vicious circle. The character of sovereignty is such that it contradicts efforts to circumvent it. But there are new social facts implicit in the endangered-planet situation that will become increasingly prominent. A collective emergency exists that cannot be dealt with by any of the traditional responses of nation-states to danger. Let us imagine the world reaction to a verified report that the earth was about to be invaded by a hostile extraterrestrial enemy and that earth defense required a pooling of energies and capabilities under a unified command. Let us also suppose that the extraterrestrial threat is likely to continue according to intelligence reports for 50 to 100 years at a minimum. What kind of response could be organized? How would it be implemented? Can we discover a way to make world leaders worried about the survival and welfare of their societies in the manner that U Thant is presently worried? By what evidence, or other form of persuasion?

Science fiction has often anticipated the technological future by its invention of plausible fantasies. A similar genre or sub-genre of fantasy-projection might be able to invent plausible political futures. In the remainder of this chapter we shall experiment with such a fantasy-projection.

Think of an old-time sailing vessel being carried toward unfamiliar and distant rocky shoals by a momentous wind, the sea torn asunder with gigantic waves and swirls. Think also of a captain and helmsman who have drugged themselves to sleep, exhausted by the routine pressures of eluding capture in pirate-infested waters and confident of the readiness of the gun crews to withstand pursuit. Such a ship is headed for shipwreck unless rescued by a miracle or a sudden arousal of unprecedented energy. Consider the plight of our world beneath the sway of such an image.

THE FUTURE AS PROJECTED FROM THE PRESENT.
We suppose in this first exercise of the imagination that nothing very startling happens to reverse the present course of disintegrating developments. The attentions of governments continue to concentrate upon competitive rivalry in world affairs and economic growth and civil order in domestic affairs. The big states maintain current levels of arms spending and smaller states gradually acquire the capability to wage mass war in the nuclear age. Pollution and population issues are often discussed

in the more industrialized societies and larger efforts at abatement and birth control are made. Alarmist predictions about the future are made by a variety of specialists and are carried along with other news of the day. Great strides forward are reported in space exploits, biomedical research, deep-sea exploration, synthetic food production, and computer sciences. The technetronic age is upon us with cold efficiency, central data banks of constricting knowledge, and the disappearance of privacy.

Poorer countries continue with their efforts to modernize their societies, which includes becoming better and better equipped to fight modern wars of mass destruction. Civil strife is present in widely separated parts of the globe. Technological and ecological disasters occur with greater frequency but continue to be reported as unfortunate accidents and isolated disasters: A half-million-ton oil tanker splits open in mid-ocean, a smog attack causes sharp jumps in urban death rates, famine and epidemic reports from the crowded countries of Asia are given prominence from time to time, some radioactive waste leaks from its underground burial place into a subterranean water system that seeps into a major river system, several countries are suspected of developing nuclear weapons in secret and several more are stockpiling nerve gas and bacteriological weapons.

The same political structures as we know today persist as do the basic forms of political conflict. Governments continue to be preoccupied with the ambitions of their rivals, with building up their own GNP, and with stamping out fires of discontent wherever they flare up. The race will continue between the bureaucratic capacities of repressive governments and the insurrectionist capacities of revolutionary groups. Political energy will be directed toward maintaining control over man-in-society by managing the dynamics of power within the nation and in the relations among nations.

Such an emphasis will resist information about the deteriorating character of the international environment. Various kinds of pollution will cross thresholds of irretrievable disaster, causing the disappearance of many species of marine and wildlife. Urban growth will press upon the open spaces and crowd still further the already heavily settled parts of the world. All kinds of ingenious proposals for alleviating environmental pressure will be advanced, but numbers will keep increasing, as will per capita demands.

The human spirit will be worn down even if no cataclysmic event such as a nuclear war or a drastic change of earth temperature, oxygen supply, or ocean level occurs. The regimentation in large societies and the inability of governments to secure a decent quality of life for their populations will lead to widespread human despair. The loss of man's confidence to shape his destiny will become almost total with the appearance of robots and hominoids, making it virtually impossible to distinguish between people and artificially created centers of "intelligent" action.

Against this tableau we expect a tense world system that moves from crisis to crisis and is unable to organize a coherent response to the underlying hazards and decay of the whole experience of mankind. Each national society will shift blame to the machinations of others and concentrate upon upholding "its interests" in a world of impending doom. In terms of sequence, we might expect the following course of events.

The 1970's— The Politics of Despair. In the 1970's there will arise a deepening sense of the inability of major governments to solve the central problems before their societies. The rich countries will continue to build new expensive weapons systems to secure, if possible, a competitive edge over their rivals and a dominant position in relation to their potential challengers. The poorer countries will continue to base their hopes upon small annual increases in their GNP, but population expansion and military spending will eliminate most of the improvement in the standard of living that might otherwise have taken place. Large-scale misery will persist.

The pressure of rising populations will everywhere push back the remaining preserves of nature. Urban sprawls will be the breeding ground of disease, distress, crime, and revolution. Electronic techniques of surveillance and propaganda will try to immobilize internal opposition groups. The struggles of government to survive will overshadow all other concerns and will inhibit the search for solutions to more underlying problems. People will be led to submit or revolt.

Poorer countries will grow more receptive to romantic schemes for the improvement of their position in the world: Threats to develop nuclear weapons, military buildups, and a variety of attacks on the citadels of affluence will send tremors across world society. Radical energies will continue to pin human hope on seizing power from oppressors. More and more, the strategy of seizure will acquire an international dimension. The poor societies will perceive themselves as sharing a condition of oppression and privation, and will challenge the propriety of the world power structure.

In advanced countries many individuals will grow dispirited with politics altogether. The old political concerns will lose their relevance to the imagination of the most enlightened members of a society. Alienation on a massive scale is likely to ensue, leading to severe symptoms of withdrawal and nervous disorder. Politics will lead men of thought and feeling to despair: not caring, not doing. There will be no reason to suppose a better future, just more of the same. *People will increasingly doubt whether life is worth living.* Religions will become a weaker source of comfort and guidance as people in smaller and more backward communities grow more able to act on their unbelief.

Nothing much is likely to change during the decade. Some achievements will be hailed, some disasters will be lamented; there will be more people, more noise, more garbage, more weapons possessed by more governments, more violence of all kinds threatened and proposed, worse accidents, and fewer trees, less wilderness, fewer usable resources. The world scene may be more or less in turmoil, but there will be a continuous effort to ward off catastrophe by monitoring dangerous conditions. Hotlines will interconnect most capitals of the world. Sovereign states will persist as the exclusive centers of political power, but will enjoy less automatic loyalty from their populations and will relinquish to multinational corporations much of their former control over economic activity. The UN will still be there, but its utterances will not matter much more than today, nor will its prestige be much enhanced whether or not, by then, China participates in its activities. National governments will remain entrapped by the values and constraints of the national setting, and no major institutional changes are likely to occur, although a small underground world government movement may be identified in various parts of the world. Some of its leaders may be put in jail as "subversives" or "detained" for long periods. The more prescient members of the various *anciens régimes* of the world will sense hard times ahead and make subtle proposals for thought control and moral conformity. There will everywhere be a drift toward the centralization of power accompanied by greater reliance on police methods and more regulation of dissent. Some concentration camps will be used, many more will be built. TV and newspapers will be increasingly subject to governmental control.

The 1980's — The Politics of Desperation. C. P. Snow asserts that "Despair is a sin. Or, if you talk in secular terms as I do, it prevents one taking such an action as one might, however small it is. I have to say that I have been nearer to despair this year, 1968, than ever in my life."

Those men of goodwill who have been concerned with working carefully and peacefully for reform will gradually drop out of sight, mainly victims of despair. Governments will come to realize their own inability to solve the principal problems of man-in-society. The pressure upon the environment will grow even more intense, strains and dislocations will occur with more frequency, on a larger scale, and will arouse greater alarm. Only drastic change will seem capable of doing anything about the situation, but "establishments" everywhere will resist pleas for drastic change, as the dynamics of change would appear to threaten prevailing social structures and produce sudden and massive transfers of wealth and status. Those that have, do not voluntarily give up their privileges to those that have not, but organize defensive efforts.

But not everyone will succumb to the politics of despair. There will be everywhere some who believe in the need for drastic change and are willing to accept the risks of working for it. The espousal of such an outlook will begin to threaten those with power, especially if a prospect for a large change-oriented constituency begins to emerge. The politics of desperation breeds the attitudes, tactics, and reactions of revolution and counterrevolution. Under modern circumstances the technology of government can effectively eliminate a potentiality for resistance and revolution. We already know that totalitarian regimes, if backed up by secret police, media control, propaganda, control over child education, can keep their populations at bay for long periods of time without even having to kill anyone. Such forms of repressive government will be likely to proliferate in the 1980's. The objective of government would be to depoliticize the mass, discourage and punish dissent, and hunt down deviant revolutionary personalities and groups. Such a pattern is to be expected, especially in the most powerful countries, where mass disaffection is likely to result from the inability of government to halt the decline in the quality of life for most of the population. The methods of repression may grow very sophisticated, relying upon subliminal manipulation and a system of perpetual, but nonintrusive, surveillance. Electronic censors might, for instance, be able to monitor all movements of political suspects, record their utterances, and photo-transmit their action. Knowledge alone of such a capability by government would intimidate all but the bravest individuals, and make the entire enterprise of political opposition seem self-destructive and futile. A revolution can take shape only if there are opportunities to persuade and organize, places to hide while building a movement, and sanctuaries for retreat during early stages of struggle. If these open political spaces disappear, then governments will succeed in maintaining regimes of law and order in the 1980's, and will be able to ignore the grievances and disaffection of despairing masses of men and women.

Perhaps the objective situation of impending environmental catastrophe will move members of the establishment to consider or even stage "inside" coups. Deteriorating world conditions may prompt a new, hitherto unknown kind and form of domestic revolution that might catch the establishment off guard, as if the radar was pointing toward the probable enemy, but the air attack came from a friendly quadrant of the skies. Such surprise had caught Arab air defenses off guard, even in a period of high tension, when Israeli planes attacked on June 5, 1967, initiating the Six-Day War. The point is that we will find a loss of confidence in traditional modes of political actions, and a sense of desperation will be felt by opponents and a mingled sense of anxiety and impotence by power-wielders. Such a setting is likely to generate a politics of desperation, if it does not induce the disappearance of politics altogether.

In the poorer countries the same kinds of developments are likely to occur, but the hold of the central government over its population is likely to be less secure and less complete during the 1980's. Misery will persist, despair and desperation will ensue. As such, we can project a continuation of cycles of turmoil, strife and repression throughout the Third World.

But an increasing portion of Third World energy will be turned outward in the form of hostility toward those actors in the world that are doing well, at least in an economic sense. Such hostility may turn to rage if the debris of a space flight falls on an Asian inhabited community or if sickness and disease are traced to the effluents of the rich industrial societies. Governments in these countries may also grow convinced (or think it expedient to explain) that their own failures are an outgrowth of a rigid system of international stratification that exploits the poor for the sake of the rich. As such, one can imagine governments in groups or singly pursuing the politics of desperation, by acquiring the capability to destroy or disrupt any society in the world. The strategy of the have-not nations could well move toward adopting a real military strike capability as the essential preliminary to hard bargaining for a new deal in international society. One might recall the sardonic wisdom of the English nineteenth-century literary personality, Rev. Sydney Smith: "From what motive but fear, I should like to know, have all the improvements in our constitution proceeded? I question if any justice has ever been done to large masses of mankind from any other motive. . . . "

Such an arousal of tensions in world affairs might lead to a frenzied realignment of powers that would decisively end the ideological dispute that produced the Soviet-American conflict in the years after World War II. Concessions in the terms of international trade, in capital loans, in economic assistance might be offered by the rich countries, but to no avail. Once the politics of desperation takes root it will not be cajoled into moderation.

The man who sees himself mutilated by another and is prepared to die fighting for a new order is not likely to settle for a 2 to 5 percent rate of readjustment, but this is the fastest pace of voluntary adjustment that can be imagined. A rate of 20 to 50 percent is deemed minimal to the extent that bargaining can go forward at all. The politics of desperation may involve a gigantic North-South build-up in preparation for World War III. Surely, the rich, challenged power-wielders would resist the demands for drastic reform as excessive and impractical. The technological gaps may have grown so large by the late 1980's that the superstates will be in a position to disarm and recolonize the poor, populous countries at small risk to themselves. There may be a movement to do this in the interest of world peace and to safeguard the international economy.

In fact, it would not be implausible to discover that a new civilizing mission in the rich countries is engendered by the urgencies of man-in-nature. The powerful governments may convince themselves that their own welfare and the future of the planet depend on imposing population and pollution curbs in Asian, African, and Latin American countries. Trusted advisers to the Soviet and American governments may conclude that the poorer countries of Asia and Africa have no capacity to provide for the welfare of their own populations or to ease up the pressure on the world system. Such a perception is especially likely if population growth continues in the 1980's at a rate of 1 percent or more and if food and environmental resources become scarce in quantity and degenerate in quality. This perception will be reinforced by demands from these troubled lands for freedom to migrate to less populated countries and for the receipt of a fixed proportion of GNP from the richer societies. Perhaps desperate strategies will be initiated by the rich countries to depopulate Asia, Africa, and Latin America to relieve the overall pressure of man upon the environment and to assure the maintenance of an ample flow of raw materials and primary commodities. We may not yet have heard the end of "the white man's burden," or perhaps, of "the yellow man's burden."

Whatever its form, the 1980's are likely to end on some note of desperation in a world fraught with tension and fear. We would still expect no changes in world structure. Sovereign states will continue to act as autonomous agents for national gain, being both abetted and impeded by a very powerful network of multinational corporations; the growth of regional institutions and special functional regimes is likely to remain marginal to mainstream world political concerns, as is the still extant, still enfeebled United Nations Organization.

The 1990's—The Politics of Catastrophe. As we retreat further into the future our political moorings are likely to grow more and more insecure. Our purpose is to identify the probable political impacts of those unchecked trends toward disintegration that have earlier been summarized as the essence of an endangered planet. The gloomy prognosis rests on the conviction that prevailing political perceptions and institutions will prove unable to cope with the challenges of the future, that these inabilities will induce a dominant mood of despair in the 1970's and a dominant mood of desperation in the 1980's, and that this dismal sequence of development will eventuate in catastrophe in the 1990's. This pattern of expectation about the future is an interpretation of emerging rhythms of action and reaction; our schedule may be too slow or too rapid, it may also be too confined by the boundaries of present awareness. We are projecting outward from our current sense of the situation. The world has been vulnerable to catastrophe since the beginning of the nuclear age and certainly since both the United States and the Soviet Union possessed deliverable high megatonnages of nuclear warheads targeted at each other's centers of population and industry. Where something of this sort is possible, it may happen. The risks cannot be calculated, but there is a finite risk of general nuclear war at any time, that appears raised to higher levels during periods of superpower crisis and confrontation, most spectacularly, one supposes, during the Cuban Missile Crisis of 1962. Therefore, throughout the period of the future there will exist at every moment some risk of catastrophe, perhaps gradually increased by the steady buildup of a danger of ecological collapse. There are also other kinds of risks of catastrophe ranging from man-initiated large-scale weather modification and from induced earthquakes to large-scale pollution of oceans, rivers, forests, and skies. The risk of disaster will rise continually in the years ahead, and barring some colossal human intervention, a disaster of catastrophic proportions is likely to occur in the 1990's. In any event, people attuned to their world will anticipate catastrophe in the closing decade of the century.

If there should be some recolonization of the poor portions of the world, then it is possible that the sense of impending catastrophe would recede from political consciousness, especially if the new colonizers were motivated by a large-scale effort to relieve pressure on the environment and reorganize the entire world system along ecologically sound lines. But basement H-bombs and the like will make it quite likely that in some weakly governed societies subnational or resistance groups will gain access to formidable military capabilities. The hypermodern state will grow ever more dependent on an interlinked network of controls and flows, and will become highly sensitive to intended and accidental disruption. The government will devote great energy and resources to prevent disruptive breakdowns and expend much ingenuity on the development of quick-detection and recovery procedures. Such a prospect of imminent breakdown will make ruling groups exceedingly nervous and encourage the deployment of surveillance and repressive technology to enable early warning, detection, and neutralization of disruptive behavior. The rising influence and expanding mission of the police will be justified, if justifications are needed, as essential vigilance given the vulnerability of technology to disruption and the presence of political unrest in the system.

Nevertheless, the mounting pressure on the environment is likely to interact with the politics of catastrophe to bring about the downfall of the world system as we have known it. The depth of the catastrophe, its character, and the prospects for rehabilitation will influence the reaction to it. In its aftermath proposals for sweeping world-order reforms that were earlier dismissed as "impractical" are likely to be studied and considered with solemn seriousness. The occurrence of a global catastrophe is likely to bring an end to the long reign of the sovereign state as the basic organizing unit of human life. What world system will be put in its place will almost certainly proceed from the premise of the wholeness of the planet and work out the ramifications of this premise in light of the dominant goals, values, and points of convergence that take shape in the bargaining and negotiating process. Obviously, it will be one kind of post-sovereign world if only the Afro-Asian societies are left intact, another if only the Euro-American societies survive, and still another if there is a more-or-less uniform impact throughout the world. The outcome will depend greatly on whether the organizing initiative is compulsory or voluntary and whether the dominant human forces are despotic or enlightened in political persuasion.

The Twenty-First Century—An Era of Annihilation. Let us suppose that the organized life of mankind is not permanently disrupted by the politics of catastrophe projected for the 1990's. Perhaps the police techniques of control and the propaganda techniques of manipulation will indeed keep dissident groups under control.

Perhaps the dominant states will be able to maintain their control over the activities of secondary states. Perhaps pollution abatement will become a major undertaking of principal governments and manage to keep the air we breathe clear enough and the food we eat pure enough. Perhaps an S-curve effect will lead to a reduced rate of population growth, so that the oxygen supply will not be seriously depleted, or the temperature of the earth rise so much as to produce drastic polar melting or disruptive weather change. Perhaps no Hitlers or Napoleons will take over the administration of a major government, no major leakage of radioactive waste will occur, no major war will break out, no major subversive plot will succeed. Perhaps, in other words, the worst will not happen, despite a context of continuing disintegration and extraordinary danger. Then what?

We would expect no major adaptation in terms of value-change or institutional innovation. Sovereign states would still run the political affairs of mankind, military power would still be the basis of national security, economic growth would still be the primary measure of social progress, and the gap between the rich countries and the poor countries would still be expanding.

How can such an outmoded world-order system hope to endure? Would it not seem certain that it would either be subverted from within, perhaps by a sect of super-robots who are instructed to take over the world for the sake of man, or crash down toward some cataclysmic end? By the next century the tensions will be so taut, the pressures and dangers so great, the lack of space so evident, the interdependencies so critical, the conflicts of interest so profound that the capacity for spontaneous adjustment is likely to be minimal, if not negative. It seems almost certain that the compulsions of the power-wielders will lead to a surge of repressive political energy reinforced by the super-technologies that lie ahead. Such a system cannot be peacefully transformed: It is taken over or it crashes down in ruin, leaving the entire system in disarray, perhaps irretrievably so. For this reason, we would suppose that the twenty-first century will inaugurate the Era of Annihilation.

THE FUTURE—AS A RESPONSE TO THE ENDANGERED PLANET. The positive line of potential development is even less predictable than its negative shadow because the stress is placed on institutional innovation and value change. Success requires a rapid erosion of national constraints on thought and action, perhaps through a quiet initial buildup at first, supplemented later on by more frontal assaults. The political premise that informs our analysis of the future is that the sovereign state cannot solve the characteristic problems of our time, and the sub-premise is that although some kind of global federalism is eventually desirable as a substitute for the sovereign state system it cannot (and perhaps should not) be brought into being until a period of transition and preparation has transpired.

To penetrate the political consciousness of governmental elites is the first task. There are frequent verbal acknowledgments at this time that problems of population pressure, the arms race, and the environment are matters of serious concern. But the acknowledgments tend to be fragmentary and to stop short of drawing appropriate economic and political conclusions. The population problem tends to be *dissociated* from other pressures on the world political and life-support systems, and the proposed solution tends to call for a little more attention, resources, and technology, but nothing more fundamental is deemed necessary. Such views are totally misleading, induce complacency, and do not encourage solution-oriented proposals. There is no way to remove the four-fold threat to mankind within the present structures of attitude, value, and institutional design. Robert McNamara, more sensitive than most eminent men of the day to the emerging crisis, himself split off population pressure from other interrelated matters as he called upon people to become responsible actors: "A rational, responsible, moral solution to the population problem must be found. You and I—all of us—share the responsibility to find and apply that solution. If we shirk that responsibility, we will have committed the crime. But it will be those who come after us who will pay the undeserved . . . and the unspeakable . . . penalties." It is questionable, at least for today's young, whether it will only be those who come after us that will suffer these penalties. Mr. McNamara, despite the ardor of his rhetoric, the depth of his concern, still seems to be a captive of the engineering mentality. The distinctive character of the endangered-planet crisis is that it emphasizes the need for *new political systems* and the inability of the present system, even if attentive to the crisis, to bring about a restored equilibrium between man and his environment. Mr. McNamara's statement by its limited responsiveness helps locate the present highwater mark of enlightenment among the most intelligent members of the power-wielding groups. Such power-wielders tend to keep the authority framework rigid in their prescriptions for the future. This authority framework is part of the obsolete machinery of our present system of world order. As such, it points up the need for a crash program in reality-testing, learning theory, and general education so that a more coherent response can be formed in a pre-catastrophe mood.

We turn now to consider a decade-by-decade account of a positive future for the earth, based on building a new system of world order.

The 1970's—The Decade of Awareness. The first task is to get a real fix on the situation presented by the fourfold threat to planetary welfare and to work toward a model for response. It will be essential to begin monitoring the critical trends immediately to avoid great hazards or points of no return. But the main effort should be to achieve a dissemination of concern by overt and covert forms of transmission throughout the entire world. The United Nations can perform useful roles by involving other countries in the discussion of the endangered-planet agenda.

The point is to convince as many people as possible that we are in the midst of an emergency, that the traditional priorities, aims, and conflicts need to be subordinated, and that there is a way out, but it involves change, sacrifice, and danger for all societies. The vested elites of the various power structures can be expected to fight back, more violently and ruthlessly as the consequences of the endangered-planet argument become more widely understood. Political leaders will resort to many forms of mystification, including giving their endorsement to the endangered-planet analysis accompanied by false assurances that the situation is not so dangerous as is contended or that it can be brought under control or that it is being exaggerated by enemies of the state.

The 1980's—The Decade of Mobilization. A general aroused awareness will produce a number of constituencies around the world demanding change. Proposals for new world-order systems will begin to be taken seriously in various political arenas and will be discussed within all major types of societies. Since the trend-lines appear quite clear there is likely to emerge a strong consensus as to the character of the endangered-planet crisis and the direction, at least, of political response. Sharp disagreements can be expected as to how to resolve the world-order crisis, by what tactics, with what priorities, toward what goals. The locus of jeopardy will be shifted in the direction of other actors, so that the rich and powerful will want their privileges mainly to survive transition, whereas the poor and weak will demand that the transition also emphasize equalization policies.

Transnational elites will play an increasingly important bargaining role in crystallizing a positive consensus upon which action can be taken. In some countries those who are committed to the old system will try to suppress this kind of world political movement, but at some point the momentum generated by the crisis-awareness will become overwhelming. The excitement caused by this mobilization process is likely to produce converging perspectives on action-proposals as well as a wave of confident expectation about a new world community that eliminates war and poverty and works to achieve harmony between man and nature. A new kind of optimism and energy will become evident at many points. New songs, symbols, heroes, movements, and myths will celebrate the transition process, heralding the new order for mankind that is being born.

Unpredictable alliances and antagonisms will reflect the decay of traditional politics in both national and international arenas.

The 1990's — The Decade of Transformation. Of course, as awareness crystallizes and as mobilization proceeds, the transformation of world order will begin to occur. But two decades, more or less, are needed to permit a favorable psychopolitical buildup to take place. At that point, a new definition of political acceptability will prevail in great portions of the world, ideas of sovereign prerogative will be muted or abandoned, and the work of community- and institution-building at a global level will begin to be the dominant political activity of the times.

A new kind of political ethos will have enlarged the idea of politics to include man-and-nature, as well as man-and-society. Several forms of ecological humanism will have arisen to replace earlier ideologies. Each principal cultural and regional setting will have stimulated at least one main ecologically grounded ideology by the 1990's. The separation of man from the rest of nature will seem less defensible than in earlier times. The capacities of supercomputers will deflate the pride of man in his own

uniqueness whereas man's sense of dependence on a
fragile environment will encourage a more conservative
and humble view of man-in-nature. In high places we will
find it standard to reflect an ecological outlook of the sort
that today we associate with exceptional and nonpolitical
men, such an outlook as is possessed by such a poetically
inclined naturalist as John Hay:

> There is a chance, in this newly defined arena
> of darkness and light, a chance to live and let live,
> a chance, paradoxically enough, to get rid of that
> terrible, isolating concept of man as the lord of
> creation. . . . We have a great deal of exploring
> to do in order to find the place where we share our
> lives with other lives, where we breathe and repro-
> duce, employ our sight, and join the breadth of
> chances not as separate, unique entities with dooms-
> day on our docket but as vessels for universal ex-
> perience.

We will be working toward a stable population that
recycles its wastes so as to assure future generations an
ample resource base. We will be working toward mass
consumption of needs and the provision of satisfactions
that are consistent with preserving the strength of diverse
forms of life that do not imperil men or other animal
species. A new political man will emerge from such a
climate of opinion and change. The planet will be governed as a system that needs to guard against relapse and
reversion, and regards the diversities within itself as a
source of vitality and vigor. Procedures and institutions
will be needed to balance the need for unity and uni-
formity against the claims for autonomy and pluralism.
There will be conflicts, disaffections, reactions, movements to go back to the earlier period of international
history. The strength of ecological humanism, however,
will be too great; its satisfactions will give man a new
sense of awe and of magic about the experience of life,

and a will to guard this new stage of human evolution against forces that would drag man back to the war-torn, sovereignty-ridden world of the 1970's and 1980's.

The 21st Century — The Era of World Harmony. By the end of the century the institutions and attitudes appropriate for the ideology of ecological humanism will be firmly established among men. Population levels will be falling toward optimum levels. There will be debates about the level and composition of the global optimum, its allocational shares, and the system of incentives attached to its maintenance. The dangers of pollution will be managed by a central international agency working in conjunction with regional and local institutions. Resource use will also be governed by a strict set of standards and responsibilities fixed for recycling. Pollution will have been largely controlled, streams will be pure, air will be clean even in and around large cities. There will, of course, be no reliance at all, or virtually none, on the internal combustion engine for private transportation. Great energy will be concentrated upon community activity, local crafts will flourish, and pride in work and affection will be much more visible.

The applications of science and other branches of knowledge will be guided by service to the maintenance of harmony on the planet. A less homocentric scale of values will underlie political decision. The welfare of plants, animals, and machines will all be considered benevolent in this kind of humanism. It is a humanism only because the whole process is conceived of and worked out by man, as if man were hired as an architect to rehabilitate the ecosphere inhabited by all that exists on earth.

These two visions of the future correspond with "the newly defined arena of darkness and light" that infiltrates the mind. These are the two poles of our political imagination that illumine the choice confronting each of us. This choice will be made knowingly or by default throughout the globe in the years ahead. The prospects for the welfare and survival of the planet will undoubtedly reflect how lucidly and effectively this choice will be made in the critical places at the crucial hours. We improve the prospects to the extent we inform ourselves about "the facts" so as to be in a position to reject or act upon the endangered-planet appraisal.

X. The Personal Imperative

The problem of personal action has been left until the end. To some extent it is a personal question, as one acts from a particular vantage point of style and belief, of hope, energy, and ability. Accordingly, I will answer mainly for myself, as an American living in the United States.

The principal point is that the argument of this book calls for action in response. Of course, it is possible— although I do not understand on what basis—to dismiss the whole argument as unfounded: Cassandra whistling in the dark. It is possible, but not very plausible, given the mounting evidence of ecological danger on so many fronts. I shall proceed on the assumption that the endangered-planet argument is basically correct. Albert Szent-Györgyi, who won the Nobel Prize in 1937 for his studies of metabolism and for the discovery of ascorbic acid (Vitamin C), expressed the prospects for the future in a moderate form: "According to very respectable scientists man's chances of survival are dropping toward 50

percent, and by the end of the century will be below that."
It is in this setting that each of us should ask himself
"What can I do?" Ideally, each reader should write his
or her own version of this chapter from here on.

What can any of us expect to do to reverse this prospect of destruction of the human species? The struggle
of peace groups through the years to stop the Vietnam
War has proved virtually fruitless. The system grinds on
like some gigantic machine, insensible to failure, without
a shred of compassion, and even unswayed by motives
of self-interest and greed. If we cannot stop the Vietnam
War, how can we expect to save the world from ecological
collapse? Although the objectives and obstacles are different, the experience of challenge in the one area illuminates the problems in the other area. There should be no
illusions about rapid and spectacular successes on the
ecological front. Large turnouts at demonstrations do not
transform structures of power.

At the same time, counsels of despair and resignation
are a costly disservice. The antiwar movement did help
place an upper ceiling on the Vietnam involvement and
induce a turnaround in official policy: far too late and
much too slow, but the machine has at least been made
to change gears, if not direction.

Furthermore, to accept the inevitability of disaster or
the impotence of individual action is to help assure the
triumph of the worst elements in our society. At the same
time, unless we act with a keen awareness of the obstacles
in our path, we shall be quickly disillusioned, and tempted
to retire back into the comforts of our private worlds.

The enormity of the minimum goals—transforming the
market economy at home and the state system abroad—
seems overwhelming. How can an individual contend
with such gigantic and remote forces? It should be clear
that only a fanatic—and a supply of effective fanatics
may be politically useful—can hope directly to challenge
major social forces and political structures. The rest

of us must find a more modest corner from which to begin, and then proceed to do what we can on the assumption that the sum of many separate efforts may erupt suddenly, at the right moment, into a potent political movement. Consider the early days of the antislavery movement or Zionism or the various struggles against colonial rule or the several successful revolutionary movements. In each instance the initial balance of forces made the radical project appear virtually hopeless even to its most ardent adherents. And then, a gathering of momentum from many separate sources converged and certain external conditions created a new, more favorable context for action. What has been hopeless from an earlier outlook in time and place all at once can become irresistible. This reality undermines such prevalent non-action postures as arise from moods of despair or moods of complacency. Situations do change, survivors of a shipwreck are sometimes rescued, smooth-flying planes do sometimes explode in flight. Neither hopelessness nor confidence can be justified as an attitude toward the future.

Movements against the tide can and do succeed, even when the nature of success involves toppling the power structure. We have no very reliable way of distinguishing between what is and is not possible in the future. Therefore, if we are convinced that certain changes are both *desirable* and *necessary*, then it is worth working for these changes up to the level of our capacities.

But to work effectively means to comprehend the special features of time, place, and circumstance. I would work differently for the same truth in the United States than I would in the Soviet Union, differently in Uganda than I would in South Africa. Similarly, there is a matter of knowing oneself, one's talents, limitations, points of access. I would work differently as a corporate executive than as an academician, differently as a poet than as a Congressman, differently as a fat cat than as a poor white.

Responsible political action depends on knowing one-self, on interpreting the scene, on selecting, and there-after sticking to, a line of action. Perseverance is a very helpful ingredient in the outlook of a radical innovator. William Blake put this point well: "If the fool would only persist he would become wise." "The fool" can be under-stood as a person who is not discouraged by mere evidence or by apparent lack of progress.

Ideology often helps sustain the spirit of an activist: a coherent account of why and wherefore. In these early years of the ecological age we are still groping for an interpretation that combines a critique with a cure. I have tried in this book to put forward one such coherent account. There is obviously a need for many others, taken from a variety of cultural, political, vocational, and generational perspectives. One clear obligation of an ecological activist, then, is to keep informed and to organize his information around the best available overall interpretation. The mass of relevant information now being spewed forth by the media makes it impossible to understand what is going on without having command of a fairly simple set of organizing concepts—what educationists call "organizers."

The problem of where to begin is a difficult one that relates to such matters as to where one is at. Every political movement requires for growth action that expresses the level of consciousness prevailing in a given community at a particular time. On ecological issues, this level of consciousness has been sharply ebbing and flowing over the last few months. What would have seemed a futile gesture a few years ago, perhaps even last year—holding an environmental teach-in—emerged as a nationwide booming success, with almost everyone clamoring to be part of the action. And then subsequently, a letdown, a failure of momentum, an ebbing, even a shift of focus to the matter of women and their status. The situation needs to be reappraised constantly in order to understand what is possible and worthwhile.

On the most modest, but concrete, level, it is certainly worthwhile to join and give financial support to such organizations as the World Law Fund, the Sierra Club, Friends of the Earth, The Wilderness Society, as well as to groups concerned with limiting population whether by birth control technology or by legalized abortion. Similarly, it is usually worthwhile to give support and time to local environmental crusades against the disposal of noxious wastes, or against building a new power plant on a river, or adding a new or larger jetport. It is certainly worth calling the attention of legislators in Washington and in state capitals to environmental issues, if possible at an early stage in their development. It is certainly worth making conservation groups aware of the links between their traditional concerns and the war system on one side, and population and resource pressure on the other. It is certainly worth questioning growth for growth's sake in town councils, chambers of commerce, and civic organizations.

In America, particularly, deeply held beliefs become newsworthy as they attain visibility. Such beliefs are becoming visible by means of slogans, bumper stickers, buttons, and placards. "Get Lead Out of Gasoline," "Ban DDT," "Don't Buy Union" slogans have quickly become characteristic of our time. Rachel Carson initiated a process of questioning in *The Silent Spring* that has not stopped to this moment. That book changed the climate of opinion so much that even its detractors—and these included many honorable men—were made sensitive to a whole new class of issues: the dreadful side effects of that miracle product DDT.

The challenge of action is to advance a set of goals that is supported by a general explanation and that can be implemented by a series of enactments. For most of us these individual acts will have social consequences only if similar efforts are made by thousands, perhaps millions, of others. It will be the cumulative impact of separate assertions of concern that form the basis of the one solid strategy for change that does not rest on the tactics of desperation.

There are two problems with the tactics of desperation as applied in the United States at this stage to ecological issues. First, recourse to these tactics plays mostly and normally into the hands of the repressive and reactionary elements in the society. The SDS Weatherman Movement is dangerous at least to the extent that it makes repression more credible to many "quiet Americans." The situation in America is not ripe at this time for violent, frontal challenge; the balance of opinion and of forces is arrayed against such convulsive strategies of change. Second, and perhaps more convincing, reliance on the tactics of desperation usually depends on counterviolence and confrontation, which are themselves destructive of life quality and inconsistent with the central earth values of harmony, wholeness, and restraint. We have had too much experience with movements that have polluted their ends by adopting repugnant means.

I would also add a note of caution. If the ecological situation is allowed to slide much longer, then there will exist no alternative to the politics of desperation. The longer we wait to act, the more we induce ecological collapse or invite some wild repressive scheme to avert it.

In planning action, it is important to grow aware of the many potentialities that all of us have. We each play many roles: at home, at work, in civic affairs, political clubs, in church, in fraternal and voluntary organizations, at social events. Each role is carried out within a politically significant arena. As a parent, I buy toys, games, and books for my child, surround his early years with images and examples, have access to his teachers and friends. In each context, I can exhibit my own sense of the endangered-planet crisis. Exhibit, not preach. To impose a set of views, rather than to bear witness and manifest their truth, is again to rely upon the brutal, spirit-diminishing weapons of the weak and insecure. The facts and values that endanger our future are capable of their own somber eloquence. There is no need to cram a creed down resistant throats.

We live in a reasonably free society. It is a rich and proud society, capable of great actions and responsible for great wrongs. Each of our lives is carried on within a complex setting with many points of contact. Each point of contact is also an opportunity for action. Also we can and should go beyond this daily show of concern, and devote as much energy as we can spare to explicit efforts to educate, persuade, and organize. I find no pretext for inaction, no grounds to believe that others, or the government, will get the job done when the danger grows *really* serious. The urgent work of rescuing our planet needs all the help it can get. Now.

Appendix

It is consistent with the spirit of this book to minimize the intrusion of scholarly apparatus. Hence the absence of footnotes to document the many assertions of the text based on evidence available elsewhere. In this back section I will acknowledge direct quotations from other sources and will indicate, in a few instances, particularly heavy debts to other authors. The page number is given for each source, and in the event that there is more than one source on a page, they are given in the order of their occurrence.

I have also included a few charts to illustrate and support the main lines of analysis contained in Chapter IV.

Despite the absence of fuller documentation, it should be evident that I have often relied on the work of others. Indeed it is precisely because I have relied on so many sources that I am reluctant to burden the argument with their full exhibition. I am also convinced that a more complete system of footnoting would not add any substance to my basic argument and would clearly make it more burdensome to follow.

To compensate for this decision I have added, following the charts, a short select bibliography built around the eight major concerns of this book. I have also added the names and addresses of several organizations whose work seems directly relevant to the challenges posed by an endangered planet. My whole purpose is to encourage awareness and induce action, and hence I suggest both further reading and relevant arenas for action.

NOTES

Chapter I

Page
 4 *Science,* Vol. 166, No. 3905, October 31, 1969, p. 591.
 8 Rolf Edberg, *On the Shred of a Cloud,* translated by Sven Ahman (Tuscaloosa, Ala.: University of Alabama Press, 1969), p. 15.
 14 Arthur C. Clarke, *Profiles of the Future* (New York: Harper & Row, 1963), pp. xi–xii.
 18 Arnold J. Toynbee, "The World View," *The American Way* (January 1970).
 19 An interview carried in *The New York Times* (December 31, 1969), p. 4.

Chapter II
 21 The orientation adopted in this chapter has been influenced by Harold and Margaret Sprout, *The Ecological Perspective on Human Affairs, With Special Reference to Inter-*

Page

national Politics (Princeton: Princeton University Press, 1965).

24 The idea of "objective consciousness" is developed by Theodore Roszak, *The Making of a Counter Culture* (New York: Anchor Books, 1969). It is significant to note that a dialectic existed even within Greek culture. The great dramatic tragedies of Aeschylus, Sophocles, and Euripides were all concerned with the sin of pride, *hubris,* which led a character of heroic stature into catastrophe. The essence of *hubris* is the overweening confidence of man in the autonomous powers of his own intellect. In Greek drama, unlike Greek philosophy, the mind of man was no sure guide, and in fact was an instrument that could easily turn upon itself; the *dualism* of mind and body and of reason and nature was repudiated, implicitly at least, in the vision of these playwrights.

25 Lynn H. White, Jr., "The Historic Roots of Our Ecologic Crisis," *Science,* Vol. 155, No. 3767 (March 10, 1967), pp. 1203–1207, at p. 1207.

28 Interview carried in *The New York Times* (December 31, 1969), p. 4.

29 "A Conversation," *Psychology Today,* Vol. III (January 1970), pp. 43–50, 58, at p. 50.

29–30 Eugene P. Odum, "The Strategy of Ecosystem Development," *Science,* Vol. 164, No. 3877 (April 18, 1969), pp. 262–270, at p. 268.

31–32 Quoted from Peter Collier, "The Red Man's Burden," *Ramparts* (February 1970), pp. 26–38, at p. 32.

32 John P. Lewis, *Wanted in India: A Relevant Radicalism,* Princeton University: Center of International Studies Policy Memorandum No. 36, December 1969, especially footnote 3, pp. 15–16.

On the Arusha Doctrine see Henry Bienen, "An Ideology for Africa," *Foreign Affairs,* Vol. 47, No. 3 (April 1969), pp. 545–559; see also Julius K. Nyerere, *Ujamaa— Essays on Socialism* (London: Oxford University Press, 1968).

Jean Mayer, "Toward a Non-Malthusian Population Policy," *Columbia Forum,* Vol. XII, No. 2 (Summer 1969), pp. 5–13, at p. 13.

Page

33 For evidence see Ted Robert Gurr, *Why Men Rebel* (Princeton: Princeton University Press, 1970).

34 On world poverty see David Simpson, "The Dimensions of World Poverty," *Scientific American,* Vol. 219, No. 5 (November 1968), pp. 27–35.

36 Gene Marine, "Politics of the Environment," *The Nation* (January 26, 1970), pp. 82–84, at p. 84.

Chapter III

37 Paul Valéry, *The Outlook for Intelligence,* translated by Denise Folliot and Jackson Mathews (New York: Harper & Row, 1962), p. 91.

40–41 For some interpretations of the world significance of the non-Western nations, see Barbara Ward, *Spaceship Earth* (New York: Columbia University Press, 1966); also F. S. C. Northrop, *The Meeting of East and West* (New York: Macmillan, 1952); and F. S. C. Northrop, *The Taming of the Nations: A Study of the Cultural Bases of International Policy* (New York: Macmillan, 1952).

42 For further exposition of ideas, see Leopold Kohr, *The Breakdown of Nations* (London: Routledge, 1957).

44 Kurt Vonnegut, *Slaughterhouse-Five* (New York: Delacorte Press, 1969), p. 123.

Henry A. Kissinger, "Central Issues of American Foreign Policy," in Kermit Gordon, ed., *Agenda for the Nation* (Garden City: Doubleday, 1968), pp. 585–614.

47 C. E. Ayers, *The Theory of Economic Progress* (New York: Schocken Books, 2nd ed., 1962), p. xxv.

48 See Garrett Hardin, "The Tragedy of the Commons," *Science,* Vol. 162 (December 13, 1968), pp. 1243–1248.

51 See Alvin M. Weinberg, "In Defense of Science," *Science,* Vol. 167, No. 3915 (January 9, 1970), pp. 141–45, at p. 145.

67 Raymond Richard Neutra, *Survival by Design* (New York: Oxford, Galaxy ed., 1969), p. ix.

69 David Braybrooke, "Private Production of Public Goods," in Kurt Baier and Nicholas Rescher, *Values and the Future* (New York: Free Press, 1969), pp. 368–388, at p. 387, fn. 31.

Page

Daniel Bell, "Notes on the Post-Industrial Society, I," *The Public Interest,* No. 6 (Winter 1967), pp. 24–35.

The Mathematical Sciences: A Report by the Committee on Support in the Mathematical Sciences of the National Research Council for the Committee on Science and Public Policy (Washington, D.C.: National Academy of Sciences, Pub. 1681), p. 92.

70–71 Harold Rosenberg, *The Anxious Object: Art Today and Its Audience* (New York: Mentor ed., 1969), pp. 16, 22.

72 Kurt Vonnegut, Jr., "Excelsior! We're Going to the Moon! Excelsior!" *The New York Times Magazine* (July 13, 1969), pp. 9–11, at p. 10.

74 Quoted in *The Futurist* (June 1969), p. 63.

75–76 Margaret Mead, "The Generation Gap," *Science,* Vol. 164, No. 3876 (April 11, 1969), p. 135.

76 John Schaar and Sheldon Wolin, "Berkeley: The Battle of People's Park," *New York Review of Books,* Vol. XII, No. 12 (June 19, 1969), pp. 25–31, at p. 31.

77–78, 83 President Nixon's message on population, *The New York Times* (July 19, 1969), p. 8.

86 Roszak, cited above, p. xiv.

88 But see Alvin Toffler, "Value Impact Forecaster—A Profession of the Future," in Baier and Rescher, cited above, pp. 1–30; p. 2: "In the last 300 years, however, the rate of value change appears to have speeded up—to the point at which major shifts in the value system of a society become apparent within the span of a single lifetime and within even shorter periods"; even this velocity of change is not comparable to the rate of technological change, nor to its fundamental character.

My discussion of long-range goals for mankind has been influenced by Gerald Feinberg, *The Prometheus Project: Mankind's Search for Long-Range Goals* (Garden City: Doubleday, 1968).

René Dubos, *So Human an Animal* (New York: Scribner's, 1968), p. 230: "While the rate of environmental change has immensely accelerated, the social and biological responses have not kept pace with the new situations thus created."

Dubos, cited above, p. 232.

Page
90 Barry Weisberg, "The Politics of Ecology," *Liberation,* Vol. 14, No. 9 (January 1970), pp. 20–25, at p. 22.
90–91 Andrei Sakharov, *Progress, Coexistence and Intellectual Freedom,* translated by *The New York Times* (New York: W. W. Norton, 1968).

Chapter IV

97 C. P. Snow, *The State of Siege* (New York: Scribner's, 1969), p. 10.
107 For careful studies of the main efforts to propose world peace systems, see F. H. Hinsley, *Power and the Pursuit of Peace* (London: Cambridge University Press, 1967); also Walter Schiffer, *The Legal Community of Mankind: A Critical Analysis of the Modern Concept of World Organization* (New York: Columbia University Press, 1954).
109 Ralph K. White, *Nobody Wanted War: Misperception in Vietnam and Other Wars* (Garden City: Doubleday, 1968).
111 Garonwy Rees, "Living Death," *Encounter,* Vol. XXXII, No. 1 (January 1969), pp. 77–82, at p. 82, in a review of Robert Jay Lifton, *Death in Life: The Survivors of Hiroshima* (New York: Random House, 1968).
113–114 Lin Piao, *Long Live the Victory of People's War* (Peking: Foreign Language Press, 1968), pp. 107–109.
124 For an assessment of the dangers of militarism by the former Marine Commandant, see David Monroe Sharp and J. S. Donovan, "New American Militarism," *Atlantic Monthly,* Vol. 223, No. 4 (April 1969), pp. 51–56.
126 The book critical of drastic disarmament is Arnold Wolfers, Robert E. Osgood, et al., *The United States in a Disarmed World* (Baltimore: John Hopkins Press, 1966).
127 United States Arms Control and Disarmament Agency, *World Military Expenditures 1966–67,* Research Report 68-52, December 1968, pp. 8; 1–2.
132 Thomas Robert Malthus, "An Essay on the Principle of Population" (1798), in Garrett Hardin, ed., *Population, Evolution, and Birth Control* (San Francisco: W. H. Freeman, 1969), pp. 4–16, at p. 7.

Page

133 Georg Borgstrom, *Too Many: A Study of Earth's Biological Limitations* (London: Macmillan, 1969), p. 317.

133–134 Colin Clark, "World Population," *Nature,* Vol. 181, No. 4618 (May 3, 1958), pp. 1235–1236, at p. 1236. A more recent, even more optimistic projection is to be found in Clark, *Population Growth and Land Use* (New York: St. Martin's Press, 1967).

134–135 Committee on Resources and Man, National Academy of Sciences–National Research Council, *Resources and Man* (San Francisco: W. H. Freeman, 1969), p. 5.

136 Undernourishment is likely to impair infant learning capacity: see Heinz F. Eichenwald and Peggy Crooke Fry, "Nutrition and Learning," *Science,* Vol. 163, No. 3868 (February 14, 1969), pp. 644–648.

Mayer, cited above, p. 10.

136–137 Robert G. Lewis, "False Focus on Food," book review in *The Progressive,* Vol. 33, No. 4 (April 1969), pp. 46–47, at p. 46.

138 Mayer, cited above, p. 5.

139 Herman Kahn and Anthony J. Wiener, "The Rich Will Grow Richer and the Poor Comparatively Poorer," in Foreign Policy Association, *Toward the Year 2018* (New York: Cowles Education Corporation, 1968), pp. 148–164, at p. 160.

139–140 Wayne H. Davis, "Overpopulated America," *New Republic* (January 10, 1970), pp. 13–15, at p. 15.

140 See Gunner Myrdal, *Asian Drama: An Inquiry into the Poverty of Nations,* 3 vols. (New York: Twentieth Century Fund, 1968), for a full inquiry into the link between population and mass poverty in the poor countries of Asia.

141 Adapted from table on rate of population increase in Hardin, ed., *Population, Evolution, and Birth Control,* cited above, p. 19.

141–142 Letter by Karl Sax, *Science,* Vol. 163, No. 3869 (February 21, 1969), pp. 763–764.

144 René Dubos, *Man Adapting* (New Haven: Yale University Press, 1965), p. 281.

Philip M. Hauser, "The World's People Will Nearly

Page

Triple in Number," in *Toward the Year 2018,* cited above, pp. 136–147, at p. 137.

145 Ward, cited above, p. 13.

"Meetings," *Science,* Vol. 163, No. 3865, p. 408.

147 "The Perfect Contraceptive Population: Extent and Implications of Unwanted Fertility in the U.S.," mimeographed, January 1970, p. 7. For other views on free family planning for the poor and population growth, see Judith Blake, "Population Policy for Americans: Is the Government Being Misled?" *Science,* Vol. 164, No. 3879 (May 2, 1969), pp. 522–529. Cf. Oscar Harkavy, Frederick S. Jaffe, and Samuel M. Wishik, "Family Planning and Public Policy: Who Is Misleading Whom?" *Science,* Vol. 165, No. 3891 (July 25, 1969), pp. 367–373.

150 Statistics taken from Simpson, cited above, p. 32.

151 For general analysis see Harold Frederiksen, "Feedbacks in Economic and Demographic Transition," *Science,* Vol. 166 (November 14, 1969), pp. 837–847; supported by letter by William B. Greenough, III, *Science,* Vol. 167 (January 16, 1970), p. 237.

153 Kenneth Boulding in Hardin, ed., *Population, Evolution, and Birth Control,* cited above, at p. 46.

160 National Academy of Sciences–National Research Council, *Natural Resources: A Summary Report to the President of the United States,* Publication 1000 (Washington, D.C., 1962), p. 2.

160–162 This discussion is based on Earl Finbar Murphy, *Governing Nature* (Chicago: Quadrangle Books, 1967); quotation is on p. 30.

162 Hans H. Landsberg, Leonard L. Fischman, and Joseph L. Fisher, *Resources in America's Future: Patterns of Requirements and Availabilities* (Baltimore: Johns Hopkins Press, 1963), pp. 43–44.

163 See, e.g., Marion Clawson, ed., *Natural Resources and International Development* (Baltimore: Johns Hopkins Press, 1964); and John H. Pincus, ed., *Reshaping the World Economy: Rich and Poor Countries* (Englewood Cliffs, N.J.: Prentice-Hall, 1968).

Page

      Joseph L. Fisher and Neal Potter, in Philip M. Hauser, ed., *The Population Dilemma: An American Assembly Book* (Englewood Cliffs, N.J.: Prentice-Hall, 1963), p. 120; see also chart on p. 102 comparing resources with projected demand levels for the year 2000.

164–165 Charles F. Park, Jr., *Affluence in Jeopardy: Minerals and the Political Economy* (San Francisco: Freeman, Cooper, 1968), pp. 15–16, 20–21.

165–166 *Resources and Man,* cited above, pp. 5, 6.

166–167 *Marine Resources Development—A National Opportunity,* Department of the Interior, not dated, p. 6.

168 Joseph L. Fisher and Neal Potter, "Resources in the United States and the World," in Philip M. Hauser, ed., *The Population Dilemma,* cited above, pp. 120–121.

168–169 Park, cited above, p. 334.

      For a table of mineral imports vital to United States industry, see Park as cited above, pp. 17, 27.

171–172 Hannah Arendt, *On Revolution* (New York: Viking Press, 1963), p. 15.

173 For a presentation of representative militant perspectives, see Tariq Ali, ed., *The New Revolutionaries: A Handbook of the International Radical Left* (New York: William Morrow, 1969).

174–175 Borgstrom, cited above, pp. 32, 325, 323. See also p. 326: "Hundreds of millions in the tropics are forced to shrink their food production to raise peanuts, cotton, bananas, coffee, tea, cacao, etc., for export, in order to accrue foreign currency."

      Paul M. Sweezy, "Toward a Critique of Economics," *Monthly Review,* Vol. 21 (January 1970), pp. 1–9, at p. 9.

176 "The Delusive Safeguard," *New York Times* editorial (August 6, 1969), p. 38.

177 Borgstrom, cited above, p. 323.

178 Murphy, cited above, p. 48.

179 This presentation relies considerably on Harrison Brown, *The Challenge of Man's Future* (New York: Viking Press, 1954).

179–180 Ansel Adams and Nancy Newhall, *This is the American Earth* (San Francisco: Sierra Club, 1960), p. 72.

180 Quoted in *Newsweek* (January 26, 1970), p. 34.

Page

182 R. Buckminster Fuller, *Operating Manual for Spaceship Earth* (Carbondale and Edwardsville, Ill.: Southern Illinois University Press, 1969), p. 39.

185 Editorial in *Life* (August 1, 1969), p. 32.

186–187 See Paul Shepard, Introduction, "Ecology and Man: A Viewpoint," in Shepard and McKinley, eds., *Subversive Science: Essays Toward an Ecology of Man* (Boston: Houghton Mifflin, 1969), p. 1.

188 Quoted in an interview published in *The New York Times* (August 10, 1969), p. 53.

190 David Ehrenfeld, *Biological Conservation* (New York: Holt, Rinehart & Winston, 1970), pp. 24–25.

191 Ernest J. Sternglass, "The Death of All Children," *Esquire,* Vol. LXXII, No. 3 (September 1969), pp. 1a–1d.

  Quoted from Margaret Gowing, *Britain and Atomic Energy, 1939–1945* (New York: St. Martin's Press, 1964), p. 385*n.*

194 John Kenneth Galbraith, *The New Industrial State* (Boston: Houghton Mifflin, 1967).

195 *New York Times* interview (August 10, 1969), p. 53.

  Quoted in *The New York Times* (August 3, 1969), Sec. 4, p. 4.

197 Quoted in *Plain Truth* (August 1969), p. 48.

200 Jaro Mayda, *Environment and Resources: From Conservation to Ecomanagement* (San Juan, P.R.: University of Puerto Rico, 1968), pp. 32–33. The analysis of Professor Mayda has influenced my approach toward this portion of the chapter.

201 For a general discussion of noise pollution see Donald F. Anthrop, "Environmental Noise Pollution: A New Threat to Sanity," *Bulletin of the Atomic Scientists,* Vol. XXV, No. 5 (May 1969), pp. 11–16; lawsuits, p. 13; or SST as "economic boondoggle," pp. 15–16.

203 Quoted from *Newsweek* (January 26, 1970), p. 31.

204 *Newsweek* (July 7, 1969), p. 58.

204–205 E. D. Goldberg, "The Chemical Invasion of the Oceans," mimeographed, pp. 1, 3, 7.

208 Reported in San Francisco *Chronicle* (August 11, 1969), p. 14.

208–209 See George R. Stewart, *Not So Rich as You Think*

Page

(Boston: Houghton Mifflin, 1968), for a very comprehensive and creative approach to the range of environmental problems covered by the term "pollution." See especially pp. 229 *et seq.*

210 Egler, "Pesticides—in Our Ecosystem," in Shepard and McKinley, cited above, p. 245.

210–211 O. M. Solandt, "The Control of Technology," editorial, *Science,* Vol. 165, No. 3892 (August 1, 1969), p. 445.

Chapter V

220 Dankwart A. Rustow, *A World of Nations* (Washington, D.C.: Brookings Institution, 1967), p. 3.

223 Mostafa Rejai and Cynthia H. Enloe, "Nation-States and State-Nations," *International Studies Quarterly,* Vol. 13, No. 2 (June 1969), pp. 140–158, at p. 143.

250–251 Robert W. Stock, "Saving the World the Ecologist's Way," *The New York Times Magazine* (Oct. 5, 1969), p. 33.

255 Samuel P. Huntington, *Political Order in Changing Societies* (New Haven: Yale University Press, 1968), p. 5.

Chapter VI

259–260 Morton H. Halperin, *Contemporary Military Strategy* (Little, Brown, 1966), p. 5.

270 February 16, 1970, p. 31.

274 Hugo Grotius, *Prolegomena to the Law of War and Peace* (Indianapolis; New York: Bobbs-Merrill, Liberal Arts Press Book, 1957), p. 21.

Chapter VII

285 René Dubos, "A Social Design for Science," *Science,* Vol. 166, No. 3907 (November 14, 1969), p. 823.

287 Hans J. Morgenthau, Introduction to David Mitrany, *A Working Peace System* (Chicago: Quadrangle Books, 1966), p. 9.

288 For a concise analysis of Soviet and American approaches to disarmament after World War II, see Richard J. Barnet, *Who Wants Disarmament?* (Boston: Beacon Press, 1960), p. 14; see also pp. 7–12.

296–297 W. Warren Wagar, "The Bankruptcy of the Peace

Page

Movement," *War/Peace Report,* August–September, 1969, pp. 3–6, at p. 4; see also Wagar, *The City of Man* (Baltimore: Penquin Books, 1967; copyright 1963).

298 Manfred Halpern, "Israel's Incoherent Response to an Incoherent Arab World," *Saturday Review* (February 14, 1970), pp. 36–38, 84.

304 For summary of FAO analysis, see *The New York Times* (November 2, 1969), p. 25.

310 For background on reversal of Japanese population policy, see Philip Baffey, "Japan: A Crowded Nation Wants to Boost Its Birthrate," *Science,* Vol. 167, No. 3920 (February 13, 1970), pp. 960–962.

312 One account of prospects for a U.S.–U.S.S.R. condominium is presented by Gene Gregory, "Political Masterplan for the Future," *Atlas,* Vol. 18, No. 4 (October 1969), pp. 19–21, at p. 21.

320–321 Mitrany, cited above, p. 17.

322 See especially Ernst B. Haas, *Beyond the Nation-State* (Stanford, Cal.: Stanford University Press, 1964); and for a related approach to the growth of international institutions see James Patrick Sewell, *Functionalism and World Politics* (Princeton: Princeton University Press, 1966).

325–326 For an analysis of major future trends see John McHale, *The Future of the Future* (New York: George Braziller, 1969).

327 Useful accounts of relevant diplomatic history are Henry A. Kissinger, *A World Restored* (New York: Grosset & Dunlap, 1964); and Richard N. Rosecrance, *Action and Reaction in World Politics: International Systems in Perspective* (Boston: Little, Brown, 1963).

Chapter VIII

353–354 Report on lead poisoning presented by *The New York Times* (February 26, 1970), p. 29.

360–361 "What Environmental Crisis?" *Concerned Demography,* Vol. I, No. 3 (February 1970), pp. 1–5, at p. 5.

364–365 Wagar, "The Bankruptcy" in *War/Peace Report,* cited above, pp. 4–6.

365 Wagar, *The City of Man,* cited above, p. 267.

Page

365–366 Wagar, *The City of Man,* cited above, p. 266.

369 Quoted in *The New York Times* (November 21, 1969), p. 14.

372 Text of Declaration printed in *The New York Times* (November 26, 1969), p. 16.

376–377 See Grenville Clark and Louis B. Sohn, *World Peace Through World Law* (Cambridge: Harvard University Press, 3rd rev. ed., 1962).

379–380 Quoted in *The New York Times* (March 1, 1970), Sec. 4, p. 16.

382–385 Richard J. Barnet, *The Economy of Death* (New York: Atheneum, 1969), pp. 126, 107, 112–113.

385–386 Goulden and Singer, "Dial-A-Bomb: AT&T and the ABM," *Ramparts* (November 1969), pp. 29–37, at pp. 30–31.

390 On the constitutional status of war-making, see Merlo J. Pusey, *The Way We Go to War* (Boston: Houghton Mifflin, 1969).

391 For historical origins of this war-declaring power see: Lawrence R. V. Velvel, "The War in Viet Nam: Unconstitutional, Justiciable, and Jurisdictionally Attackable," pp. 651–710, and also Francis D. Wormuth, "The Vietnam War: The President Versus the Constitution," pp. 711–807, in Richard A. Falk, ed., *Vietnam and International Law,* Vol. 2 (Princeton, Princeton University Press, 1969).

406 See Hearings held before the Conservation and Natural Resources Subcommittee of the House Committee on Government Operations, 91st Cong., 1st Sess., September 15–16, 1969 (Washington, D.C.: Government Printing Office).

409 See *The New York Times* (November 27, 1969), p. 28.

412 Edberg, cited above, p. x.

Chapter IX

415 U Thant, "Ten Critical Years," *U.N. Monthly Chronicle,* Vol. VI, No. 7 (July 1969), pp. i–v, at p. ii.

417 Harold D. Lasswell, "The Social and Political Framework of War and Peace," in Carmine D. Clemente and Donald B. Lindsley, eds., *Aggression and Defense,* UCLA Forum in

Page

   Medical Sciences, No. 7 (Berkeley; University of California, 1967).

422 C. P. Snow, cited above, p. 31.

425 Quoted in E. S. Turner, *Roads to Ruin* (London: Michael Joseph, 1950), p. 10.

432 Robert S. McNamara, speech at University of Notre Dame, 1969.

435 John Hay, *In Defense of Nature* (Boston: Little, Brown, 1969), p. x.

Table of values

World summary: constant price figures   **MILITARY EXPENDITURE**

	1948	1949	1950	1951	1952	1953	1954	1955	1956	1957	1958
USA		16 629	17 733	37 781	52 992	54 409	46 915	44 428	45 307	46 843	46 432
Other NATO		7 276	8 959	12 450	15 495	15 878	14 796	14 557	15 375	15 539	14 379
Total Nato		**23 905**	**26 692**	**50 231**	**68 487**	**70 287**	**61 711**	**58 985**	**60 682**	**62 382**	**60 811**
Other European	677	723	726	828	1 280	1 260	1 243	1 243	1 240	1 335	1 368
Middle East	210	270	300	330	320	350	390	500	640	670	790
South Asia[a]	610	620	650	680	740	680	690	740	830	750	810
Far East (excl. China)	470	650	1 120	1 400	1 420	1 650	1 670	1 580	1 590	1 790	2 050
China	[2 000]	[2 500]	[2 750]	[3 500]	[3 000]	[2 500]	[2 500]	[2 500]	[2 500]	[2 750]	[2 500]
Oceania	369	281	342	496	595	596	536	547	535	496	491
Africa	50	50	50	90	90	80	80	90	130	150	170
Central America	270	270	270	270	270	280	260	270	280	300	300
South America	850	790	710	760	760	830	810	870	1 030	990	1 100
World total excl. Warsaw pact		**30 059**	**33 610**	**58 585**	**78 513**	**76 962**	**69 890**	**67 325**	**69 457**	**71 613**	**70 390**
Warsaw pact (A): USSR	7 366	8 800	9 208	10 709	12 111	11 978	11 144	11 888	10 811	10 747	10 400
Other Warsaw pact		[4 800]	[4 800]	[4 800]	[4 800]	[4 800]	[4 800]	[4 800]	[5 250]	5 488	5 773
Total Warsaw pact		**13 600**	**14 008**	**15 509**	**16 911**	**16 778**	**15 944**	**16 688**	**16 061**	**16 235**	**16 173**
World total incl. Warsaw pact (A)		**43 659**	**47 618**	**74 094**	**93 873**	**95 291**	**85 834**	**84 013**	**85 518**	**87 848**	**86 563**
Warsaw pact (B): USSR	15 783	18 857	19 731	22 948	25 952	25 666	23 881	25 476	23 167	23 029	22 286
Other Warsaw pact		[2 500]	[2 500]	[2 500]	[2 500]	[2 500]	[2 500]	[2 500]	[2 750]	2 827	2 918
Total Warsaw pact		**21 357**	**22 231**	**25 448**	**28 452**	**28 166**	**26 381**	**27 976**	**25 917**	**25 856**	**25 204**
World total incl. Warsaw pact (B)		**51 416**	**55 841**	**84 033**	**105 414**	**106 679**	**96 271**	**95 301**	**95 374**	**97 469**	**95 594**

US $ mn, at 1960 prices and 1960 exchange-rates (Final column, X, at current prices and exchange-rates)

	1959	1960	1961	1962	1963	1964	1965	1966	1967	1968	1969	1968X
USA	47 085	45 380	47 335	51 203	50 527	48 821	48 618	57 951	66 889	68 213	(67 770)	79 605
Other NATO	15 342	15 955	16 354	17 898	18 408	18 752	18 662	18 825	19 719	19 542	(19 673)	24 365
Total NATO	**62 427**	**61 335**	**63 689**	**69 101**	**68 935**	**67 573**	**67 280**	**76 776**	**86 608**	**87 755**	**(87 443)**	**103 970**
Other European	1 412	1 397	1 510	1 637	1 677	1 772	1 785	1 840	1 834	1 892	(1 897)	2 527
Middle East	870	890	950	1 060	1 180	1 390	1 565	1 695	2 250	2 699	. . .	2 748
South Asia^a	800	812	854	1 080	1 640	1 638	1 735	1 769	1 563	1 610	. . .	1 860
Far East (excl. China)	2 180	2 290	2 440	2 525	2 315	2 535	2 800	2 820	3 110	3 570	. . .	3 970
China	[2 800]	[2 800]	[3 300]	[3 800]	[4 300]	[4 800]	[5 500]	[6 000]	[6 000]	[6 000]	. . .	[7 000]
Oceania	498	496	498	512	536	605	735	874	1 033	1 199	. . .	1 401
Africa	210	320	390	555	610	750	880	985	[1 000]	[1 100]	. . .	[1 220]
Central America	310	330	340	380	380	395	415	455	475	[480]	. . .	515
South America	960	970	940	1 010	1 030	1 080	1 250	1 130	1 280	1 390	. . .	2 120
World total excl. Warsaw pact	**72 467**	**71 640**	**74 911**	**81 660**	**82 603**	**82 538**	**83 945**	**94 344**	**105 153**	**107 695**	**. . .**	**127 331**
Warsaw pact (A):												
USSR	10 411	10 333	12 889	14 111	15 444	14 778	14 222	14 889	16 111	18 556	(19 667)	18 556
Other Warsaw pact	6 665	6 991	7 823	8 540	9 231	9 386	9 616	10 259	10 871	12 600	(14 224)	13 394
Total Warsaw pact (A)	**17 076**	**17 324**	**20 712**	**22 651**	**24 675**	**24 164**	**23 838**	**25 148**	**26 892**	**31 156**	**(33 891)**	**31 950**
World total incl. Warsaw pact (A)	**89 543**	**88 964**	**95 623**	**104 311**	**107 278**	**106 702**	**107 783**	**119 492**	**132 045**	**138 851**	. . .	**159 281**
Warsaw pact (B):												
USSR	22 310	22 143	27 619	30 238	33 095	31 667	30 476	31 905	34 450	39 780	(42 143)	39 780
Other Warsaw pact	3 198	3 379	3 752	4 186	4 445	4 439	4 416	4 733	5 082	6 023	(6 795)	6 294
Total Warsaw pact (B)	**25 508**	**25 522**	**31 371**	**34 424**	**37 540**	**36 106**	**34 892**	**36 638**	**39 532**	**45 803**	**(48 938)**	**46 074**
World total incl. Warsaw pact (B)	97 975	97 162	106 282	116 084	120 143	118 644	118 837	130 982	145 045	153 498	. . .	173 405

(A) = At official exchange-rates   (B) = At Benoit-Lubell exchange-rates
SIPRI Yearbook of World Armaments and Disarmament, 1968–69, pp. 200–201.

^a India, Pakistan, Afghanistan, Ceylon

1969 WORLD POPULATION DATA SHEET

POPULATION INFORMATION

Region or Country	Population Estimates Mid-1969 (Millions)	Birth Rate per 1,000 population	Death Rate per 1,000 population	Current Rate of Population Growth †	Number of Years to Double Population ‡	Infant Mortality Rate (Deaths under one year per 1,000 live births)	Population Under 15 years (percent)	Population Projections to 1980 (millions) ‡	Per Capita Gross National Product (US$) §
WORLD[1]	**3551**	**34**	**15**	**1.9**	**37**		**37**	**4368**	**589**
AFRICA[1]	**344**	**46**	**22**	**2.4**	**28**		**43**	**456**	**140**
NORTHERN AFRICA									
Algeria	13.3	44	11–14	2.9	24	86	47	19.5	220
Libya	1.9			3.6	19		44		640
Morocco	15.0	46	15–19	3.0	23	149	46	22.4	170
Sudan	15.2	52	18–22	3.0	23		47	21.0	100
Tunisia	4.8	45	17	2.8	25	110	41	6.4	200
UAR	32.5	43	15	2.9	24	120	43	46.7	160
WESTERN AFRICA									
Dahomey	2.7	54	26–31	2.9	24	110	46	3.4	80
Gambia	0.4	39	19	2.1	33		38	0.5	90
Ghana	8.6	47	20	2.5	28	156	45	12.2	230
Guinea	3.9	55	35	2.0	35	216	44	5.1	80
Ivory Coast	4.2	56	33	2.3	31		43	5.5	220
Liberia	1.2	40		1.8	39		37	1.5	210
Mali	4.9	52	30–32	2.0	35	123	49	6.5	60
Mauritania	1.1	45	25–28	2.0	35	187			130

—POPULATION REFERENCE BUREAU
FOR 137 COUNTRIES

WESTERN AFRICA Cont.									
Niger	3.7	52	25–27	2.7	26	200	46	4.8	80
Nigeria	53.7²	50	25	2.5	28		43		80
Senegal	3.9	43	17	2.5	28	93	42	4.9	210
Sierra Leone	2.5	44	22	2.2	32	146	37	3.3	150
Togo	1.8	55	29	2.6	27	127	48	2.3	100
Upper Volta	5.3	53	35	2.0	35	182	42	6.3	50
EASTERN AFRICA									
Burundi	3.5	46	26	2.0	35	150	47	4.7	50
Ethiopia	24.4	50		2.0	35			30.1	60
Kenya	10.6	46	20	3.0	23	132	46	14.6	90
Madagascar	6.7		22–25	2.4	29	102	46	8.5	90
Malawi	4.3	30		2.5	28		45	6.1	50
Mauritius	0.8	42	9	2.0	35	65	44	1.1	210
Mozambique*	7.3	52	31	1.2	58			9.0	100
Rwanda	3.5			2.7	26	137		5.0	40
Somalia	2.8			3.1	23			4.1	50
Southern Rhodesia*	4.8	48	14–18	3.0	23	122	47	7.1	210
Tanzania	12.9	45	23	2.9	24	189	42		80
Uganda	8.3	42	20	2.5	28	160	41		100
Zambia	4.2	51	20	3.1	23	259	45		180
MIDDLE AFRICA									
Angola*	5.4	50	26–28	1.4	50		42		170
Cameroon (Western)	5.7	48	30	2.2	32	137	39		110
Central African Republic	1.5	45	31	1.7	41	190	42		110
Chad	3.5	41	24	1.5	47	160	46		70
Congo (Brazzaville)	0.9			1.7	41			1.1	120

1969 WORLD POPULATION DATA SHEET

POPULATION INFORMATION

Region or Country	Population Estimates Mid-1969 (Millions)	Birth Rate per 1,000 population	Death Rate per 1,000 population	Current Rate of Population Growth†	Number of Years to Double Population‡	Infant Mortality Rate (Deaths under one year per 1,000 live births)	Population Under 15 years (percent)	Population Projections to 1980 (millions)‡	Per Capita Gross National Product (US$)§
MIDDLE AFRICA Cont.									
Congo (Democratic Rep.)	17.1	43	20	2.3	31	104	39		60
Gabon	0.5	35	30	0.9	78	229	36		400
SOUTHERN AFRICA									
Botswana	0.6	40		2.0	35		43		60
Lesotho	0.9	46		1.8	39		43		60
South Africa	19.6		23	2.4	29	181	40	26.8	550
Southwest Africa (Namibia)	0.6			1.7	41			0.9	
Swaziland	0.4	36		2.9	24		40	0.6	290
ASIA[1]	1990	38	18	2.0	35		40	2472	184
SOUTH WEST ASIA									
Cyprus	0.6	25	7	1.8	39	28	35	0.7	690
Iraq	8.9	48		2.5	28		45	13.8	270
Israel	2.8	25	6.6	2.9	24	25	33		1160
Jordon	2.3	47	16	4.1	17		46	3.3	220
Kuwait	0.6	52	6	7.6	9	37	38		3410
Lebanon	2.6			2.5	28			3.6	480

—POPULATION REFERENCE BUREAU
FOR 137 COUNTRIES

SOUTH WEST ASIA Cont.									
Saudi Arabia	7.2	37		1.8	39			9.4	240
Southern Yemen	1.3		8	2.2	32	80		1.6	
Syria	6.0			2.9	24		46	9.2	180
Turkey	34.4	46	18	2.5	28	161	44	48.5	280
Yemen	5.0							6.9	90
MIDDLE SOUTH ASIA									
Afghanistan	16.5	32	8	2.3	31		41	22.1	70
Bhutan	0.8			2.7	26			1.0	
Ceylon	12.3	43	18	2.4	29	56	41	16.3	150
India	536.9	50	20	2.5	28	139	46		90
Iran	27.9	41	21	3.1	23		40	38.0	250
Nepal	10.9			2.0	35			14.1	70
Pakistan	131.6	52	19	3.3	21	142	45	183.0	90
SOUTH EAST ASIA									
Burma	27.0	50	25–31	2.2	32		40	35.0	60
Cambodia	6.7	41	20	2.2	32	127	44	9.8	120
Indonesia	115.4	43	21	2.4	29	125	42	152.8	100
Laos	2.9	47	23	2.6	27				70
Malaysia (East & West)	10.7	36	7	3.1	23	49	44	14.9	280
Philippines	37.1	50	10–15	3.5	20	73	47	55.8	160
Singapore	2.1	27	5	2.5	28	26	43	3.2	570
Thailand	34.7	46	13	3.1	23	31	43	47.5	130
North Vietnam	21.4			3.1	23		38		
South Vietnam	17.9			2.6	27				120

1969 WORLD POPULATION DATA SHEET

POPULATION INFORMATION

Region or Country	Population Estimates Mid-1969 (Millions)	Birth Rate per 1,000 population	Death Rate per 1,000 population	Current Rate of Population Growth†	Number of Years to Double Population‡	Infant Mortality Rate (Deaths under one year per 1,000 live births)	Population Under 15 years (percent)	Population Projections to 1980 (millions)‡	Per Capita Gross National Product (US$)§
EAST ASIA									
China (Mainland)	740.3[3]	34	11	1.4	50			843.0	
China (Taiwan)	13.8	29	6	2.6	24	20	44	17.6	230
Hong Kong*	4.0	23	5	2.3	31	26	40	5.5	560
Japan	102.1	19	6.8	1.1	63	15	25	112.9	860
North Korea	13.3	38	10–14	2.4	29			17.5	
South Korea	31.2	41	10–14	2.8	25		42	43.4	150
Mongolia	1.2	40	10	3.0	23		30	1.7	
AMERICA[1]									
NORTHERN AMERICA	225	18	9	1.1	63	22	30	264	3399
Canada	21.3	18.0	7.3	2.0	35	23.1	33	22.3	2240
United States	203.1	17.4	9.6	1.0	70	22.1	30	240.1	3520
LATIN AMERICA[1]	**276**[4]	**39**	**10**	**2.9**	**24**		**43**	**376**	**385**
MIDDLE AMERICA									
Costa Rica	1.7	45	7	3.8	18	70.0	38	2.7	400
El Salvador	3.3	47	13	3.3	21	62.0	45	4.9	270

—POPULATION REFERENCE BUREAU
FOR 137 COUNTRIES

MIDDLE AMERICA Cont.									
Honduras	2.5	49	17	3.4	21		51	3.7	220
Mexico	49.0	43	9	3.4	21	63.0	46	71.4	470
Nicaragua	2.0	46	16	3.0	23		48	2.8	330
Panama	1.4	41	8	3.2	22	45.0	43	1.9	500
CARIBBEAN									
Barbados	0.3	30	9	0.9	78	48.0	38	0.3	400
Cuba	8.2	27	8	2.0	35	40.0	37	10.1	320
Dominican Republic	4.2	49	15	3.4	21	80.0	47	6.2	250
Haiti	5.1	44	20	2.4	29		38	6.8	70
Jamaica	1.8	40	8–9	2.5	28	35.0	41	2.1	460
Puerto Rico*	2.7	26	6	1.1	63	33.0	39	3.1	1090
Trinidad & Tobago	1.1	38	8	2.4	29	42.0	43	1.6	630
TROPICAL SOUTH AMERICA									
Bolivia	4.5	44	19	2.4	29	99.0	44	6.0	160
Brazil	90.6	38	10	2.8	25	79.0	43	124.0	240
Colombia	21.4	45	11	3.4	21	80.0	47	31.4	280
Ecuador	5.8	45	11	3.4	21	90.0	48	8.4	190
Guyana	0.7	40	9–10	2.7	26	40.0	46	1.0	300
Peru	13.2	42	11	3.1	23	63.0	45	18.5	320
Venezuela	10.4	41	8	3.3	21	46.0	46	15.0	850
TEMPERATE SOUTH AMERICA									
Argentina	24.0	23	9	1.5	47	58.0	29	28.2	780
Chile	9.6	33	10	2.3	31	108.0	40	12.2	510
Paraguay	2.3	45	11	3.4	21	80.0	45	3.5	200
Uruguay	2.9	21	9	1.2	58	43.0	28	3.3	570

1969 WORLD POPULATION DATA SHEET

POPULATION INFORMATION

Region or Country	Population Estimates Mid-1969 (Millions)	Birth Rate per 1,000 population	Death Rate per 1,000 population	Current Rate of Population Growth†	Number of Years to Double Population‡	Infant Mortality Rate (Deaths under one year per 1,000 live births)	Population Under 15 years (percent)	Population Projections to 1980 (millions)‡	Per Capita Gross National Product (US$)§§
EUROPE[1]	456	18	10	0.8	88		25	499	1230
NORTHERN EUROPE									
Denmark	4.9	18.4	10.3	0.9	78	16.9	24	5.3	1830
Finland	4.7	16.5	9.4	0.6	117	14.2	27	5.2	1600
Iceland	0.2	22.4	7.0	2.0	35	13.7	34	0.3	1740
Ireland	2.9	21.1	10.7	0.5	140	24.4	31	3.5	850
Norway	3.8	18.0	9.2	0.8	88	16.8	25	4.3	1710
Sweden	8.0	15.4	10.1	0.8	88	12.6	21	8.6	2270
United Kingdom	55.7	17.5	11.2	0.6	117	18.8	23	60.2	1620
WESTERN EUROPE									
Austria	7.4	17.4	13.0	0.5	140	26.4	24	7.7	1150
Belgium	9.7	15.2	12.2	0.1	700	23.7	24	10.2	1630
France	50.0	16.9	10.9	1.0	70	20.6	25	53.8	1730
West Germany	58.1	17.3	11.2	0.4	175	23.5	23	61.0	1700
Luxembourg	0.3	14.8	12.3	0.1	700	20.4	22	0.4	1920
Netherlands	12.9	18.9	7.9	1.1	63	13.4	28	15.3	1420
Switzerland	6.2	17.7	9.0	0.9	78	17.5	23	5.9⁵	2250

—POPULATION REFERENCE BUREAU
FOR 137 COUNTRIES

EASTERN EUROPE									
Bulgaria	8.4	15.0	9.0	0.6	117	33.1	24	9.2	620
Czechoslovakia	14.4	15.1	10.1	0.5	140	23.7	25	15.8	1010
East Germany	16.0	14.8	13.2	0.1	700	21.2	22	17.7	1220
Hungary	10.3	14.6	10.7	0.3	233	38.4	23	10.7	800
Poland	32.5	16.3	7.7	0.8	88	38.1	30	36.6	730
Romania	20.0	27.1	9.3	1.8	39	46.8	26	22.4	650
SOUTHERN EUROPE									
Albania	2.1	34.0	8.6	2.7	26	86.8		3.0	300
Greece	8.9	18.5	8.3	1.2	58	34.3	25	9.3	660
Italy	53.1	18.1	9.7	0.7	100	34.3	24	58.8	1030
Malta	0.3	16.6	9.4	0.6	117	27.5	32	0.4	510
Portugal	9.6	21.1	10.0	1.1	63	59.3	29	10.9	380
Spain	32.7	21.1	8.7	0.8	88	33.2	27	34.8	640
Yugoslavia	20.4	19.5	8.7	1.1	63	61.3	30	22.8	510
U.S.S.R.	**241**	**18**	**8**	**1.0**	**70**	**26**	**32**	**277.8**	**890**
OCEANIA[1]	**19**	**24**	**11**	**1.8**	**41**	**18**	**30**	**23**	**1857**
Australia	12.2	19.4	8.7	1.8	39	18.2	29	15.2	1840
New Zealand	2.8	22.4	8.4	1.9	37	17.7	33	3.6	1930

1969 WORLD POPULATION DATA SHEET —POPULATION REFERENCE BUREAU

WORLD AND REGIONAL POPULATION (Millions)

	WORLD	AFRICA	ASIA	NORTH AMERICA	LATIN AMERICA	EUROPE	OCEANIA	U.S.S.R.
MID-1969	3,551	344	1,990	225	276	456	19	241
2000 PROJECTIONS, U.N. CONSTANT FERTILITY	7,522	860	4,513	388	756	571	33	402
PERCENT INCREASE	112%	150%	127%	72%	174%	25%	74%	67%
2000 PROJECTIONS U.N. MED. EST.	6,130	768	3,458	354	638	527	32	353
PERCENT INCREASE	73%	123%	74%	57%	131%	16%	68%	46%

1969 DATA SHEET FOOTNOTES

* Non-sovereign country.

† Latest available year.

‡ Assuming continued growth at current annual rate.

§ 1966 data supplied by the International Bank for Reconstruction and Development.

[1] Population totals take into account small areas not listed on Data Sheet.

[2] Official government estimate of 64.8, based on 1963 census, is now considered high.

[3] U.N. estimate. Other estimates range from 800–950 million.

[4] Mid-1969 population estimates for the Latin America countries are taken from the latest figures of the Latin American Demographic Center of the United Nations. These figures are more recent than those of the 1967 U.N. Demographic Yearbook on which most of this Data Sheet is based.

[5] Foreigners with resident permits not taken into account.

UNITED STATES POPULATION IN MILLIONS

	1965	1980	2000
Actual	194		
Substantial decline in birth rate		232	297
Moderate decline in birth rate		244	334

INDICES OF NATIONAL ANNUAL REQUIREMENTS OF NATURAL RESOURCES 1950 TO 2000

	1950	1965	1980 (Percent Increases over 1965)		2000 (Percent Increases over 1965)	
Population $\times 10^6$	152	194	238[1]	(23)	315[1]	(62)
GNP $\times 10^9$ (1964$)	458	666	1,254	(88)	2,900	(335)
Disposable Personal Income $\times 10^9$ (1964$)	267	457	756	(65)	1,473	(222)
Energy Demand $\times 10^5$ BTU[2]	34.5	50.6	82.4	(63)	136	(169)
Metals Consumption $\times 10^6$ (1964$)	8,100	8,805	13,058	(48)	22,085	(151)
Water $\times 10^9$ gals/day	203	372	509	(38)	770	(107)
Fishery Products $\times 10^6$ (1964$)	669	966	1,345	(39)	2,092	(116)
Recreation (Use-days)	2×10^9[3]	4×10^9	6×10^9	(50)	9×10^9	(125)

[1]Average of moderate and substantial declines in fertility rates, Bureau of the Census, Series P-25.

[2]Fossil fuels, nuclear and hydro; nuclear is non-renewable until about 2000 when breeders will have been developed.

[3]Estimate, no survey data available.

FURTHER READING

The main purpose of this bibliography is to encourage and facilitate further inquiry into the main concerns of the book. A second purpose is to provide a list of books that have had a major influence on my understanding of these issues. In some instances, I disagree with the main position taken by an author, but nevertheless recommend his book. The bibliography is organized around eight principal topics. Some books concern more than one; these I have placed in relation to their main contribution.

I. The Future: Conjecture, Prediction, and Prophecy

Fritz Baade, *The Race to the Year 2000* (Garden City: Doubleday, 1962).

Harrison Brown, James Bonner, and John Weir, *The Next Hundred Years* (New York: Viking Press, 1957).

Zbigniew Brzezinski, *Between Two Ages: America's Role in the Technetronic Age* (New York: Viking, 1970).

Theodosius Dobzhansky, *Mankind Evolving* (New Haven: Yale University Press, 1962).

Jacques Ellul, *The Technological Society* (New York: Knopf, 1964).

Gerald Feinberg, *The Prometheus Project: Mankind's Search for Long-Range Goals* (Garden City: Doubleday, 1968).

R. Buckminster Fuller, *Utopia or Oblivion: The Prospects for Humanity* (New York: Bantam, 1969).

Olaf Helmer, *Social Technology* (New York: Basic Books, 1966).

Edward Higbee, *A Question of Priorities: New Strategies for Our Urbanized World* (New York: Morrow, 1970).

Erich Jantsch, *Technological Forecasting in Perspective* (Paris: Organisation for Economic Cooperation and Development, 1967).

Karl Jaspers, *The Future of Mankind* (Chicago: University of Chicago Press, 1961).

Bertrand de Jouvenel, *The Art of Conjecture* (New York: Basic Books, 1967).

Robert Jungk and Johan Galtung, eds., *Mankind 2000* (London: Allen and Unwin, 1969).

Herman Kahn and Anthony J. Wiener, *The Year 2000: A framework for Speculation on the Next Thirty-Three Years* (New York: Macmillan, 1967).

Martin E. Marty, *The Search for a Usable Future* (New York: Harper and Row, 1969).

Donald N. Michael, *The Unprepared Society: Planning for a Precarious Future* (New York: Basic Books, 1968).

John McHale, *The Future of the Future* (New York: Braziller, 1969).

The Next Ninety Years, Proceedings of a Conference held at the California Institute of Technology, March 1967 (Pasadena, Calif.: Institute of Technology, 1967).

John R. Platt, *The Step to Man* (New York: Wiley, 1966).

Fred L. Pollak, *The Image of the Future,* 2 vol. (New York: Oceana, 1961).

Bruce M. Russett and others, *World Handbook of Political and Social Indicators* (New Haven: Yale University Press, 1964).

A. T. W. Simeons, *Man's Presumptuous Brain: An Evolutionary Interpretation of Psychosomatic Disease* (New York: Dutton, 1961).

Gordon Rattray Taylor, *The Biological Time Bomb* (Cleveland: World, 1968).

Alvin Toffler, *Future Shock* (New York: Random House, 1970).

II. The War System

Raymond Aron, *The Century of Total War* (Garden City: Doubleday, 1954).

Richard J. Barnet, *The Economy of Death* (New York: Atheneum, 1969).

Paul Bohannan, ed., *Law and Warfare: Studies in the Anthropology of Conflict* (Garden City: Natural History Press, 1967).

Kenneth Boulding, *Conflict and Defense* (New York: Harper Bros., 1962).

Alastair Buchan, *War in Modern Society* (New York: Harper & Row, 1968).

Hedley Bull, *The Control of the Arms Race* (New York: Praeger, 1965).

Frantz Fanon, *The Wretched of the Earth* (New York: Grove, 1963).

James Finn, ed., *A Conflict of Loyalties* (New York: Pegasus, 1968).

Morton Fried, Marvin Harris, and Robert Murphy, eds., *War: The Anthropology of Armed Conflict and Aggression* (Garden City: Natural History Press, 1967).

J. F. C. Fuller, *The Conduct of War, 1789–1961* (New York: Funk & Wagnalls, 1961).

J. Glenn Gray, *The Warriors: Reflections on Men in Battle* (New York: Harcourt, Brace, 1959).

Philip Green, *Deadly Logic: The Theory of Nuclear Deterrence* (Columbus, Ohio: Ohio State University Press, 1966).

Joseph Heller, *Catch-22* (New York: Simon and Schuster, 1961).

Stanley Hoffmann, *The State of War* (New York: Praeger, 1965).

Herman Kahn, *On Thermonuclear War* (Princeton: Princeton University Press, 1960).

Erwin Knoll and Judith Nies McFadden, *American Militarism 1970* (New York: Viking, 1969).

Klaus Knorr, *On the Uses of Military Power in the Nuclear Age* (Princeton: Princeton University Press, 1966).

Robert A. Levine, *The Arms Debate* (Cambridge: Harvard University Press, 1963).

Robert Jay Lifton, *Death in Life: Survivors of Hiroshima* (New York: Random House, 1967).

Peter Mayer, ed., *The Pacifist Conscience* (New York: Holt, Rinehart, 1966).

Marshall McLuhan and Quentin Fiore, *War and Peace in the Global Village* (New York: Bantam, 1968).

Robert S. McNamara, *The Essence of Security: Reflections in Office* (New York: Harper & Row, 1968).

Seymour Melman, *Pentagon Capitalism: The Political Economy of War* (New York: McGraw-Hill, 1970).

C. Wright Mills, *The Causes of World War Three* (New York: Ballantine, 1960).

H. L. Nieburg, *Political Violence: The Behavioral Process* (New York: St. Martin's, 1969).

Robert E. Osgood and Robert W. Tucker, *Force, Order, and Justice* (Baltimore: Johns Hopkins Press, 1967).

Anatol Rapoport, *Strategy and Conscience* (New York: Harper & Row, 1964).

Theodore Ropp, *War in the Modern World* (Durham, N.C.: Duke University Press, 1959).

Thomas C. Schelling, *Arms and Influence* (New Haven: Yale University Press, 1966).

George Thayer, *The War Business: The International Trade in Armaments* (New York: Simon and Schuster, 1969).

Kurt Vonnegut, Jr., *Slaughterhouse-Five or The Children's Crusade* (New York: Delacorte, 1969).

Kenneth N. Waltz, *Man, the State, and War* (New York: Columbia University Press, 1959).

Arthur and Lila Weinberg, eds., *Instead of Violence* (New York: Grossman, 1963).

Donald A. Wells, *The War Myth* (New York: Pegasus, 1967).

Ralph K. White, *Nobody Wanted War: Misperception in Vietnam and Other Wars* (Garden City: Doubleday, Anchor edition, 1970).

Quincy Wright, *A Study of War* (Chicago: University of Chicago Press, 2nd ed., 1965).

Herbert York, *Race to Oblivion: A Participant's View of the Arms Race* (New York: Harper & Row, 1970).

III. Population Policy

Marston Bates, *The Prevalence of People* (New York: Charles Scribner's, 1962).

Georg Borgstrom, *The Hungry Planet: The Modern World at the Edge of Famine* (New York: Macmillan, 1965).

Lester R. Brown, *Seeds of Change: The Green Revolution and Development in the 1970's* (New York: Praeger, 1970).

Paul Ehrlich, *The Population Bomb* (San Francisco: Ballantine, 1968).

Garrett Hardin, ed., *Population, Evolution, and Birth Control: A Collage of Controversial Readings* (San Francisco: W. H. Freeman, 2nd ed., 1969).

Philip M. Hauser, ed., *The Population Dilemma* (Englewood Cliffs, N.J.: Prentice-Hall, 2nd ed., 1969).

International Action to Avert the Impending Protein Crisis. Report to the Economic and Social Council of the Advisory Committee on the Application of Science and Technology to Development (New York: United Nations, 1968).

Katherine and A. F. K. Organski, *Population and World Power* (New York: Knopf, 1961).

William and Paul Paddock, *Famine—1975! America's Decision: Who Will Survive* (Boston: Little, Brown, 1967).

William Vogt, *Road to Survival* (New York: William Sloan Assoc., 1948).

IV. Resource Policy

P. T. Flawn, *Mineral Resources* (Chicago: Rand McNally, 1968).

Hans H. Landsberg and others, *Natural Resources for U.S. Growth: A Look Ahead to the Year 2000* (Baltimore: Johns Hopkins Press, 1964).

J. F. McDivitt, *Minerals and Men* (Baltimore: Johns Hopkins Press, 1964).

Charles F. Park, Jr., *Affluence in Jeopardy: Minerals and the Political Economy* (San Francisco: Freeman, 1968).

N. W. Pirie, *Food Resources: Conventional and Novel* (Baltimore: Penguin, 1969).

Resources and Man. Committee on Resources and Man, National Academy of Science–National Research Council (San Francisco: Freeman, 1969).

Brian J. Skinner, *Earth Resources* (Englewood Cliffs, N.J.: Prentice-Hall, 1969).

V. Environmental Policy

Harrison Brown, *The Challenge of Man's Future* (New York: Viking, 1954).

Lynton K. Caldwell, *Environment: A Challenge for Modern Society* (New York: Natural History, 1970).

Rachel Carson, *Silent Spring* (Boston: Houghton Mifflin, 1962).

Richard Curtis and Elizabeth Hogan, *Perils of the Peaceful Atom* (Garden City: Doubleday, 1969).

J. Clarence Davies, *The Politics of Pollution* (New York: Pegasus, 1970).

Garrett De Bell, ed., *The Environmental Handbook* (New York: Ballantine, 1970).

Marshall I. Goldman, ed., *Controlling Pollution: The Economics of a Cleaner America* (Englewood Cliffs, N.J.: Prentice-Hall, 1967).

Frank Graham, Jr., *Since Silent Spring* (Boston: Houghton Mifflin, 1970).

John Hay, *In Defense of Nature* (Boston: Little, Brown, 1969).

Henry Jarrett, ed., *Environmental Quality in a Growing Economy* (Baltimore: Johns Hopkins Press, 1966).

Wesley Marx, *The Frail Ocean* (New York: Ballantine, 1969).

Emanuel F. Mesthene, *Technological Change: Its Impact on Man and Society* (New York: Mentor, 1970).

Earl Finbar Murphy, *Governing Nature* (Chicago: Quadrangle, 1967).

Max Nicholson, *The Environmental Revolution* (New York: McGraw-Hill, 1970).

Sheldon Novick, *The Careless Atom* (Boston: Houghton Mifflin, 1969).

Fairfield Osborn, *Our Plundered Planet* (Boston: Little, Brown, 1948).

Peter J. Schmitt, *Back to Nature: The Arcadian Myth in Urban America* (New York: Oxford, 1969).

William A. Shurcliff, *SST and Sonic Boom Handbook* (New York: Ballantine, 1970).

George R. Stewart, *Not So Rich As You Think* (Boston: Houghton Mifflin, 1967).

Henry Still, *The Dirty Animal* (New York: Hawthorn, 1967).

Thomas Whiteside, *Defoliation* (New York: Ballantine, 1970).

VI. The Ecological Perspective

Kenneth Boulding, *The Meaning of the 20th Century* (New York: Harper & Row, 1964).

Harrison Brown, *The Challenge of Man's Future* (New York: Viking, 1954).

Barry Commoner, *Science and Survival* (New York: Viking, 1968).

Fred Cottrell, *Energy and Society: The Relation Between Energy, Social Change, and Economic Development* (New York: McGraw-Hill, 1955).

René Dubos, *Man Adapting* (New Haven: Yale University Press, 1965).

Rolf Edberg, *On the Shred of a Cloud* (Tuscaloosa, Ala.: University of Alabama Press, 1969).

David W. Ehrenfeld, *Biological Conservation* (New York: Holt, Rinehart, 1970).

Paul R. and Anne H. Ehrlich, *Population, Resources, Environment: Issues in Human Ecology* (San Francisco: Freeman, 1970).

Amitai Etzioni, *The Active Society: A Theory of Societal and Political Processes* (New York: The Free Press, 1968).

Erich Kahler, *The Tower and the Abyss: An Inquiry into the Transformation of Man* (New York: Braziller, 1957).

E. Kormondy, *Concepts of Ecology* (New York: Prentice-Hall, 1969).

Jaro Mayda, *Environment and Resources: From Conservation to Ecomanagement* (San Juan, P.R.: University of Puerto Rico, 1968).

Ian L. McHarg, *Design with Nature* (Garden City: Natural History Press, 1969).

Eugene Odum, *Ecology* (New York: Holt, Rinehart, 1969).

Theodore Roszak, *The Making of a Counter Culture* (Garden City: Doubleday, 1969).

Paul Shepard and David McKinley, eds., *The Subversive Science: Essays toward an Ecology of Man* (Boston: Houghton Mifflin, 1969).

Harold and Margaret Sprout, *The Ecological Perspective on Human Affairs, with Special Reference to International Politics* (Princeton: Princeton University Press, 1965).

VII. World-Order Perspectives

Gar Alperowitz, *Atomic Diplomacy: Hiroshima and Potsdam* (New York: Vintage, 1965).

Robert C. Angell, *Peace on the March: Transnational Participation* (New York: Van Nostrand Reinhold, 1969).

Raymond Aron, *Peace and War: A Theory of International Relations* (Garden City: Doubleday, 1966).

Cyril E. Black, *The Dynamics of Modernization: A Study in Comparative History* (New York: Harper & Row, 1966).

Cyril E. Black and Richard A. Falk, eds., *The Future of the International Legal Order, Vol. I: Trends and Patterns* (Princeton: Princeton University Press, 1970).

Kenneth E. Boulding, *The Organizational Revolution: A Study in the Ethics of Economic Organization* (Chicago: Quadrangle, 1968).

Adda B. Bozeman, *Politics and Culture in International History* (Princeton: Princeton University Press, 1960).

Ernst Cassirer, *The Myth of the State* (New Haven: Yale University Press, 1946).

Richard N. Cooper, *The Economics of Interdependence: Economic Policy in the American Community* (New York: McGraw-Hill, 1968).

Amitai Etzioni, *Political Unification: A Comparative Study of Leaders and Forces* (New York: Holt, Rinehart, 1965).

Richard A. Falk, *Legal Order in a Violent World* (Princeton: Princeton University Press, 1968).

Richard A. Falk, *The Status of Law in International Society* (Princeton: Princeton University Press, 1970).

Richard A. Falk and Saul H. Mendlovitz, eds., *The Strategy of World Order,* 4 vols. (New York: World Law Fund, 1966, 1967).

Roger Fisher, *International Conflict for Beginners* (New York: Harper & Row, 1969).

Wolfgang Friedmann, *The Changing Structure of International Law* (New York: Columbia University Press, 1964).

Richard N. Gardner, *In Pursuit of World Order* (New York: Praeger, rev. ed., 1966).

Robert W. Gregg and Michael Barkun, eds., *The United Nations and Its Functions* (Princeton: Van Nostrand, 1968).

Ernst B. Haas, *The Tangle of Hopes: American Commitments and World Order* (Englewood Cliffs, N.J.: Prentice-Hall, 1969).

Louis Henkin, *How Nations Behave: Law and Foreign Policy* (New York: Praeger, 1968).

John H. Herz, *International Politics in the Atomic Age* (New York: Columbia University Press, 1959).

F. H. Hinsley, *Power and the Pursuit of Peace* (London: Cambridge University Press, 1963).

Arthur N. Holcolme, *A Strategy of Peace in a Changing World* (Cambridge: Harvard University Press, 1967).

Philip E. Jacob and James U. Toscano, eds., *The Integration of Political Communities* (Philadelphia: Lippincott, 1964).

Harry Magdoff, *The Age of Imperialism: The Economics of U.S. Foreign Policy* (New York: Monthly Review, 1969).

Arno J. Mayer, *Wilson vs. Lenin: Political Origins of the New Diplomacy 1917–1918* (Cleveland: World, Meridian edition, 1964).

Myres S. McDougal and Florentino P. Feliciano, *Law and Minimum World Public Order* (New Haven: Yale University Press, 1961).

William H. McNeill, *The Rise of the West: A History of the Human Community* (Chicago: Chicago University Press, 1963).

Gunnar Myrdal, *Asian Drama: An Inquiry into the Poverty of Nations,* 3 vols. (New York: Twentieth Century Fund and Pantheon, 1968).

Gunnar Myrdal, *Beyond the Welfare State* (New Haven: Yale University Press, 1960).

Gunnar Myrdal, *The Challenge of World Poverty* (New York: Pantheon, 1970).

Joseph H. Nye, Jr., ed., *International Regionalism* (Boston: Little, Brown, 1968).

F. S. C. Northrop, *The Meeting of East and West* (New York: Macmillan, 1946).

Hans F. Petersson, *Power and International Order* (Lund, Sweden: C. W. K. Gleerup, 1964).

John A. Pincus, ed., *Reshaping the World Economy: Rich Countries and Poor Countries* (Englewood Cliffs, N.J.: Prentice-Hall, 1968).

Emery Reves, *The Anatomy of Peace* (New York: Viking, 1963).

Richard N. Rosecrance, *Action and Reaction in World Politics* (Boston: Little, Brown, 1963).

Andrei D. Sakharov, *Progress, Coexistence and Intellectual Freedom* (New York: Norton, 1968).

Julius Stone, *Legal Control of International Conflict* (Sydney, Australia: Maitland, 1954).

F. P. Walters, *A History of the League of Nations,* 2 vols. (London: Oxford University Press, 1952).

Jan Tinbergen, *Shaping the World Economy: Suggestions for an International Economic Policy* (New York: Twentieth Century Fund, 1962).

William Appleman Williams, *The Roots of the Modern American Empire* (New York: Random House, 1969).

Charles Yost, *The Insecurity of Nations: International Relations in the Twentieth Century* (New York: Praeger, 1968).

Oran R. Young, *The Intermediaries: Third Parties in International Crises* (Princeton: Princeton University Press, 1967).

VIII. Utopian Perspectives

Robert Boguslaw, *The New Utopians: A Study of System Design and Social Change* (Englewood Cliffs, N.J.: Prentice-Hall, 1965).

Martin Buber, *Paths in Utopia* (Boston: Beacon, 1949).

Grenville Clark and Louis B. Sohn, *World Peace through World Law* (Cambridge: Harvard University Press, 3rd rev. ed., 1966).

Harvey Cox, *The Feast of Fools* (Cambridge: Harvard University Press, 1969).

Robert C. Elliott, *The Shape of Utopia: Studies in a Literary Genre* (Chicago: University of Chicago Press, 1970).

Peter Farb, *Man's Rise to Civilization as Shown by the Indians of North America from Primeval Times to the Coming of the Industrial State* (New York: Dutton, 1968).

Errol E. Harris, *Annihilation and Utopia: The Principles of International Politics* (London: Allen & Unwin, 1966).

Karl Mannheim, *Ideology and Utopia* (New York: Harcourt, Brace, Harvest edition, no date; original edition, 1936).

Frank E. Manuel, *Utopias and Utopian Thought* (Boston: Beacon, 1967).

David Mitrany, *A Working Peace System* (Chicago: Quadrangle, 1966).

Lewis Mumford, *The Story of Utopias* (New York: Viking, Compass edition, 1962).

Glenn Negley and J. Max Patrick, eds., *The Quest for Utopia* (Garden City: Doubleday-Anchor, 1952).

Judith N. Shklar, *After Utopia* (Princeton: Princeton University Press, 1957).

W. Warren Wagar, *The City of Man* (Baltimore: Penguin, 1967).

Frank Waters, *Book of the Hopi* (New York: Viking, 1963).

Robert Paul Wolff, *In Defense of Anarchism* (New York: Harper & Row, 1970).

CONTINUING ACTION

Prospects for human survival depend upon many kinds of action undertaken by many people in many societies. Within the setting of the United States in the early 1970's I have found that the following activities are especially worthy of support (i.e., time, energy, and resources):

The World Law Fund
11 West 42nd Street
New York, N.Y. 10036
(the leading private organization concerned with education and world order)

Fund for Peace
1865 Broadway
New York, N.Y. 10023
(an ambitious effort to support major research and educational activities in the world-order area)

Friends of the Earth
30 East 42nd Street
New York, N.Y. 10017
(an ecological approach to conservation and environmental quality organized to influence public opinion and government action)

The Sierra Club
1050 Mills Tower
San Francisco, Calif. 94104
(the most effective organization in America with respect to the traditional agenda of conservation concerns)

Planned Parenthood/World Population
515 Madison Ave.
New York, N.Y. 10022
(a highly respected organization devoted to moderating population growth by education and family planning)

John Muir Institute for Environmental Studies
451 Pacific Avenue
San Francisco, Calif. 94133
(a research institute sharing the outlook and orientation
of the more activist Friends of the Earth)

The Institute for Policy Studies
1520 New Hampshire Ave., N.W.
Washington, D.C. 20036
(center of creative research and education on major
policy issues of American society)

Environmental Action
2000 P St., N.W.
Washington, D.C. 20006
(a continuing operation by the organizers of Earth Day
1970)

The Wilderness Society
729 15th Street, N.W.
Washington, D.C. 20005
(an organization concerned with the defense of the
quality of nature against human encroachment)

Environmental Defense Fund
P.O. Drawer 740
Stony Brook, N.Y. 11790
(a highly successful organization of mainly lawyers and
scientists that concentrate on specific environmental
issues with great intensity)

Index

ABOUT THE AUTHOR

RICHARD A. FALK is the Albert G. Milbank Professor of International Law and Practice at Princeton University, where he has been teaching since 1961. Before that he taught for six years in the College of Law at Ohio State University.

Professor Falk was born in New York City in 1930 and has received academic degrees from the University of Pennsylvania (B.S.), Yale Law School (LL.B.), and Harvard Law School (J.S.D.). He is a member of the New York Bar, but has not practiced except to serve as one of the Counsel on behalf of Ethiopia and Liberia in a case against South Africa in the International Court of Justice in The Hague.

Involved in various professional activities, Professor Falk has been acting as the Director of the North American Section of the World Order Models Project, World Law Fund. This project represents a global effort of scholars to develop feasible strategies for improving the quality of world order by the end of the century. He is also a Vice President of the American Society of International Law and a member of the Board of Directors of the Fund for Peace and the Foreign Policy Association.

Professor Falk lives in Princeton with his wife, Florence, and their son, Dimitri.